Seismology and Wave Propagation

Seismology and Wave Propagation

Editor: Daniel Galea

www.callistoreference.com

Callisto Reference,
118-35 Queens Blvd., Suite 400,
Forest Hills, NY 11375, USA

Visit us on the World Wide Web at:
www.callistoreference.com

ISBN: 978-1-64116-090-2 (Hardback)

Cataloging-in-Publication Data

Seismology and wave propagation / edited by Daniel Galea.
p. cm.
Includes bibliographical references and index.
ISBN 978-1-64116-090-2
1. Seismology. 2. Seismic waves. 3. Seismic wave propagation. 4. Waves.
I. Galea, Daniel.
QE534.3 .S45 2019
551.22--dc23

Table of Contents

Preface

Earthquakes are disasters caused by seismic waves which are formed due to a sudden release of energy from the Earth's lithosphere. This can develop into disturbances of massive magnitude resulting in substantial damage to life and property. Seismology is the study of earthquakes. It encompasses the study of diverse aspects related to earthquakes including seismicity analysis, frequency, wave nature, rupture dynamics, wave propagation, etc. It also involves evaluating earthquake aftershocks and effects using principles of subfields like paleoseismology. This book provides significant information of this discipline to help develop a good understanding of seismology and wave propagation. It also unravels the recent studies in this field and highlights the modern tools of earthquake prediction, evaluation and analysis. The book will be helpful to a broad spectrum of readers such as seismologists, geologists, engineers, researchers, academicians, experts and students associated with this discipline.

This book is the end result of constructive efforts and intensive research done by experts in this field. The aim of this book is to enlighten the readers with recent information in this area of research. The information provided in this profound book would serve as a valuable reference to students and researchers in this field.

At the end, I would like to thank all the authors for devoting their precious time and providing their valuable contribution to this book. I would also like to express my gratitude to my fellow colleagues who encouraged me throughout the process.

Editor

1

An IBEM solution to the scattering of plane SH-waves by a lined tunnel in elastic wedge space

Zhongxian Liu · Lei Liu

Abstract The indirect boundary element method (IBEM) is developed to solve the scattering of plane SH-waves by a lined tunnel in elastic wedge space. According to the theory of single-layer potential, the scattered-wave field can be constructed by applying virtual uniform loads on the surface of lined tunnel and the nearby wedge surface. The densities of virtual loads can be solved by establishing equations through the continuity conditions on the interface and zero-traction conditions on free surfaces. The total wave field is obtained by the superposition of free field and scattered-wave field in elastic wedge space. Numerical results indicate that the IBEM can solve the diffraction of elastic wave in elastic wedge space accurately and efficiently. The wave motion feature strongly depends on the wedge angle, the angle of incidence, incident frequency, the location of lined tunnel, and material parameters. The waves interference and amplification effect around the tunnel in wedge space is more significant, causing the dynamic stress concentration factor on rigid tunnel and the displacement amplitude of flexible tunnel up to 50.0 and 17.0, respectively, more than double that of the case of half-space. Hence, considerable attention should be paid to seismic resistant or anti-explosion design of the tunnel built on a slope or hillside.

Z. Liu (✉) · L. Liu
Tianjin Chengjian University, Tianjin 300384, China
e-mail: zhongxian1212@163.com

Z. Liu
Key Laboratory of Soft Soils and Engineering Environmental of Tianjin, Tianjin Chengjian University, Tianjin 300384, China

L. Liu
Earthquake Engineering Research Institute of Tianjin, Tianjin 300384, China

Keywords Wedge space · Scattering · Lined tunnel · Plane SH-waves · Indirect boundary element method (IBEM) · Dynamic stress concentration

1 Introduction

The scattering of elastic waves by the underground structure and the phenomenon of the dynamic stress concentration is an interesting and important topic in many fields, e.g., in earthquake engineering, non-destructive detection, etc. In general,the solution methods can be divided into the analytical method and numerical method. The analytical methods include the wave function expansion (Lee and Trifunac 1979; Liang et al. 2010; Li et al. 2009), earthquake coefficient method and response displacement method, etc. The numerical methods include the finite element method (Yang and Liu 1994), boundary element method (boundary integral equation method) (Du et al. 1993; Stamos and Beskos 1996; Liang et al. 2013; Chen et al. 2011) and the hybrid method (Datta et al. 1984), etc.

Note that above studies are mainly restricted to the full-space or half-space model at present. However, in practical engineering, many tunnels or underground pipes are built on sloping topography or cliffs, and then wedge space model would be more appropriate for preliminary quantitative analysis in such sites. Compared with the half-space model, more difficulties will arise for exactly satisfying the free boundary conditions of the wedge space. Achenbach (Achenbach 1970) studied the transient wave propagation problem in wedge space, considering spatially uniform shear tractions applied to one or both faces of the wedge. Knopoff (1969), Budaev and Bogy (1995) and Gautesen (2002) have investigated the wave reflection and transmission coefficients of the free field in elastic wedge space by numerical or experimental method. Moreover, Li and Gong studied the reflection and transmission of obliquely incident Rayleigh wave in two adjacent rectangular space (Li and Gong 1998).

As for the wave scattered by obstacle in wedge space, available results are rarely seen due to the complicated characteristics of wave propagation and scattering. Lee and Sherif studied the scattering of SH-waves by a canyon in wedge-shape space (Lee and Sherif 1996); Shi et al. studied the scattering solutions with a fixed circular inclusion or a hole in Cartesian space by the method of complex variable function (Shi et al. 2006, 2007). However, up to date, there is no results published for the wave scattering around a tunnel in wedge space of arbitrary angle, only Liu et al. (2009) presented some results around a cavity in wedge space. Zhang et al. (2013) studied the scattering of SH-wave by circular cavity in a right-angle plane.

This paper aims to study the scattering of SH-waves by a lined tunnel in elastic wedge space of any angle by the indirect boundary element method. It is illustrated that this method has several advantages such as reducing dimensions of problems, automatic satisfaction of radiation condition, and high calculation precision. Moreover, in this method, the virtual loads can directly act on the boundary surface, which can be recognized as a direct implementation of the Huygens' principle.

This paper will be arranged as follows. Firstly, the numerical procedure for IBEM solution to SH-waves diffraction in wedge space is presented. Then, the accuracy of this method is verified by the comparison between the degenerated solutions and available solutions. Finally, the effects of key parameters, such as the wedge angle, excitation frequency and the incident angle on dynamic response of tunnel are investigated in detail through numerical examples, and some important conclusions have been obtained.

2 Model

Figure 1 shows a lined tunnel of infinite length and constant cross-section located at arbitrary position of the wedge space. Define the vertex of the wedge as o, the angle between the inclined and horizontal surface as $v\pi$, the geometric center of the lined tunnel as o'. H and D are the vertically and horizontal distance between o and o', respectively. Assume that the material in the wedge space and lined tunnel is homogeneous and linearly elastic. Define the outer and inner surface of tunnel as Γ_1 and Γ_2, the region of the wedge space and tunnel as D_1 and D_2, the ground and wedge surface as S and B, respectively. The shear wave velocity of the wedge space is defined as

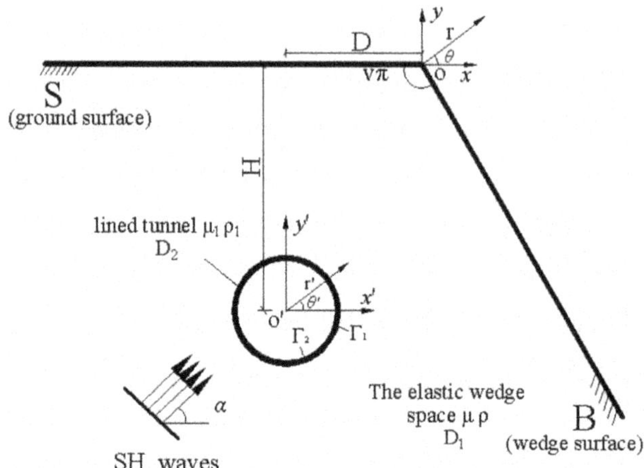

Fig. 1 Model for calculation

$\beta = \sqrt{\mu/\rho}$, here μ and ρ are the shear modulus and the mass density, respectively. Accordingly, β_1, μ_1, ρ_1 are corresponding parameters in the tunnel lining. Considering plane SH-waves incidence with angle α to the horizontal surface, the two dimensional anti-plane scattering problem needs to be solved. For simplicity, only the cylindrical tunnel is considered in this paper. Note that the tunnel of arbitrary shape can be treated by the IBEM.

3 Method and solution

The wave scattering problem can be solved by IBEM (Sanchez-Sesma and Campillo 1991) in the following way: based on the theory of single-layer potential, the scattered field can be formed by applying virtual uniform loads on the surface of the scatterers. Then the density of virtual loads can be solved by boundary conditions. The total wave field can be obtained by the superstition of free-wave field and scattered-wave field.

The displacement wave field $u^{(t)}$ in wedge-shaped space using the polar coordinate system $o{-}\theta$ satisfies the wave equation as follows:

$$\frac{\partial^2 u^{(t)}}{\partial r^2} + \frac{1}{r}\frac{\partial u^{(t)}}{\partial r} + \frac{1}{r^2}\frac{\partial^2 u^{(t)}}{\partial \theta^2} = \frac{1}{\beta^2}\frac{\partial^2 u^{(t)}}{\partial t^2} \tag{1}$$

The displacement at any location of the elastic solid, subject to a time harmonic excitation, can be written by the Somigliana integral representation:

$$cu(\xi) = \int_V G(y,\xi)f(y)\mathrm{d}V + \int_S [G(x,\xi)t(x) - T(x,\xi)u(x)]\mathrm{d}S \tag{2}$$

In which, $u(x)$ and $t(x)$ are the z-direction component of displacement and stress at arbitrary point x, respectively. $f(y)$ is the body force, $G(x,\xi)$ and $T(x,\xi)$ are the corresponding displacement and stress Green's functions, respectively.

$$u = \int_V G(x,\xi)f(\xi)\mathrm{d}V + \int_S G(x,\xi)\phi(\xi)\mathrm{d}S. \tag{3}$$

Similarly, the stresses can be derived according to Hooke's law:

$$t = \int_V T(x,\xi)f(\xi)\mathrm{d}V + \int_S T(x,\xi)\phi(\xi)\mathrm{d}S. \tag{4}$$

Equations (3) and (4) form the basis of the boundary element formulation. Numerical computation requires a discretization of the boundary surface.

3.1 Wave field construction

According to linear elastic theory, the total wave field can be decomposed into the free field and the scattered-wave field. Herein the free field is the solution of SH-waves in elastic wedge space without the lined tunnel. The scattered-wave field, based on the theory of single-layer potential, can be constructed by applying virtual uniform loads on the surface of lined tunnel and near ground surface. Then, the free field and the scattered-wave field add up to the total wave field:

$$u^{(t)} = u^{(f)} + u^{(s)} \tag{5}$$

3.1.1 *Free field*

Sanchez presented the expression for the displacement generated by plane SH-waves incidence in wedge space with wedge angle $v\pi$ (Sanchez-Sesma 1985). The resulting formulation is:

$$u^{(f)} = u_0 \frac{2}{v} \sum_{n=0}^{\infty} \varepsilon_n e^{-\frac{in\pi}{2v}} J_{n/v}(kr) \cos\frac{n\alpha}{v} \cos\frac{n\theta}{v} \tag{6}$$

where u_0 is the displacement amplitude, ε_n is the Neumann factor ($n = 0$, $\varepsilon_n = 1$; $n \geq 1$, $\varepsilon_n = 2$), α is the incident angle with the horizontal direction, and $J_{n/v}$ is the Bessel function of the first kind of order n/v.

In the polar coordinate system o–θ and the Cartesian coordinate system x–y, the stress function can be expressed by a simple derivation as follows:

$$\tau_{rz} = -\frac{2\mu u_0}{v} \sum_{n=0}^{\infty} \varepsilon_n e^{\frac{in\pi}{2v}} \left[kJ_{n/v+1}(kr) - \frac{n}{rv} J_{n/v}(kr)\right] \cos\frac{\theta n}{v} \cos\frac{n\alpha}{v} \tag{7}$$

$$\tau_{\theta z} = -\frac{2\mu u_0}{v} \sum_{n=0}^{\infty} \varepsilon_n e^{\frac{in\pi}{2v}} \frac{n}{vr} J_{n/v}(kr) \sin\frac{n\theta}{v} \cos\frac{n\alpha}{v}, \tag{8}$$

$$\tau_{yz} = \tau_{rz} \sin\theta + \tau_{\theta z} \cos\theta, \tag{9}$$

$$\tau_{xz} = \tau_{rz} \cos\theta - \tau_{\theta z} \sin\theta. \tag{10}$$

Note that the tangential stress by the free field on the boundary can be written as:

$$\tau_{nz}^{(f)} = \tau_{xz}\frac{\partial x}{\partial n} + \tau_{yz}\frac{\partial y}{\partial n}, \tag{11}$$

where $k = \omega/\beta$ is the wave number of the SH-waves in wedge–shaped space (β is the shear wave velocity), and the time factor $\exp(i\omega t)$ has been omitted.

3.1.2 *Scattered-wave field*

According to the previous discussion, the diffracted field is given by Eqs. (3) and (4), which, in the absence of body forces, can be written as:

$$u^{(s)}(x) = \int_S \phi(\xi)G(x,\xi)\mathrm{d}S_\xi, \tag{12}$$

$$t^{(s)}(x) = \int_S \phi(\xi)T(x,\xi)\mathrm{d}S_\xi. \tag{13}$$

ϕ_j denotes the densities of virtual uniform loads on the boundary. The anti-plane line source Green's function in full-space can be expressed in the following form:

$$G(x,\xi) = -\frac{\mathrm{i}}{4\mu}\mathrm{H}_0^{(2)}(kr), \tag{14}$$

$$T(x,\xi) = -\frac{\mathrm{i}}{4r}\mathrm{H}_1^{(2)}(kr)(\gamma_x n_x + \gamma_y n_y), \tag{15}$$

where $r = \sqrt{(x-x_0)^2 + (y-y_0)^2}$, $\gamma_x = (x - x_0)/r$, $\gamma_y = (y - y_0)/r$, (x, y) and (x_0, y_0) are the coordinate of the field point and source point, respectively. $\mathrm{H}_n^{(2)}(\bullet)$ is the Hankel function of the second kind of integer order n·(n_x, n_y) denotes unit normal vector on boundary surface.

3.2 Boundary conditions and solution

The boundary conditions of this problem include the zero-traction condition on the surface of wedge space and inner surface of the tunnel, the continuity of displacements and stresses on the interface between the tunnel and the wedge space.

The zero-traction conditions are expressed as follows:

$$\tau_{\theta z,\mathrm{I}}^{(t)} = \mu\frac{\partial u_\mathrm{I}^{(t)}}{r\partial r} = 0 \quad (\theta = \pi), \tag{16}$$

$$\tau^{(t)}_{\theta z,\mathrm{I}} = \mu\frac{\partial u^{(t)}_{\mathrm{I}}}{r\partial r} = 0 \quad (\theta = \pi + v\pi), \tag{17}$$

$$\tau^{(t)}_{r'z,\mathrm{II}} = \mu\frac{\partial u^{(t)}_{\mathrm{II}}}{\partial r'} = 0 \quad (on\ surface\ \Gamma_2). \tag{18}$$

The continuity conditions of displacements and stresses can be written as:

$$u^{(f)}_{\mathrm{I}} + u^{(s)}_{\mathrm{I}} = u^{(s)}_{\mathrm{II}} \quad (on\ surface\ \Gamma_1), \tag{19}$$

$$\tau^{(f)}_{\mathrm{I}} + \tau^{(s)}_{\mathrm{I}} = \tau^{(s)}_{\mathrm{II}} \quad (on\ surface\ \Gamma_1). \tag{20}$$

According to the free boundary and continuity condition, integral equations can be expressed as:

$$\int_S \phi^{\mathrm{I}}(\xi)T^{\mathrm{I}}(x,\xi)\mathrm{d}S_\xi = -t^{(f)}(x), \tag{21}$$

$$\int_S \phi^{\mathrm{II}}(\xi)T^{\mathrm{II}}(x,\xi)\mathrm{d}S_\xi = 0, \tag{22}$$

$$\int_S \phi^{\mathrm{I}}(\xi)G^{\mathrm{I}}(x,\xi)\mathrm{d}S_\xi - \int_S \phi^{\mathrm{II}}(\xi)G^{\mathrm{II}}(x,\xi)\mathrm{d}S_\xi = -u^{(f)}_{\mathrm{I}}, \tag{23}$$

$$\int_S \phi^{\mathrm{I}}(\xi)T^{\mathrm{I}}(x,\xi)\mathrm{d}S_\xi - \int_S \phi^{\mathrm{II}}(\xi)T^{\mathrm{II}}(x,\xi)\mathrm{d}S_\xi = -t^{(f)}. \tag{24}$$

It's a singular Fredholm integral equation of the second kind for the boundary sources. We need to discretize the inner and outer surface of tunnel and the nearby wedge surface, then apply virtual uniform loads on each element. Due to the attenuation characteristics of the scattered-wave, the computational accuracy can easily reach to 10^{-3} as the discretization range of the surface of wedge space reaches to 8 times the wavelength near the tunnel. $\phi_j(\xi)$ is assumed to be constant on each boundary element. Define the discretization numbers of the wedge surface, the outer and inner surface of tunnel are N_1, N_2, and N_3, respectively, then the linear Eqs. (21–24) can be rewritten as:

$$\sum_{l=1}^{N_1+N_2} \phi^{\mathrm{I}}(\xi_l)t^{\mathrm{I}}(x_{n_1},\xi_l) = -t^{(f)}(x_{n_1}), \quad n_1 = 1, N_1, \tag{25}$$

$$\sum_{l=1}^{N_2+N_3} \phi^{\mathrm{II}}(\xi_l)t^{\mathrm{II}}(x_{n_3},\xi_l) = 0, \quad n_3 = 1, N_3, \tag{26}$$

$$\sum_{l_1=1}^{N_1+N_2} \phi^{\mathrm{I}}(\xi_{l_1})g^{\mathrm{I}}(x_{n_2},\xi_{l_1}) - \sum_{l_2=1}^{N_2+N_3} \phi^{\mathrm{II}}(\xi_{l_2})g^{\mathrm{II}}(x_{n_2},\xi_{l_2}) = -u^{(f)}, \quad n_2 = 1, N_2, \tag{27}$$

$$\sum_{l_1=1}^{N_1+N_2} \phi^{\mathrm{I}}(\xi_{l_1})t^{\mathrm{I}}(x_{n_2},\xi_{l_1}) - \sum_{l_2=1}^{N_2+N_3} \phi^{\mathrm{II}}(\xi_{l_2})t^{\mathrm{II}}(x_{n_2},\xi_{l_2}) = -t^{(f)}, \quad n_2 = 1, N_2. \tag{28}$$

Namely, constructing matrix equation: $[H]\ [\phi] = [B]$. ϕ (densities of virtual loads) can be solved by the equation, $[\phi] = [H]^{-1}[B]$.

In which, dynamic influence functions can be expressed:

$$t(x_n,\xi_l) = \int_{\xi_l-\frac{\Delta S}{2}}^{\xi_l+\frac{\Delta S}{2}} T(x_n,\xi)\mathrm{d}S_\xi, \tag{29}$$

$$g(x_n,\xi_l) = \int_{\xi_l-\frac{\Delta S}{2}}^{\xi_l+\frac{\Delta S}{2}} G(x_n,\xi)\mathrm{d}S_\xi. \tag{30}$$

Equations (12) and (13) can be calculated directly by using two or three-point Gauss quadrature rules when $x \neq \xi$. Analytical expressions can be obtained by the series expansion of Green's functions when x is in the neighborhood of ξ. It can be expressed as:

$$t(x_n,\xi_n) = 0.5, \tag{31}$$

$$g(x_n,\xi_n) = \int_{-\Delta s/2}^{\Delta s/2} -\frac{\mathrm{i}}{4\mu}\mathrm{H}^{(2)}_0(ks)\mathrm{d}s = -\frac{i\Delta S}{4\mu}[1 + i\frac{2}{\pi}(1 - \gamma - \lg(\frac{k\Delta S}{4}))], \tag{32}$$

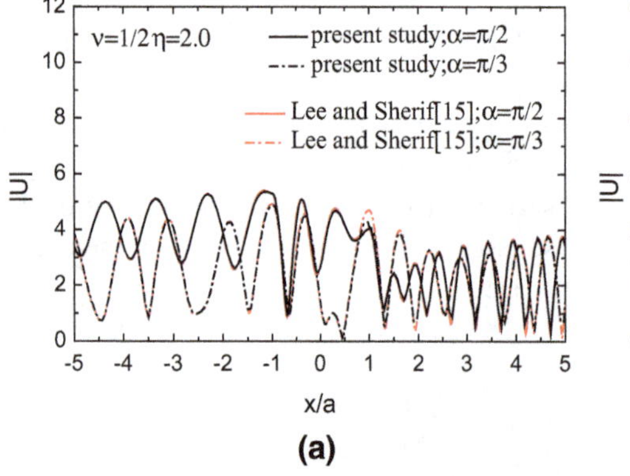

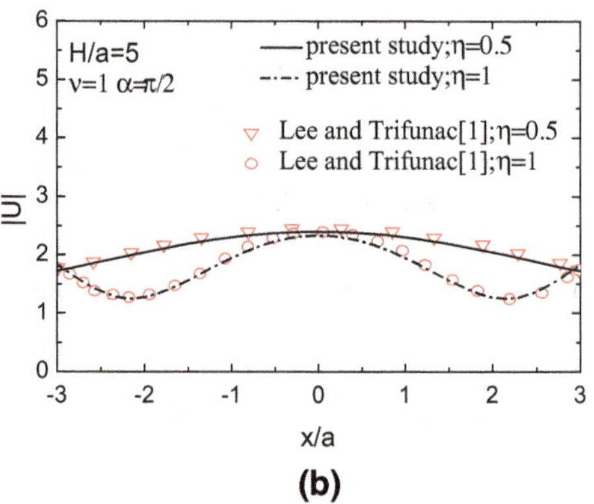

Fig. 2 Degenerated solutions compared with references (Lee and Trifunac 1979; Lee and Sherif 1996), **a** displacement amplitude around the surface of the canyon in wedge space, **b** displacement amplitude of the ground surface above the tunnel in half-space

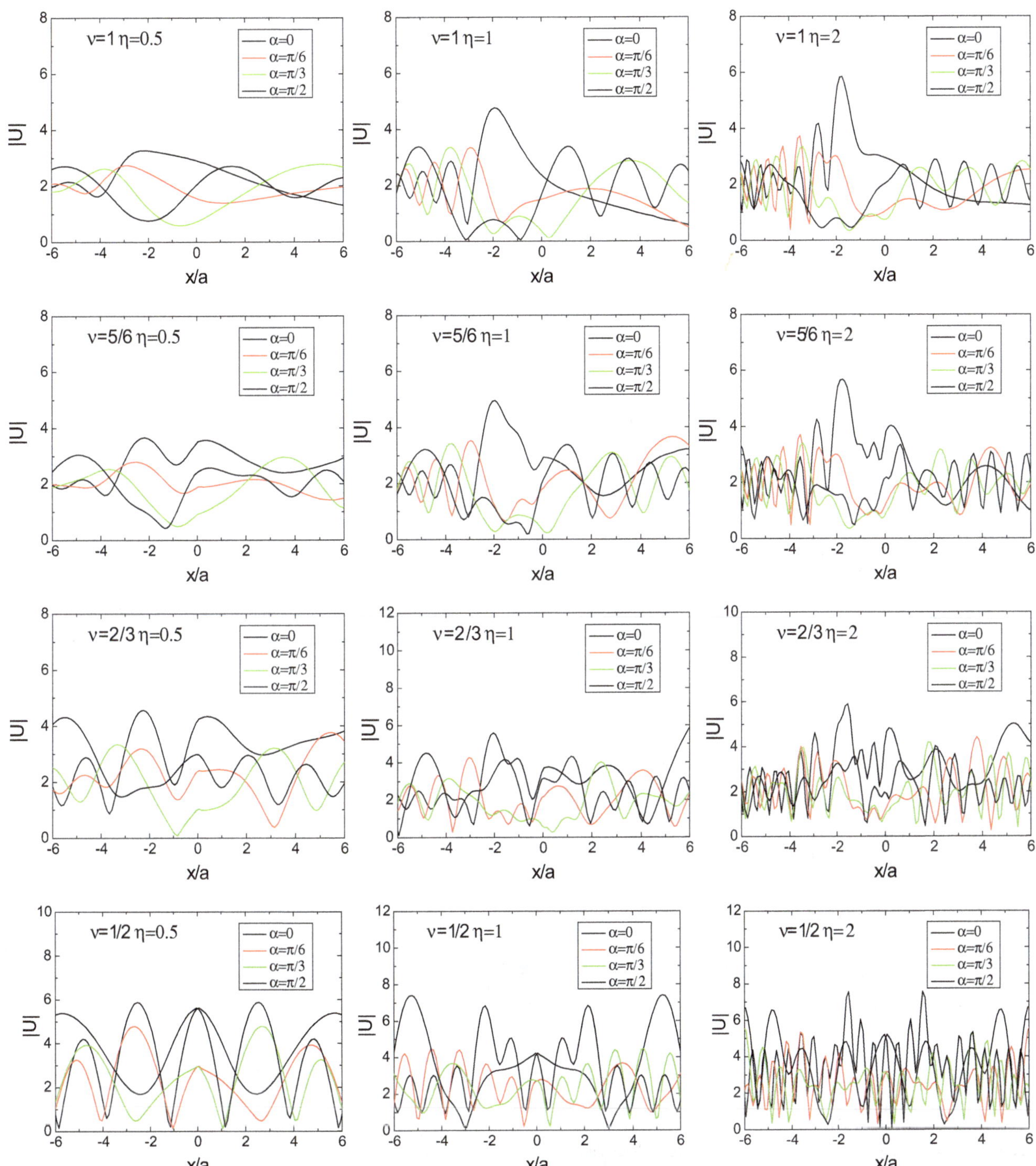

Fig. 3 Displacement amplitude around the surface of the tunnel in wedge space (v = 1/2, 2/3, 5/6, 1; rigid tunnel)

where γ is the Euler constant (0.5772), lg is the signs for logarithms, and ΔS is the length of element.

In summary, the introduction of Green's function for full-space leads to extra discretization of wedge space surface, but has an advantage of analytically treating the singular integration on each element, which is fairly beneficial for improving the calculation accuracy. Then, the density of virtual loads on each element can be solved through the Eqs. (25)–(28). The total wave field is obtained by the superposition of free field and scattered-wave field. Besides, above calculations are performed in frequency domain, and the time domain solution can be obtained by Fourier transform.

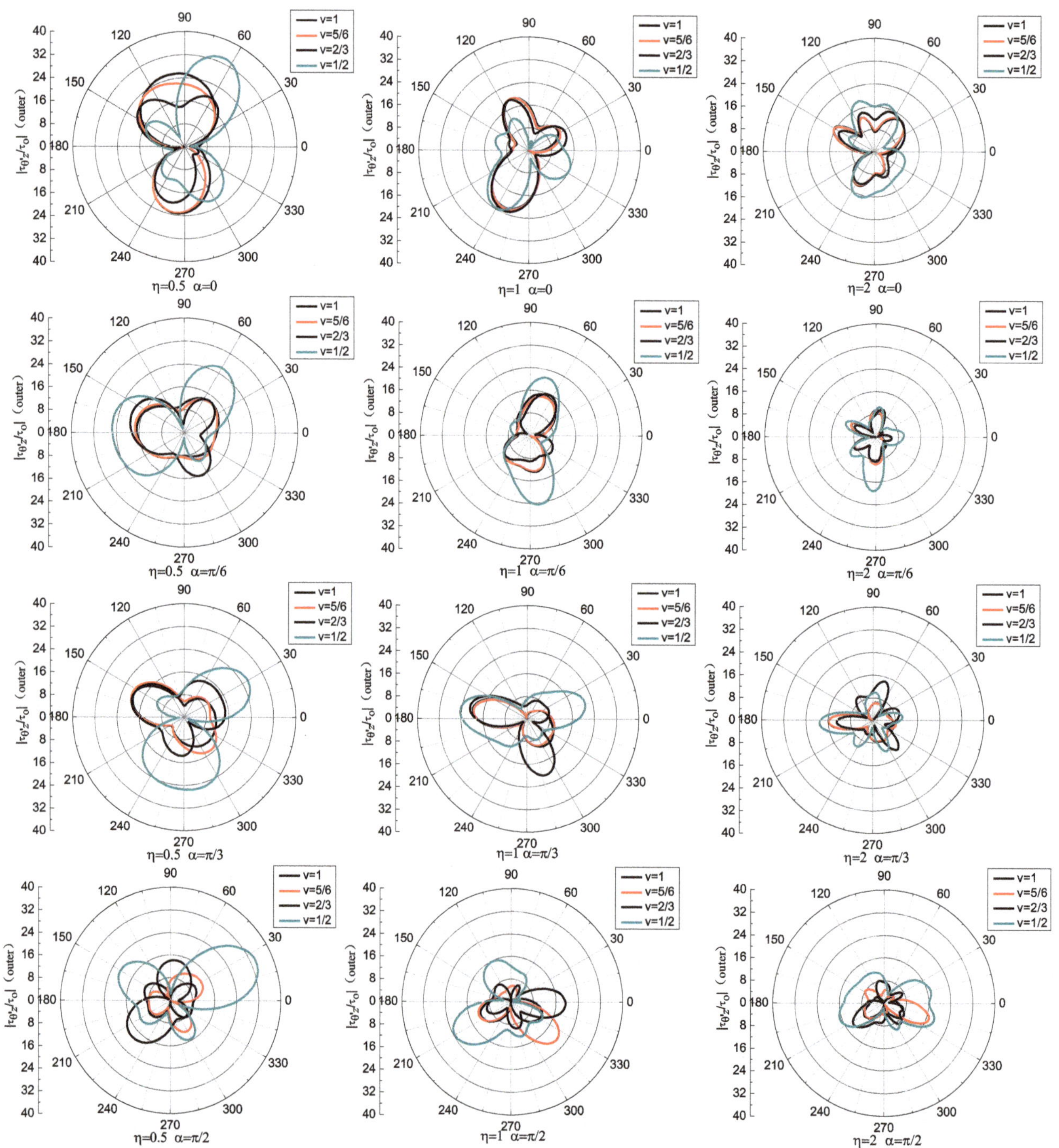

Fig. 4 Dynamic stress concentration factor at the tunnel surface in wedge space (the outer surface of the tunnel) ($\eta = 0.5, 1.0, 2.0$; rigid tunnel)

4 Accuracy verification

Firstly, define non-dimensional frequency as the ratio of the equivalent diameter of scatterer to the wavelength of the incident waves:

$$\eta = \frac{2a}{\lambda} = \frac{ka}{\pi} = \frac{\omega a}{\pi\beta}. \tag{33}$$

The degenerated solution in elastic wedge space can be calculated using this method. The result of displacement amplitudes by this method compared with the references (Lee and Sherif 1996) and (Lee and Trifunac 1979) are shown in Fig. 2. In Fig. 2(a), for the model of a canyon in wedge space, the following parameters are set: incident frequency $\eta = 2.0$ and wedge angle $v = 1/2$. In Fig. 2(b), for the model of a tunnel in half-space, $\rho_1/\rho = 1/3$, $\mu_1/$

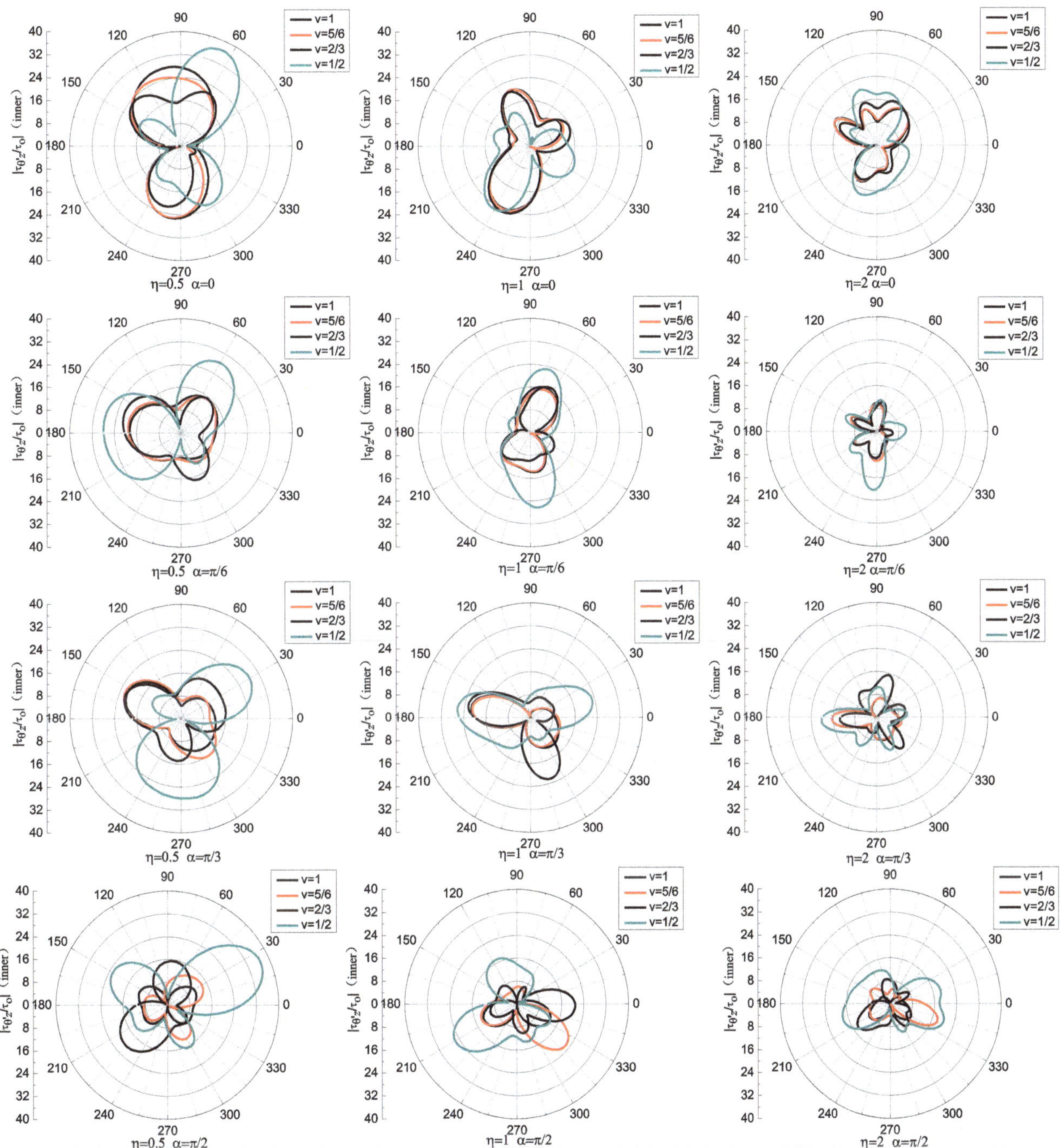

Fig. 5 Dynamic stress concentration factor at the tunnel surface in wedge space (the inner surface of the tunnel) (η = 0.5, 1.0, 2.0; rigid tunnel)

$\mu = 0.35$, $\eta = 0.5$, 1. It shows that our results by IBEM are in good agreement with the analytical method (Lee and Trifunac 1979). Note that, the scattering of elastic wave by tunnel of arbitrary shape in elastic wedge space can be solved by present method.

5 Numerical examples

In this part, detailed parameters analysis will be presented considering various wedge angles, the variety of tunnel stiffness, incident frequency and angle of incidence. For

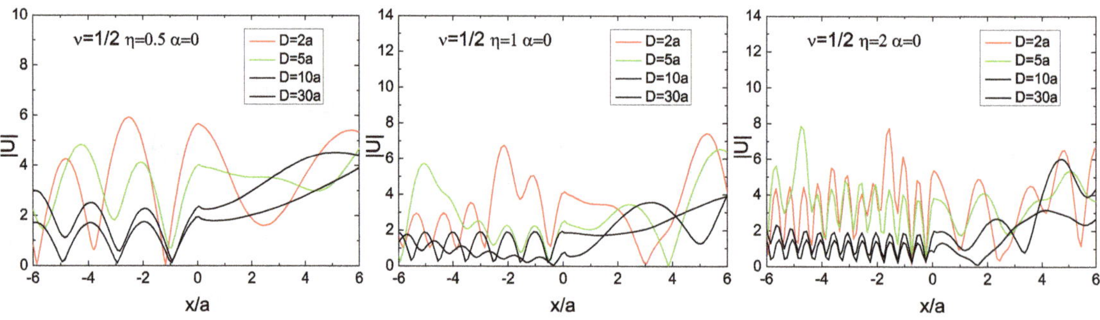

Fig. 6 The effect of the location of the tunnel in wedge space on the displacement response (D/a = 2.0, 5.0, 10, 30; rigid tunnel)

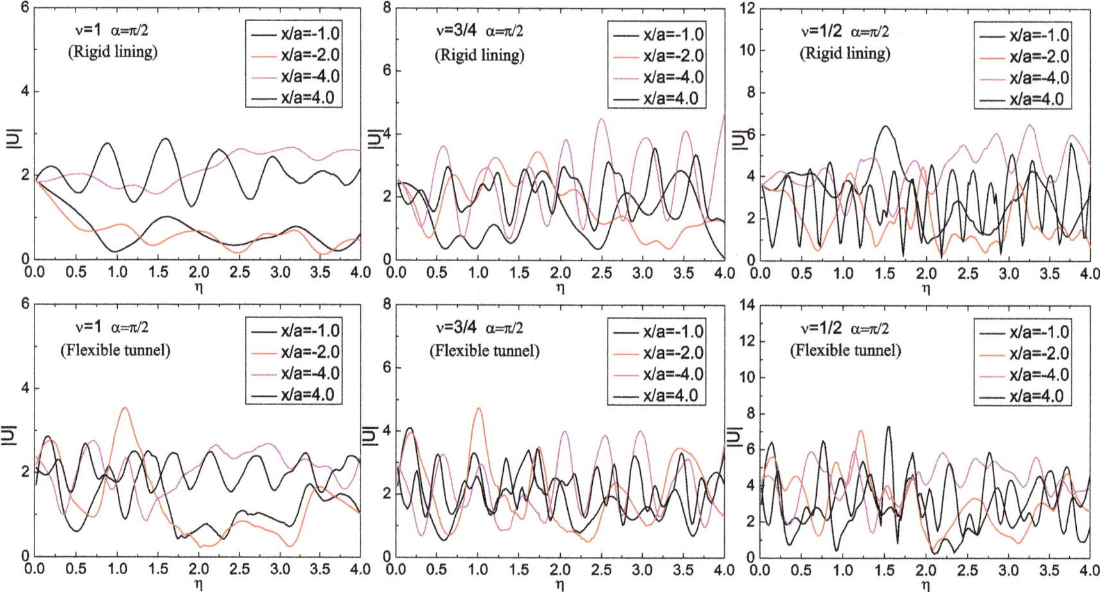

Fig. 7 Displacement amplitude spectrum around the surface of the tunnel in wedge space

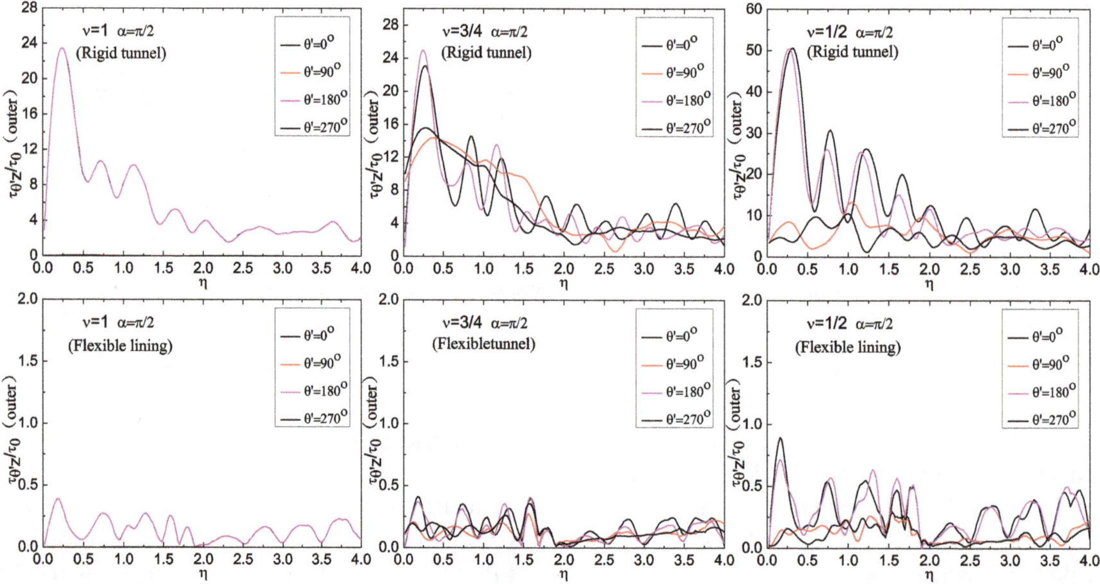

Fig. 8 Dynamic stress concentration factor spectrum of the tunnel in wedge space (the outer surface of the tunnel)

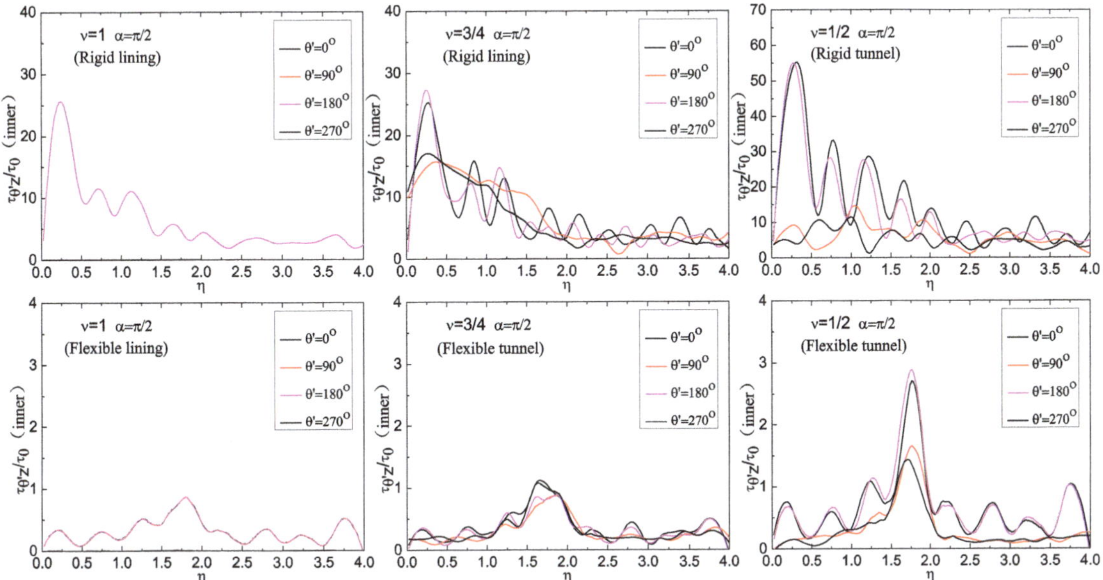

Fig. 9 Dynamic stress concentration factor spectrum of the tunnel in wedge space (the inner surface of the tunnel)

the rigid tunnel, the material parameters are: $\rho_1/\rho = 5/4$, $\beta_1/\beta = 5/1$. For the flexible tunnel, the material parameters are: $\rho_1/\rho = 4/5$, $\beta_1/\beta = 1/3$. The ratio of the inner and outer radius of tunnel is set to be $r_1/r_2 = 10/11$.

Figure 3 illustrates the surface displacement amplitudes in wedge space for different incident frequencies and angles. Set the angles of wedge to be 90°,120°,150°,180°, the incident frequencies $\eta = 0.5$, 1.0, 2.0, and the incident angles $\alpha = 0$, $\pi/6$, $\pi/3$, $\pi/2$, respectively. The location of the lined tunnel is $D = 2a$,$H = 2a$ (a is the inner radius of the tunnel). In the figure, the x-axis represents the ratio of the distance between the point on the ground surface and the vertex of the wedge to the inner radius of the tunnel (the negative axis represents the horizontal plane; the positive axis represents the inclined plane). The y-axis represents the displacement amplitude $|u^{(t)}|$. Obviously, there is large difference between the results of half-space ($v = 1$) and that of the wedge space. Due to the multiple scattering and interface effect of SH-waves between the surfaces of the wedge and the tunnel, the response characteristic is more complicated and the amplification effect is more significant. As the wedge angle decreases, the displacement amplitude increases significantly. For an example, for $v = 1/2$ (90° wedge space), the displacement amplitude can reach up to about 6.0 for $\eta = 0.5$ (which is about 5.0 in 120° wedge space, and about 3.7 in half-space). As the frequency increases, the displacement amplitude oscillates more quickly in space and the amplification effect appears to be more obvious, up to about 8.0 for $\eta = 2.0$ in 90° wedge space. Note that for horizontally incident waves, the max displacement response usually appears just above the tunnel.

Figures 4 and 5 illustrate the dynamic stress concentration factor (DSCF), defined as the ratio of the total shear stress to the stress of incident waves $\tau_{\theta' z}/\tau_0$, at the inner and outer surface of the tunnel for different incident frequencies. The calculation parameters are the same as Fig. 3. It can be seen that as the wedge angle decreases, the wave energy centralization becomes more significant and the peak value of DSCF increases gradually. There is large difference between the 90° wedge space ($v = 1/2$) and the half-space ($v = 1$). For an example, for $\eta = 0.5$ and vertical incident waves, the DSCF at the outer surface can reach up to about 34.0 for $v = 1/2$ (90° wedge space), but that is about 10.0 in half-space. Therefore, the phenomenon of the dynamic stress concentration near the tunnel in wedge space cannot be analyzed quantitatively using the model of a half-space. It can also be found that spatial characteristics of DSCF strongly depend on the incident frequency, and the stress oscillates more rapidly for high frequency. In addition, comparing Figs. 4 and 5, the peak of the DSCF at inner surface is a bit bigger than that at outer surface, but the spatial distribution characteristics are similar. From the application perspective, we should pay more attention to seismic design for the tunnel located at steep slope.

Figure 6 illustrates the effect of the location of the lined tunnel in wedge space on the displacement response, with incident frequency $\eta = 0.5$, wedge angle $v = 1/2$, incident angle $\alpha = 0$, and the buried depth of tunnel $H = 2a$. The horizontal distance from tunnel center to the vertex of wedge space takes: $D = 2a$, $5a$, $10a$ and $30a$, respectively. Obviously, when $D = 2a$, the wave interference effect between the tunnel and the wedge space is more

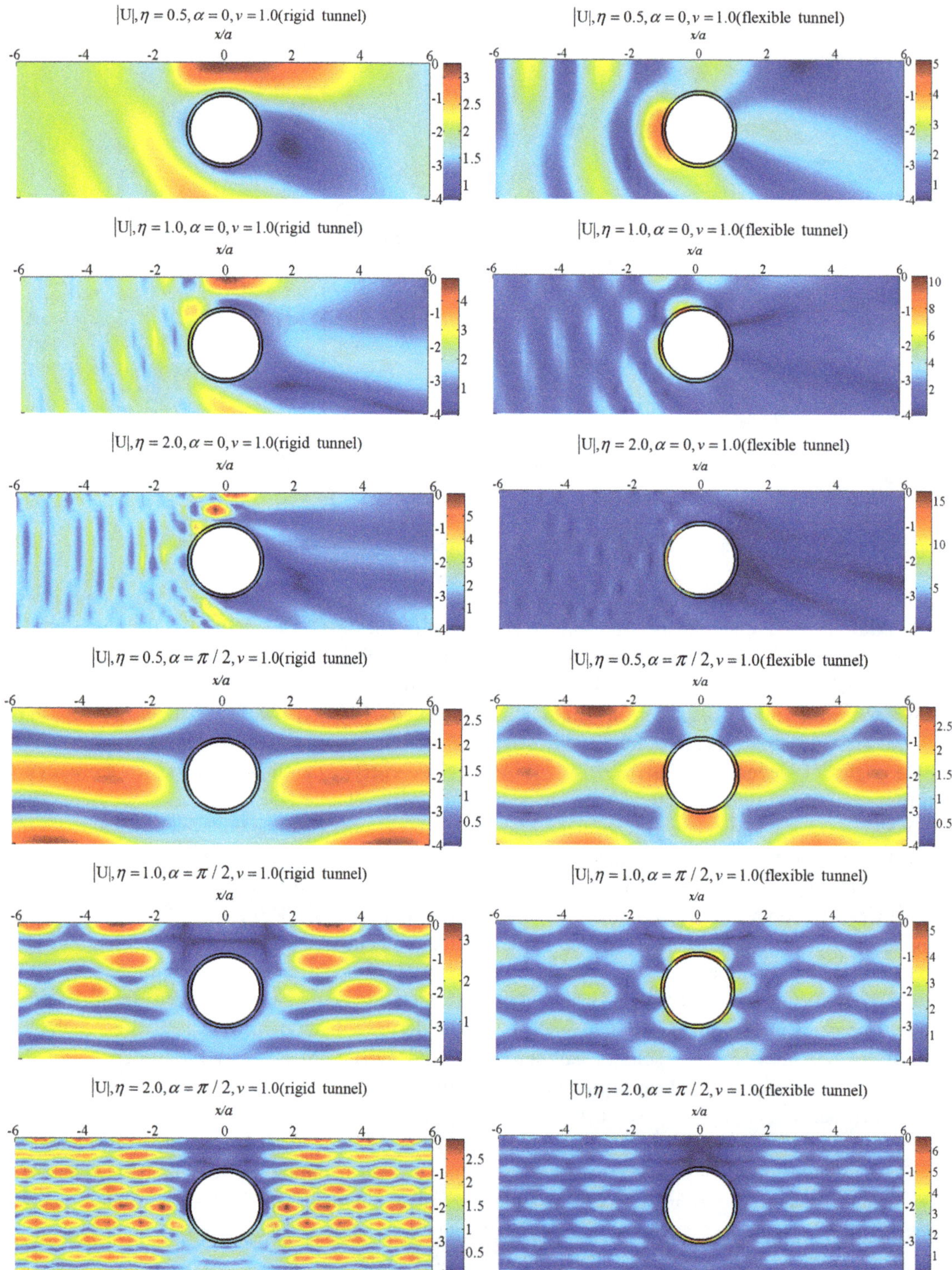

Fig. 10 Surface displacement amplitude around the tunnel in 180° wedge space (half-space) (η = 0.5, 1.0, 2.0)

significant. In general, as the distance D increases, the peak of displacement amplitude decrease gradually, which clearly indicates the great influence of vertical plane on wave scattering in a 90° wedge space.

Figures 7, 8 and 9 illustrate the displacement amplitude spectrum around the surface of the tunnel and the dynamic stress concentration factor (DSCF) spectrum on the surface of the tunnel to the incident SH-waves for different wedge angles. The incident frequency takes $\eta \in (0,4.0)$, and observation points for displacement amplitude are located at $x/a = -1, -2$ (right above the tunnel), -4, and 4 on ground surface, respectively; for DSCF, the points take

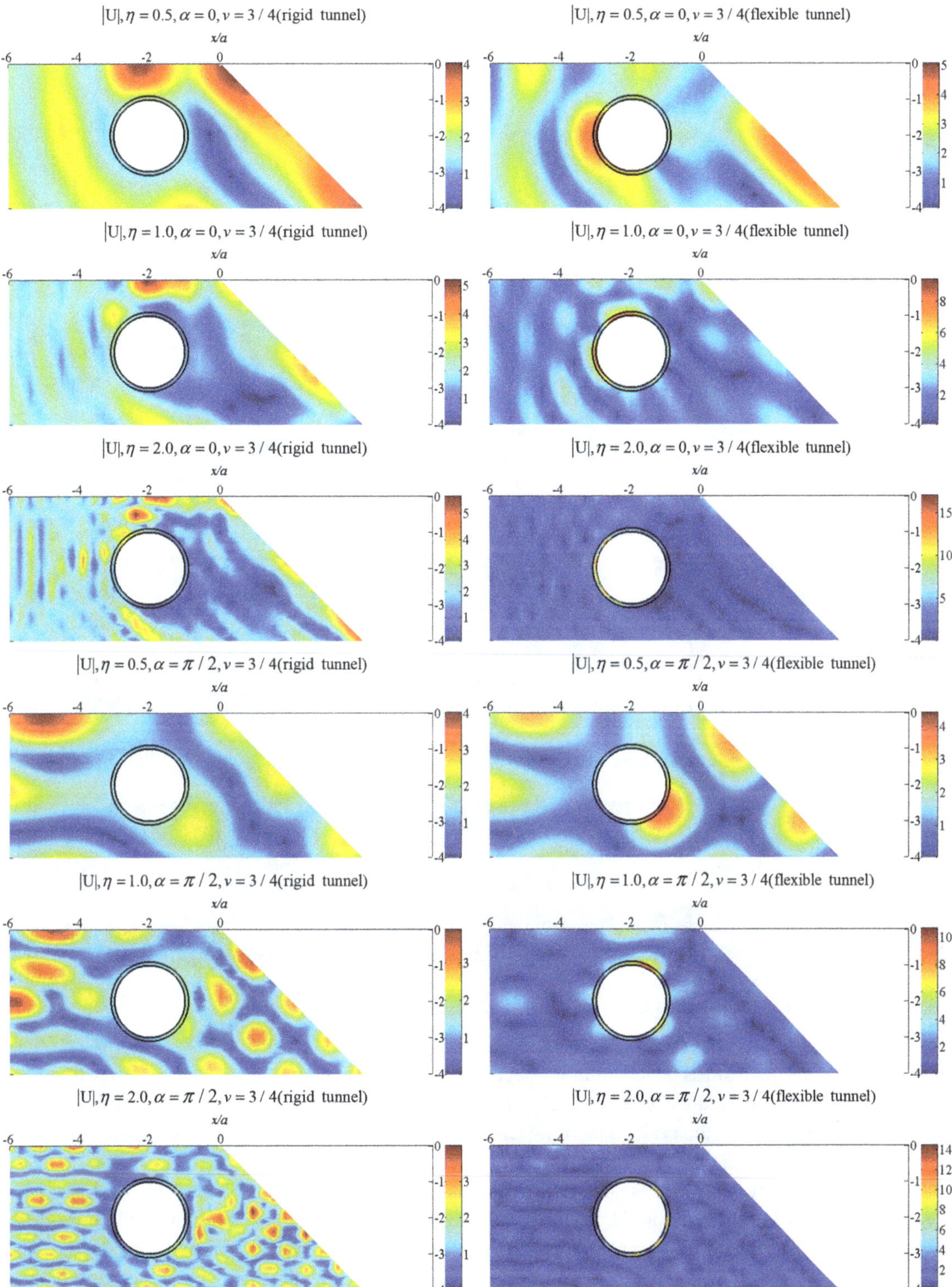

Fig. 11 Surface displacement amplitude around the tunnel in 135° wedge space (η = 0.5, 1.0, 2.0)

$\theta' = 0°$, 90°,180°, and 270° on the surface of the tunnel, respectively. It shows that as the wedge angle decreases, the peak value of the frequency spectrum increases, and the displacement amplitude can reach to 4.3 in the 90° wedge space but that is 1.8 in the half-space. In addition, compared with the half-space, the spectrum curve oscillates more rapidly in the wedge space due to the complex interference effect near the tunnel. As for DSCF spectrum

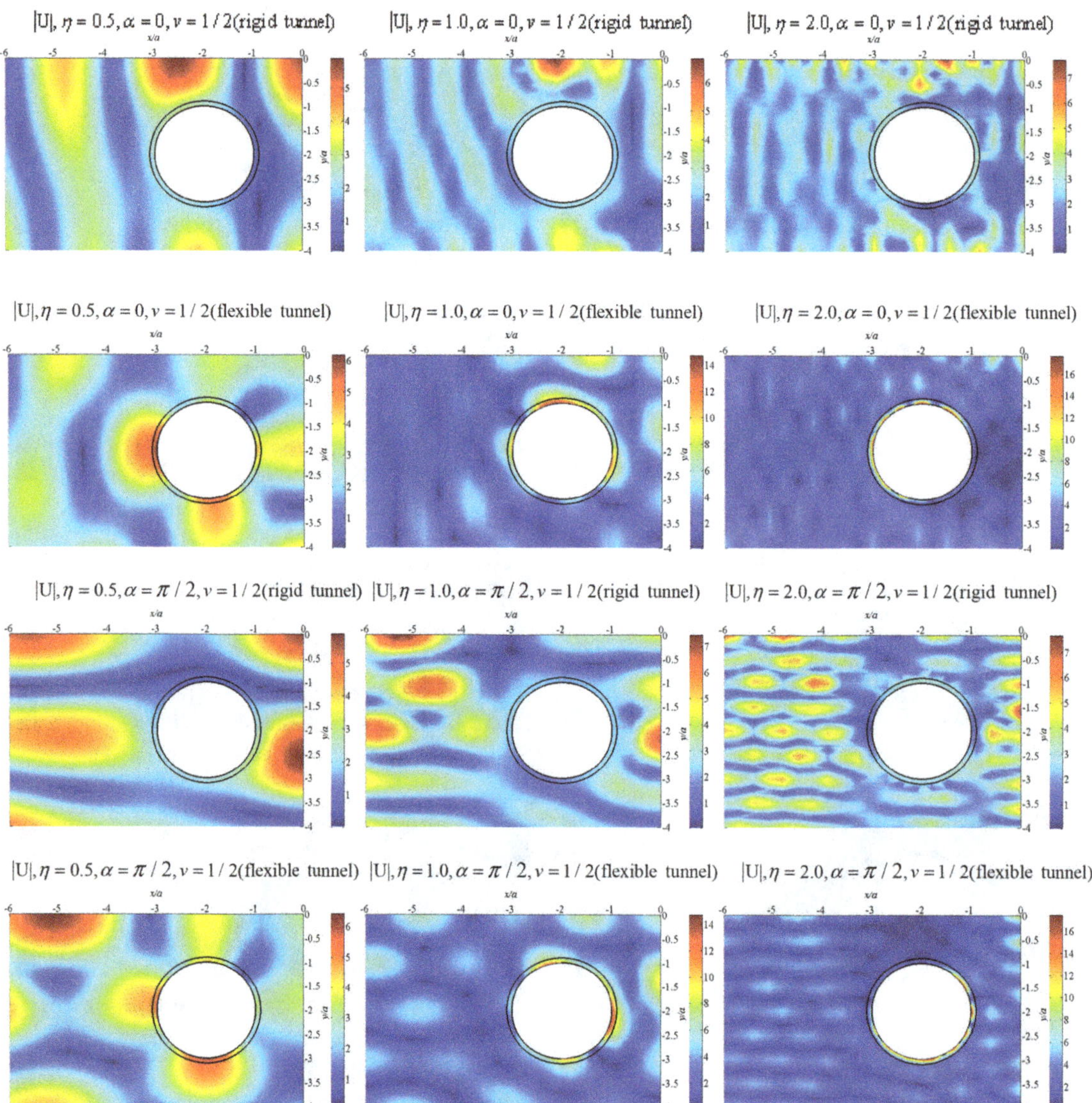

Fig. 12 Surface displacement amplitude around the tunnel in 90° wedge space (η = 0.5, 1.0, 2.0)

curve, it shows that the dynamic stress concentration effect seems more significant for low frequency waves, and the wedge angle has large influence on the peak value of DSCF spectrum. For an example, in the 90° wedge space, the peak of the stress spectrum inner the tunnel can reach to 55 ($\eta = 0.3$) at the point $\theta' = 0°$ (the point at the right of the tunnel). Accordingly, that are just 26 ($\eta = 0.25$) in half-space.

Considering different material properties of the lined tunnel, Figs. 7, 8, and 9 also show the displacement amplitude and DSCF spectrum of the flexible tunnel case. Except for the material parameters, other parameters remain the same as those of the rigid tunnel. Compared with the rigid tunnel, the displacement amplitude on wedge space surface near the tunnel shows more significant amplification effect, which can reach to 7.5 at $x/a = -2$ (just above the tunnel), but that is 4.3 for the rigid tunnel in 90° wedge space. Additionally, the DSCF spectrums show little amplification effect inside the tunnel for the flexible case.

Figures 10, 11, and 12 show the contour pictures of the displacement amplitude both around the rigid and flexible lined tunnel in wedge space. Consider the wedge angle v = 1, 3/4, 1/2, the incident angles α = 0, $\pi/2$, and the frequencies η = 0.5, 1.0, 2.0, respectively.

As for the rigid case, in wedge space, the displacement amplification effect becomes more significant, e.g., the displacement amplitude becomes 7.0 in 90° wedge space, and that is just 3.0 in half-space, 4.0 in 135° wedge space. It can be seen that as the frequency increases, the interference effect between the tunnel and the wedge space becomes more notable. There are more "focusing points"

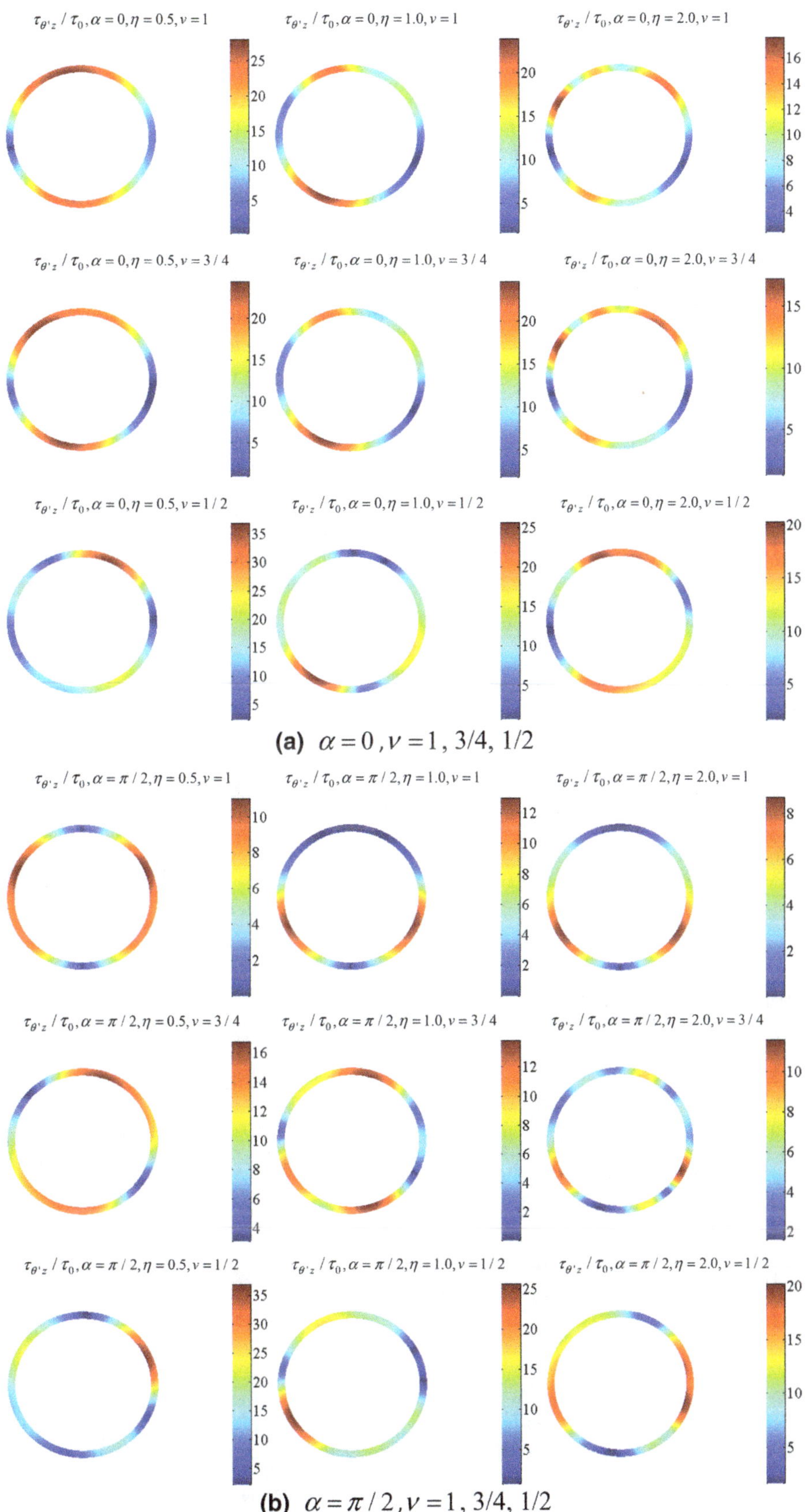

Fig. 13 Dynamic stress concentration factor of the tunnel in wedge space (η = 0.5, 1.0, 2.0; v = 1, 3/4, 1/2 rigid tunnel), **a** α = 0, v = 1, 3/4, 1/2, **b** α = $\pi/2$, v = 1, 3/4, 1/2

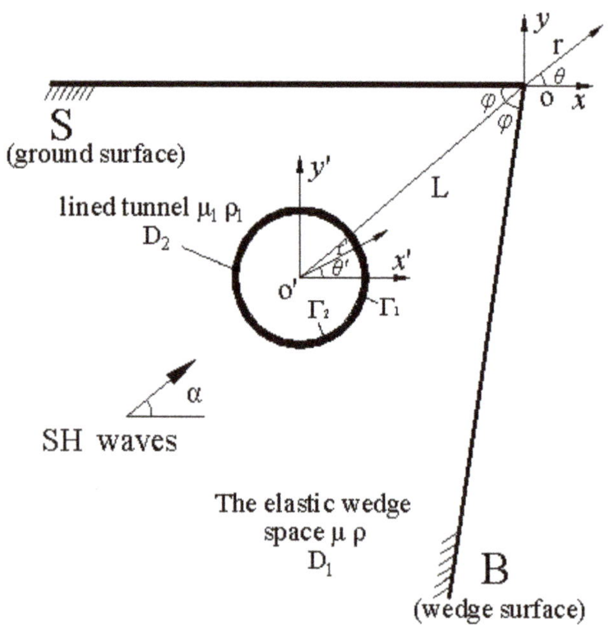

Fig. 14 Calculation model for a tunnel embedded in acute-angled wedge space

of wave energy for high incident frequencies. In addition, for the horizontal incident waves in half-space, due to the existence of the lined tunnel, the shielding effect on the displacement response can be seen clearly behind the tunnel, while this phenomenon is not so obvious for the 135° and 90° wedge space due to the existence of the inclined surface of wedge space.

As for the flexible tunnel, the displacement amplification effect inner the tunnel should be paid more attention. For example, when $\eta = 1.0$ and $\alpha = 0$ the peak of the displacement amplitude inner the flexible tunnel can reach up to 12, while that of the rigid tunnel is about 2.5. Besides, for the flexible tunnel, in the 90° wedge space, the peak of the displacement amplitude inner the tunnel can reach up to 17.0 for $\eta = 2.0$ and $\alpha = \pi/2$, while that is about 6.4 for the half-space case. In general, the peak of displacement appears in the side facing the incoming waves or most close to the wedge space surfaces. Additionally, as the incident frequency increases, the amplification effect seems more significant. Hence, the seismic resistant design of the

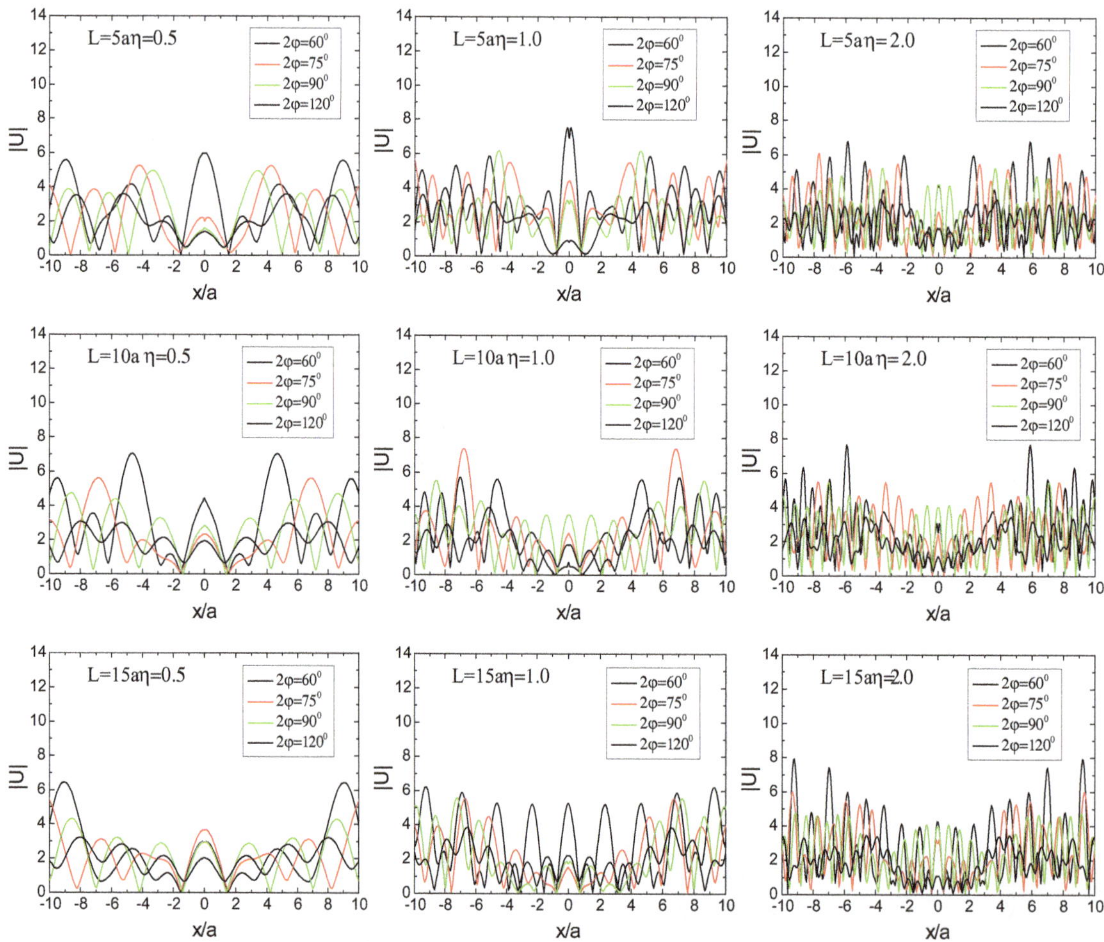

Fig. 15 Displacement amplitude around the surface of the tunnel in arbitrary-angled wedge space ($2\varphi = 60°, 75°, 90°, 120°$; $\eta = 0.5, 1.0, 2.0$)

flexible tunnel should try to control the large displacement response of the tunnels.

Figure 13 shows the contour picture of dynamic stress concentration factor (DSCF) of the rigid tunnel in wedge space. It can be seen that spatial characteristics of DSCF is more complicated for high frequency, but the peak value increases gradually. Besides, as the wedge angle decreases, the peak value of DSCF increases clearly, e.g., for $\eta = 2.0$, $\alpha = \pi/2$, the peak value can reach up to about 21.0 for $v = 1/2$ (90° wedge space), but that is about 8.6 in half-space. In general, the DSCF decreases from the inner surface to the outer surface gradually.

To study the seismic response of tunnel built in cliffy mountains, Fig. 14 shows a sharply angular wedge space. Define the apex angle of the wedge as 2φ, the distance between o and o' as L. oo' is the bisector of the wedge space, and the incident SH-wave comes along the bisector line oo'.

Figure 15 illustrates the surface displacement amplitudes both for acute and obtuse angle wedge space with different incident frequencies, apex angles and the location of the lined tunnel. Set the incident frequencies to be $\eta =$ 0.5, 1.0, 2.0, the apex angle of wedge $2\varphi = 60°$, 75°, 90°, 120°, the location of the lined tunnel $L = 5a$, $10a$, $15a$, the ratio of mass density and shear wave velocity $\rho_1/\rho = 5/4$, $\beta_1/\beta = 5/1$, respectively. Obviously, when the apex angle of wedge is an acute angle, the seismic response is more significant. For example, when $\eta = 0.5$, $L = 5a$, the displacement amplitude can reach up to about 6.0 for $2\varphi = 60°$, but that is about 1.4 for $2\varphi = 120°$ at the apex of wedge. Besides, as the distance L increases, the peak of displacement amplitude decreases gradually.

6 Conclusions

This paper presents an indirect boundary element method (IBEM) for the scattering of plane SH-waves by a lined tunnel in elastic wedge space based on the theory of single-layer potential. Compared with the exact analytical solution, the accuracy of this method has been verified. Through detailed parameter analysis, several important conclusions can be drawn as follows:

The reflection and diffraction of elastic waves are changed substantially by the inclined surface of wedge space. The wave scattering features become more complicated, and strongly depend on the wedge angle, the angle of incidence, the incident frequency, the location of lined tunnel and material parameters. As the wedge angle decreases, the displacement amplification and dynamic stress concentration effects around the tunnel become more significant. As for the rigid tunnel in 90° wedge space, compared with the half-space case, the peak values of the displacement amplitude on ground surface and the dynamic stress concentration factor inside the tunnel increase more than 100 %. As for the flexible tunnel in wedge space, the dynamic stress concentration effect is not so significant, but the displacement amplitude of the tunnel can reach up to 17 times that of the incident waves. Therefore, the seismic or anti-explosion design of the tunnel or pipes located in slope or hillside should adopt the wedge space model to improve the accuracy.

It is worth mentioning that present method is applicable to the tunnel of arbitrary shape and in arbitrary-angle wedge space. In addition, the solution technique can also be expanded to solve the scattering of P and SV waves in elastic wedge space. The specific solution procedure will be given in another paper.

Acknowledgments The authors gratefully acknowledge the support from National Natural Science Foundation of China under Grants (51278327) and the Tianjin Research Program of Application Foundation and Advanced Technology (14JCYBJC21900).

References

Achenbach JD (1970) Shear waves in an elastic wedge. Int J Solids Struct 4:379–388

Budaev BV, Bogy DB (1995) Rayleigh wave scattering by a wedge. Wave Motion 22:239–257

Chen JT, Chou KH, Lee YT (2011) A novel method for solving the displacement and stress fields of an infinite domain with circular holes and/or inclusions subject to a screw dislocation. Acta Mech 218:115–132

Datta SK, Shah AH, Wong KC (1984) Dynamic stresses and displacements in buried pipe. J Eng Mech 110:1451–1466

Du X, Xiong J, Guan H (1993) The boundary integral equation method solution to the scattering of plane SH waves. Acta Seismol Sin 15(3):331–338

Gautesen AK (2002) Scattering of a Rayleigh wave by an elastic quarter space–revisited. Wave Motion 35:91–98

Knopoff L (1969) Elastic wave propagation in a wedge. In: Miklowitz J (ed) Wave propagation in solids. ASME, New York, pp 3–42

Lee VW, Sherif RI (1996) Diffraction around circular canyon in elastic wedge space by plane SH-waves. J Eng Mech 122:539–544

Lee VW, Trifunac MD (1979) Response of tunnels to incident SH-waves. J Eng Mech, ASCE 105:643–659

Li Z, Gong P (1998) Reflection and transmission of obliquely incident rayleigh surface waves by a Visco—elastic interphase. Acta Mech Solid Sin 11:229–240

Li W, Shen Q, Zhao C (2009) Scattering of plane SV waves by a underground circular lined tunnel in half space saturated soil. J Disaster Prev Mitig Eng 29(2):172–178 (In Chinese with English abstract)

Liang J, Luo H, Lee VW (2010) Diffraction of plane SH waves by a semi-circular cavity in half-space. Earthq Sci 23:5–12

Liang J, Chen J, Ba Z (2013) 3D scattering of obliquely incident SH waves by a cylindrical cavity in layered elastic half-space (II): numerical results and analysis. Acta Seismol Sin 35:173–183 (in Chinese with English abstract)

Liu G, Chen H, Li D, Ji B (2009) Antiplane harmonic elastodynamic stress analysis of an infinite wedge with a circular cavity. J Appl Mech 76:061008

Sanchez-Sesma FJ (1985) Diffraction of elastic SH-waves by wedges. Bull Seismol Soc Am 75:1435–1446

Sanchez-Sesma FJ, Campillo M (1991) Diffraction of P, SV, and Rayleigh waves by topographic features: a boundary integral formulation. Bull Seismol Soc Am 81:2234–2253

Shi W, Liu D, Song Y, Chu J, Hu A (2006) Scattering of circular cavity in right-angular planar space to steady SH-wave. Appl Math Mech 27:1417–1423

Shi W, Liu D, Chu J, Gong H, Guo S (2007) Scattering of fixed circular inclusion in right-angled plane to steady incident planar shearing horizontal wave. Explos Shock Waves 27:57–62

Stamos AA, Beskos DE (1996) 3-D seismic response analysis of long lined tunnels in half-space. Soil Dyn Earthq Eng 16:111–118

Yang G, Liu Z (1994) A finite element-artificial transmitting boundary method for analyzing the earthquake motions in the underground tunnel structures. Eng Mech 11:122–130

Zhang GC, Qi H, LIU PA (2013) Scattering of sh-wave by circular cavity in right-angle plane and seismic ground motion. Mech Eng 35(1):60–66

2

Synthetic seismograms for finite sources in spherically symmetric Earth using normal-mode summation

Tianshi Liu · Haiming Zhang

Abstract Normal-mode summation is the most rapidly used method in calculating synthetic seismograms. However, normal-mode summation is mostly applied to point sources. For earthquakes triggered by faults extending for as long as several 100 km, the seismic waves are usually simulated by point source summation. In this paper, we attempt to follow a different route, i.e., directly calculate the excitation of each mode, and use normal-mode summation to obtain the seismogram. Furthermore, we assume the finite source to be a "line source" and numerically calculate the transverse component of synthetic seismograms for vertical strike-slip faults. Finally, we analyze the features in the Love waves excited by finite faults.

Keywords Normal-mode summation · Synthetic seismogram · Finite fault · Surface waves

1 Introduction

The calculation of synthetic seismograms is one of the most important topics in seismology, because on the one hand, synthetic seismogram is the bridge that connects the theory of seismology and the observational data, and on the other hand, it is crucial to structure inversion and rupture process inversion. Roughly speaking there are three types of methods to calculate synthetic seismograms. The first type is numerical method, e.g., the finite difference method (Boore 1972), the finite element method (Bielak et al. 2003) and the spectral element method (Komatitsch and Tromp 1999, 2002). Numerical methods are usually quite flexible. They are available for very complex structures. But the computational costs are usually high, and the compromise between accuracy and efficiency is often inevitable. The second type is asymptotic method, e.g., the generalized ray method (Gilbert and Helmberger 1972) and the WKBJ method (Chapman 1978). The asymptotic methods typically have high efficiency, but only effective for computing the high-frequency component. The third type is semi-analytical method, e.g., the discrete wavenumber method (Bouchon and Aki 1977) and the R/T coefficient method (Luco and Apsel 1983; Kennett and Kerry 1979) for stratified half-space, and the normal-mode summation method (Dahlen and Tromp 1998) for spherically symmetric Earth model.

Normal-mode summation is one of the most widely used methods in simulating teleseismic waves. After the 1960 Chile earthquake, seismologists did a great amount of studies on normal modes and gradually developed the normal-mode summation method. Gilbert (1971) explicitly showed the normal-mode summation representation of the displacement in elastic media. Singh and Ben-Menahem (1969a, b) calculated the excitation of each mode by a point source in spherically symmetric Earth. Takeuchi and Saito (1972) derived the equations that govern the normal modes. Tanimoto (1984) obtained the formulae to calculate long-period synthetic seismograms using normal-mode summation. Woodhouse (1988) developed the numerical method to calculate the radial eigenfunction, which made it possible to calculate synthetic seismograms using normal-mode summation. Dahlen and Tromp (1998) integrated the previous work and constructed a comprehensive and

T. Liu · H. Zhang (✉)
Department of Geophysics, School of Earth and Space Sciences, Peking University, Beijing 100871, People's Republic of China
e-mail: zhanghm@pku.edu.cn

T. Liu
e-mail: dash2007@163.com

compact framework for theoretical global seismology, which includes the normal-mode theory, normal-mode summation method and other related topics.

However, normal-mode summation is mostly used for point sources. In order to calculate the synthetic seismogram for a finite source, point source summation is usually used (Bouchon 1980a, b; Song and Helmberger 1996): the fault is discretized into small sub-faults which can be treated as point sources; the seismogram for each sub-fault is calculated using normal-mode summation and then the seismogram for the finite fault is obtained by adding the seismograms of the sub-faults together. Analytical or semi-analytical methods are sometimes applied to calculating the synthetic seismograms of finite faults (Ben-Menahem and Singh 1987; Israel and Kovach 1977; Saikia and Helmberger 1997; Stump and Johnson 1982).

In this work, we expand the normal-mode summation method to the finite fault case. We first derive the excitation of each mode and then add them together to obtain the seismogram. We assume that the fault is a "line source": it expand transversely but concentrate at a certain depth radially, which take into account the effect of propagation in the transverse direction, but ignore that of the propagation in the radial direction. We represent the normal mode in the form of generalized spherical harmonics (Phinney and Burridge 1973; Yang et al. 2010). Next, we use the radial eigenfunction of MINEOS (Woodhouse 1988) and calculate the numerically results of the transverse component in the case of the vertical strike-slip fault as example. We then observe some features in the Love wave of the finite source, and we use an intuitive model to explain these features based on Chap.10 of Aki and Richards (2002).

2 Normal-mode summation for finite source

2.1 From point source to finite source

Suppose that the displacement at $\boldsymbol{x}$ excited by a point source at $\boldsymbol{x}'$ can be written in the form of normal-mode summation as

$$\boldsymbol{g}(\boldsymbol{x},t;\boldsymbol{x}') = \sum_k A_k(\boldsymbol{x}',t)\boldsymbol{s}_k(\boldsymbol{x}), \tag{1}$$

in which $\boldsymbol{s}_k$ is the eigenfunction with index k, and A_k represents the excitation of mode with index k by the point source at $\boldsymbol{x}'$. Note that here the source does not have to be a body force, so $\boldsymbol{g}$ is not necessarily the Green's function. For a finite source with a spatial and temporal amplitude distribution $D(x',\tau)$, the displacement at $\boldsymbol{x}$ is

$$\begin{aligned}\boldsymbol{u}(\boldsymbol{x},t) &= \int_{\Sigma_f}\int_0^{T_f} \boldsymbol{g}(\boldsymbol{x},t-\tau;\boldsymbol{x}')D(\boldsymbol{x}',\tau)\mathrm{d}\tau\mathrm{d}S \\ &= \sum_k\left(\int_{\Sigma_f}\int_0^{T_f} A_k(\boldsymbol{x}',t-\tau)D(\boldsymbol{x}',\tau)\mathrm{d}\tau\mathrm{d}S\right)\boldsymbol{s}_k(\boldsymbol{x}),\end{aligned} \tag{2}$$

where Σ_f is the fault plane, T_f is the total rupture time. Thus, the seismogram of the finite fault can also be written in the normal-mode summation form

$$\boldsymbol{u}(\boldsymbol{x},t) = \sum_k \mathscr{A}_k(t)\boldsymbol{s}_k(\boldsymbol{x}), \tag{3}$$

where the excitation of mode with index k is

$$\mathscr{A}_k(t) = \int_{\Sigma_f}\int_0^{T_f} A_k(\boldsymbol{x}',t-\tau)D(\boldsymbol{x}',\tau)\mathrm{d}\tau\mathrm{d}S. \tag{4}$$

The excitation of mode with index k by a double-couple $\boldsymbol{M}$ is well known (Dahlen and Tromp 1998; Yang et al. 2010),

$$A_k(\boldsymbol{x}',t) = \left(\boldsymbol{M}:\boldsymbol{\varepsilon}_k^*(\boldsymbol{x}')\right)\frac{1-\mathrm{e}^{-\sigma_k t}\cos\omega_k t}{\omega_k^2}. \tag{5}$$

Substitute Eq. (5) in Eq. (4), and let

$$\dot{\boldsymbol{m}}(\boldsymbol{x}',\tau) = \boldsymbol{M}D(\boldsymbol{x}',\tau), \tag{6}$$

we obtain

$$\begin{aligned}\mathscr{A}_k(t) = \int_{\Sigma_f}\int_0^{T_f} &\left(\dot{\boldsymbol{m}}(\boldsymbol{x}',\tau):\boldsymbol{\varepsilon}_k^*(\boldsymbol{x}')\right) \\ &\times\frac{1-\mathrm{e}^{-\sigma_k(t-\tau)}\cos\omega_k(t-\tau)}{\omega_k^2}\mathrm{d}\tau\mathrm{d}S.\end{aligned} \tag{7}$$

2.2 GSH representation for spherically symmetric Earth

Any tensor can be decomposed in generalized spherical coordinate, and the components can be represented using generalized spherical harmonics (GSH). We assume that the Earth structure is spherically uniform, and we can write the normal modes using GSH. Then, we use Eqs. (7) and (2) to obtain the final solution of the displacement excited by finite source.

The normal mode of the spherical symmetric Earth is (Dahlen and Tromp 1998)

$$\boldsymbol{s}_k(r,\theta,\phi) = \sum_{\alpha=0,\pm1} s_{nl}^{\alpha}(r)Y_{lm}^{\alpha}(\theta,\phi)\hat{\boldsymbol{e}}_{\alpha}, \tag{8}$$

where Y_{lm}^{α} is the generalized scalar harmonics, $\hat{\boldsymbol{e}}_{\alpha}$ is the base of generalized spherical coordinate. For a spheroidal or a toroidal mode, the index k is the same as n, l, m.

$$s_{nl}^{\pm 1} = \frac{1}{\sqrt{2}} V_{nl}, \qquad s_{nl}^{0} = U_{nl} \tag{9}$$

for spheroidal mode, and

$$s_{nl}^{\pm 1} = \pm \frac{1}{\sqrt{2}} W_{nl}, \qquad s_{nl}^{0} = 0 \tag{10}$$

for toroidal mode, where U_{nl}, V_{nl} and W_{nl} are radial eigenfunctions.

Substituting Eq. (8) into Eqs. (7) and (2), we can derive the representation of seismogram using GSH. But before this, we make some simplifications. First, without loss of generality, we can put our spherical coordinate such that the line connecting the pole and the origin is perpendicular to the fault plane (see Fig. 1). In such coordinate, the coordinate of the receiver is $\{r, \theta, \phi\}$, the coordinate of some point on the fault plane is $\{r', \theta', \varphi\}$. We set up a local coordinate $\{\hat{\xi}, \hat{\boldsymbol{\phi}}, \hat{\boldsymbol{v}}\}$ on the fault plane. Using this coordinate system,

$$\mathrm{d}S = r' \sin\theta' \mathrm{d}\xi \mathrm{d}\varphi. \tag{11}$$

Next, we assume that the fault only expands transversely, but concentrates at a certain depth (line source). Unlike the point source approximation, which concentrates the seismic moment at a single point, the line source approximation assumes that the seismic moment concentrates on a line. It is a more general assumption than the point source approximation, and it captures the effect of the lateral propagation of the rupture, but it is still far from enough to capture the full finite fault feature of the earthquake. The line source approximation is valid because the impact of the transverse propagation of rupture is far more significant compared with the radial propagation. Thus, the integral on the fault can be simplified as

$$\mathrm{d}S = r_f w \sin\theta_d \mathrm{d}\varphi, \tag{12}$$

where w is the width of the fault, θ_d is the dip-angle of the fault plane, and r_f is the distance between the fault and the origin.

Using the aforementioned spherical coordinate and the line source approximation, we obtain the GSH representation of displacement triggered by the finite fault. We use ${}^{K}_{\gamma}u^{q}$ to represent the displacement in q direction ($q = Z, R, T$ for each direction of the ZRT coordinate), related to K type mode ($K = S, T$ for spheroidal and toroidal mode) triggered by γ component of slip ($\gamma = \varphi, \xi$ for strike-slip and dip-slip component). For simplicity, we show only the result of displacement that related to toroidal mode and triggered by strike-slip component

$$\begin{aligned}
{}^{T}_{\varphi}u^{Z}_{nl} &= 0, \\
{}^{T}_{\varphi}u^{R}_{nl} &= \sum_{n,l} -2\sin\theta_d W_{nl}(r)\, {}^{T}\mathscr{S}^{2}_{nl} \Re\left\{ \mathrm{e}^{\mathrm{i}\phi_b} \frac{{}^{T}_{\varphi}\mathscr{K}^{-2,+1}_{nl} + {}^{T}_{\varphi}\mathscr{K}^{+2,+1}_{nl}}{2} \right\} \\
&\quad + \cos\theta_d W_{nl}(r)\, {}^{T}\mathscr{S}^{1}_{nl} \Re\left\{ \mathrm{e}^{\mathrm{i}\phi_b} \frac{{}^{T}_{\varphi}\mathscr{K}^{-1,+1}_{nl} - {}^{T}_{\varphi}\mathscr{K}^{+1,+1}_{nl}}{2} \right\}, \\
{}^{T}_{\varphi}u^{T}_{nl} &= \sum_{n,l} -2\sin\theta_d W_{nl}(r)\, {}^{T}\mathscr{S}^{2}_{nl} \Im\left\{ \mathrm{e}^{\mathrm{i}\phi_b} \frac{{}^{T}_{\varphi}\mathscr{K}^{-2,+1}_{nl} + {}^{T}_{\varphi}\mathscr{K}^{+2,+1}_{nl}}{2} \right\} \\
&\quad + \cos 2\theta_d W_{nl}(r)\, {}^{T}\mathscr{S}^{1}_{nl} \Im\left\{ \mathrm{e}^{\mathrm{i}\phi_b} \frac{{}^{T}_{\varphi}\mathscr{K}^{-1,+1}_{nl} - {}^{T}_{\varphi}\mathscr{K}^{+1,+1}_{nl}}{2} \right\},
\end{aligned} \tag{13}$$

where ${}^{T}\mathscr{S}^{1}_{nl}$, ${}^{T}\mathscr{S}^{2}_{nl}$ and ${}^{K}_{\gamma}\mathscr{K}^{\alpha,\beta}_{nl}$ are given in "Appendix." ϕ_b is the back-azimuth angle (see Fig. 1). Other parts of displacement are shown in "Appendix." Note that as the length of the fault goes to zero, the formulae for the displacements due to a finite source become exactly the same as the point source case (Yang et al. 2010).

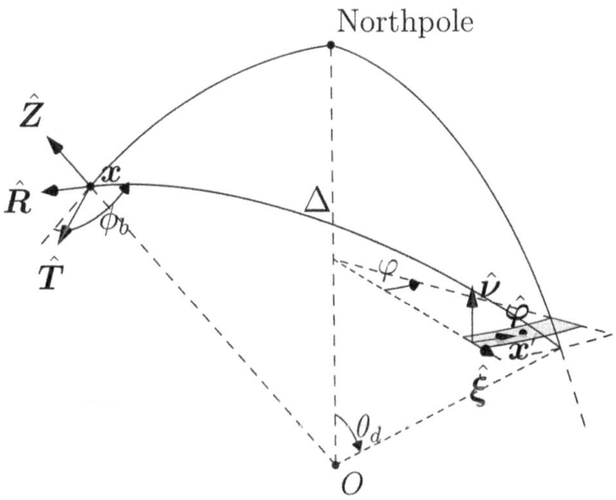

Fig. 1 Geometry of the receiver and the finite source. The global spherical coordinate is selected such that the line connecting the origin and the "Northpole" is perpendicular to the fault plane. The local coordinate at the receiver ($\boldsymbol{x}$) is the "ZRT" coordinate. The basis of local coordinate at the receiver is $\{\hat{\xi}, \hat{\boldsymbol{\phi}}, \hat{\boldsymbol{v}}\}$. The reference point of the source is $\boldsymbol{x}'$, which can be an arbitrary point near the fault (e.g., epicenter). The angular distance between the reference point and the receiver is defined as the epicentral distance Δ. The back-azimuth is ϕ_b

3 Numerical results

To show the impact of the transverse propagation of rupture on seismograms, we continue to make following simplifications:

(i) The fault is vertical and the slip only has strike-slip component, i.e., $\theta_d = \frac{\pi}{2}$,

(ii) The rupture velocity is a pulse propagating along the fault with a constant velocity, which is

Table 1 Parameters in the numerical experiment

Source depth	15 km
Seismic moment	1.06×10^{27} dyne cm
Earth structure	Continental PREM model
Normal mode	$n \leq 300$, $l \leq 2000$
Filter	1–100 mHz

$$\Delta\dot{s}_{\varphi}(\varphi, \tau) = \Delta\dot{s}_0 \delta\left(\tau - \frac{\varphi}{v_f}\right), \quad (14)$$

where the rupture velocity v_f is constant, and with unit rad/s.

(iii) The receiver coplanar to the fault plane.

We calculate the synthetic seismograms in different scenarios (different rupture length and rupture velocity, rupture toward and away from the receiver, respectively), show the features of seismograms of finite fault and compare them with the seismograms of point source, which is exactly the same as the result of MINEOS. Table 1 shows the source and structure parameters we use for our numerical calculation. Note that the source parameters are chose according to the study of the 1992 Landers earthquake (Wald and Heaton 1994).

Table 2 shows the four cases for our numerical experiment. The rupture velocity is represented in deg/s. Note that the rupture length is about 70 km and the rupture velocity is approximately 0.024 deg/s for the 1992 Landers earthquake (Wald and Heaton 1994).

Our numerical results are shown in Figs. 2, 3, 4 and 5. We can observe that the seismograms for finite sources have smaller amplitudes than those for point sources. The amplitudes are in general smaller for sources with larger rupture lengths, lower rupture velocities and when the receivers are on the back of the rupture. Moreover, the zero-amplitude "knots" can be observed in the surface waves excited by finite sources. In the next section, we give a very intuitive explanation for these phenomenons and attribute them to the interference of the waves emitted by different parts of the fault.

4 An intuitive explanation for the numerical results

In this section, we attempt to explain the features in the seismograms of finite faults shown in our numerical results with a simple and intuitive model. Aki and Richards (2002) used the unidirectional rectangular fault model (Haskell

Table 2 Parameters for different cases in the numerical experiment

	Rupture length (km)	Rupture velocity (deg/s)	Direction
Case 1	5, 10, 20, 70	0.024	Front
Case 2	5, 10, 20, 70	0.024	Back
Case 3	70	0.028, 0.024, 0.022, 0.020	Front
Case 4	70	0.028, 0.024, 0.022, 0.020	Back

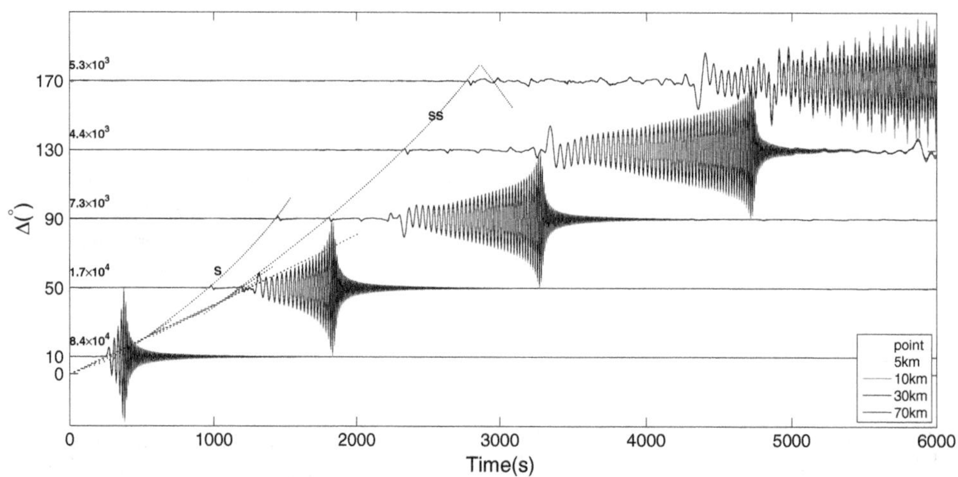

Fig. 2 Transverse component of the synthetic seismograms *in front* of the rupture for sources with the same rupture velocity, the same seismic moment but different rupture length. The *dotted line* is travel time curves of S and SS phase generated by TauP Toolkit. The rupture velocity is 0.024 deg/s, and the rupture length is 5 km (*light blue*), 10 km (*red*), 30 km (*dark blue*) and 70 km (*black*). The *yellow line* is the synthetic seismogram for point source. The waveforms are normalized according to the maximum amplitude in the point source seismogram with the same epicentral distance. The number on the *left* is the maximum amplitude in the point source seismogram. Note that we plot the seismograms throughout this paper if not stated otherwise. We can observe from this figure that the synthetic seismograms in front of the rupture for every rupture length are all very similar to those of the point sources

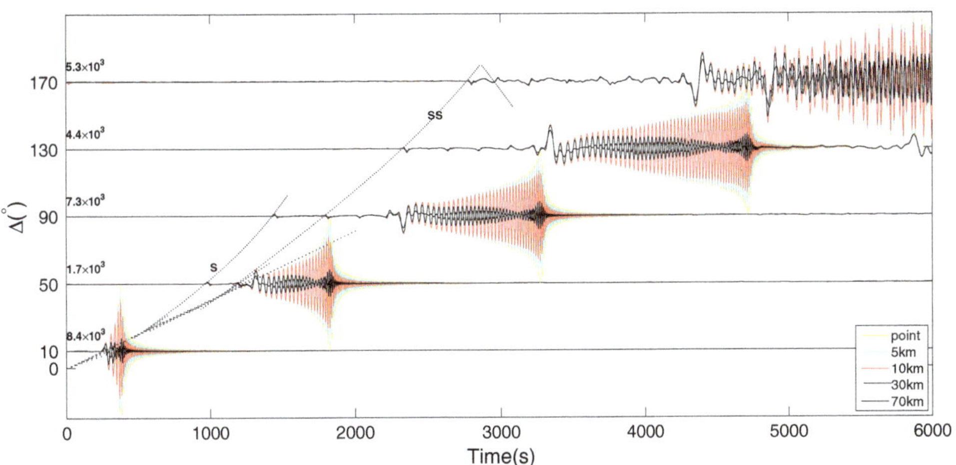

Fig. 3 Transverse component of the synthetic seismograms on the back of the rupture for sources with the same rupture velocity, the same seismic moment but different rupture length. The *dotted line* is travel time curves of S and SS phase generated by TauP Toolkit. The rupture velocity is 0.024 deg/s, and the rupture length is 5 km (*light blue*), 10 km (*red*), 30 km (*dark blue*) and 70 km (*black*). The *yellow line* is the synthetic seismogram for point source. Note that the synthetic seismograms on the back of the rupture have smaller amplitudes than those of point sources. Moreover, for large rupture length, zero-amplitude "knots" can be observed

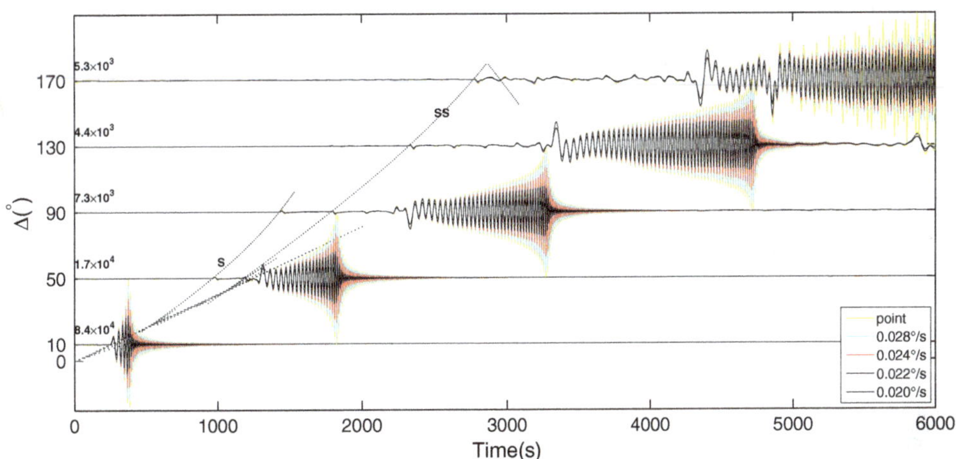

Fig. 4 Transverse component of the synthetic seismograms in front of the rupture for sources with the same rupture length, the same seismic moment but different rupture velocity. The *dotted line* is travel time curves of S and SS phase generated by TauP Toolkit. The rupture length is 70 km, and the rupture velocity is 0.028 deg/s (*light blue*), 0.024 deg/s (*red*), 0.022 deg/s (*dark blue*) and 0.020 deg/s (*black*). The *yellow line* is the synthetic seismogram for point source. Note that the seismograms of sources that propagate slowly have small amplitudes, especially for the high-frequency component

model) to explain the spectral features of the surface waves. Using this idea, we can qualitatively explain the variation of amplitudes of the Love waves due to the rupture length and velocity of the finite faults in time domain. Suppose that we have a 1-D plane dispersive wave excited by a point source

$$f(x,t) = \frac{1}{\sqrt{2\pi}} \int_{-\infty}^{+\infty} F(\omega) \mathrm{e}^{\mathrm{i}\omega\left(\frac{x}{v(\omega)} - t\right)} \mathrm{d}\omega, \tag{15}$$

in which $F(\omega)$ denotes the spectrum, and $v(\omega)$ is the phase velocity. For the finite source case, the wave field can be represented as the integral over the source

$$f(x,t) = \frac{1}{\sqrt{2\pi}} \int_{-l}^{l} \int_{-\infty}^{+\infty} \frac{F(\omega)}{2l} \mathrm{e}^{i\omega\left(\frac{x-x'}{v(\omega)} - (t - t'(x'))\right)} \mathrm{d}\omega \mathrm{d}x', \tag{16}$$

where t' is the time that the rupture arrives at x'. If we further assume that the rupture propagates at a constant velocity v_r, then Eq. (16) becomes

$$f(x,t) = \frac{1}{\sqrt{2\pi}} \int_{-\infty}^{+\infty} F(\omega) \gamma(\omega) \mathrm{e}^{i\omega\left(\frac{x}{v(\omega)} - t\right)} \mathrm{d}\omega, \tag{17}$$

where

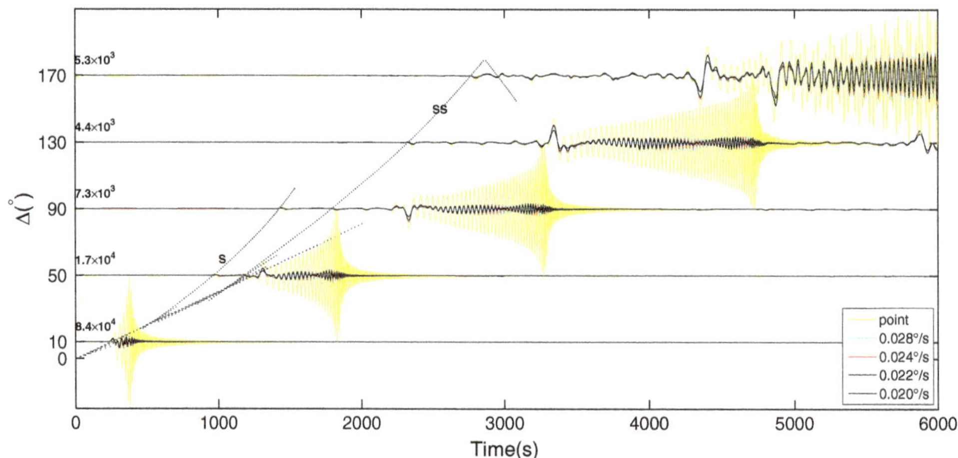

Fig. 5 Transverse component of the synthetic seismograms in front of the rupture for sources with the same rupture length, the same seismic moment but different rupture velocity. The *dotted line* is travel time curves of S and SS phase generated by TauP Toolkit. The rupture length is 70 km, and the rupture velocity is 0.028 deg/s (*light blue*), 0.024 deg/s (*red*), 0.022 deg/s (*dark blue*) and 0.020 deg/s (*black*). The *yellow line* is the synthetic seismogram for point source. Note that the synthetic seismograms on the back of the rupture have smaller amplitudes than those of point sources. Moreover, for large rupture length, zero-amplitude "knots" can be observed

$$\gamma(\omega) = \mathrm{sinc}\left[\omega l\left(\frac{1}{v_r} - \frac{1}{v(\omega)}\right)\right] = \mathrm{sinc}[\pi(n_t - n_l)]. \quad (18)$$

n_t and n_l are ratios of rupture time over period of wave and rupture length over wavelength,

$$n_t = \frac{T_r}{T}, \qquad n_l = \frac{2l}{\lambda}. \quad (19)$$

Here we define v_r and T_r to be negative if the rupture propagates away from the receiver. The "sinc" function in Eq. (17) is

$$\mathrm{sinc}(x) = \frac{\sin x}{x}, \quad (20)$$

the graph of which is shown in Fig. 6.

Note that the only difference between the wave field excited by point source Eq. (15) and that by finite source Eq. (17) is the existence of γ, which is the amplification coefficient due to the effect of finite source. Then, we can discuss the amplitude of wave in frequency domain.

Table 3 shows the value of $n_t - n_l$ in different scenarios. The dispersion relation for PREM model is given by Widmer-Schnidrig and Laske (2009). According to our model, if $|n_t - n_l| \approx 0$ then $\gamma \approx 1$, the wave emitted by all parts of the source arrives approximately at the same time, the amplitude is almost as large in the wave field produced by finite source as by point source; if $|n_t - n_l| \ll 1$, then $\gamma \approx 0$, which means that the amplitude is greatly diminished due to the destructive interference of the wave emitted by different parts of the source. The idea here is essentially the same as in Vallée and Dunham (2012). Moreover, if $n_t - n_l$ is a nonzero integer, $\gamma = 0$ which indicates the position of the zero-amplitude "knot." From Table 3, we can see that

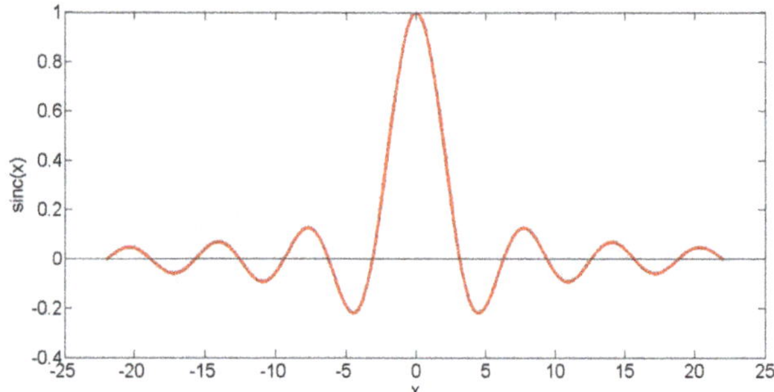

Fig. 6 Graph of the "sinc" function. $\mathrm{sinc}(0) = 1$ is the maximum of the function, and the value rapidly decays away from 0. Moreover, $\mathrm{sinc}(n\pi) = 0$ for n is nonzero integer

(i) For the same rupture length and velocity, the finiteness of the fault impacts more on high-frequency component than on low-frequency component;
(ii) For the same frequency, the waves produced by longer faults have smaller amplitudes than those produced by shorter faults;
(iii) The waves have larger amplitude in front of the rupture than in the back and have more zero-amplitude knots,

which correspond exactly to what we observe in the numerical results. Moreover, note in Table 3 that when rupture length 111.2 km, rupture velocity 0.020 deg/s and frequency $f = 50$ mHz, the $n_t - n_l$ value is very close to 1, which predicts the zero-amplitude knot. Actually, we can observe the knot in the synthetic seismogram exactly correspond to the 50 mHz group velocity (Fig. 7).

Table 3 Values of $n_t - n_l$ for some combinations of rupture length, rupture velocity and frequency

Rupture length (km)	Rupture velocity (s/deg)	3 mHz	10 mHz	20 mHz	50 mHz	100 mHz
70	0.024	0.0276	0.0744	0.1199	0.2120	0.2662
70	0.020	0.0540	0.1626	0.2962	0.6527	1.1475
111.2	0.020	0.0858	0.2583	0.4705	1.0368	1.8229
10	0.020	0.0077	0.0232	0.0423	0.0932	0.1639
70	−0.024	−0.1084	−0.3788	−0.7866	−2.0541	−4.2662
70	−0.020	−0.1348	−0.4669	−0.9628	−2.4948	−5.1475

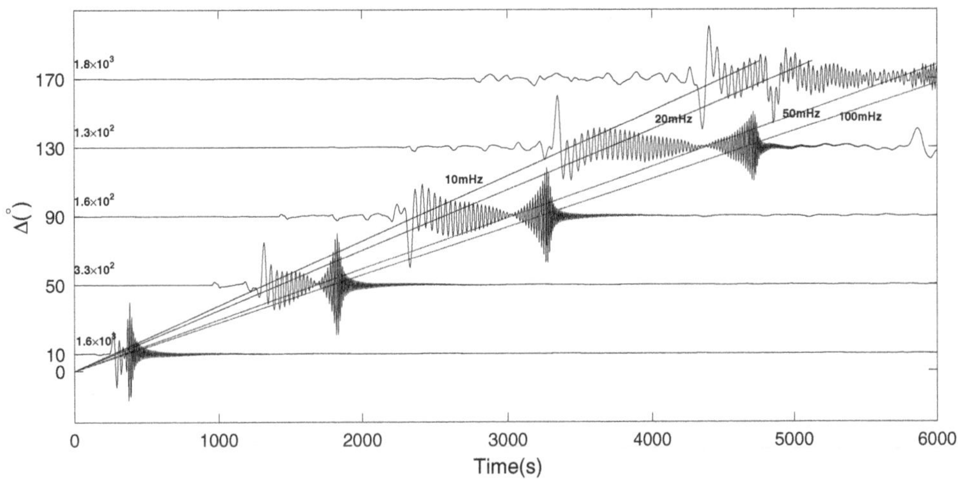

Fig. 7 Zero-amplitude "knots" in synthetic seismograms. Note that they exist near the position that correspond to the group velocity of 50 mHz Love wave

5 Conclusion

In this paper, we expand the normal-mode method to the finite source case. Instead of using point source summation, we directly calculate the excitation of each mode by the finite fault and add together to obtain the synthetic seismogram. We derive the solution of displacement produced by a "line source" and carry out numerical experiments and discuss the impact of rupture length and rupture velocity on the wave form of synthetic seismogram, especially the wave form of Love wave. We observe that

(i) The amplitude is smaller in the seismogram of finite fault than that of point source with the same seismic moment;
(ii) The finiteness of fault has more significant impact on high-frequency component than on low-frequency component;
(iii) The amplitude in the seismogram of finite fault is diminished more for the receiver that in the back of the rupture;
(iv) Zero-amplitude "knots" exist, and there are more "knots" in the back of the rupture.

Moreover, we use the Haskell model and the interference of waves emitted by different parts of the fault to provide a very intuitive explanations for all these phenomenons. Although this model is too simple to explain all the features in different cases accurately and qualitatively, it describes the basic characters of the seismogram of finite faults.

However, there are still some problems remain to be dealt with. The first one is the inclusion of radial propagation of the rupture. In this paper, we treat the fault as the "line source," which is far from enough. The line source model can only capture the effect of rupture propagation in the transverse direction. Although this is a step forward compared with the point source model, it still cannot take into account all the features on the fault. The second one is the efficiency of the method. The finiteness of the source causes the unavailability of the additional theorem of GSH, which means that the summation over m needs to be numerically calculated. Such summation is quite inefficient numerically. These two problems must be solved in order for this method to be usable in the calculation of synthetic seismograms.

Acknowledgements This work was supported by the National Natural Science Foundation of China (Grant No. 41674050) and MOST grant (2012CB417301). We thank the two anonymous reviewers for their constructive comments and suggestions which are crucial to the improvement of our manuscript.

Appendix: Formulae for the synthetic seismograms of finite faults

The displacement triggered by the finite fault is

$$
\begin{aligned}
{}^{S}_{\xi}u^{Z} &= \sum_{n,l} \sin 2\theta_d U_{nl}(r)\left({}^{S}\mathscr{S}^{0}_{nl}{}^{S}_{\xi}\mathscr{K}^{0,0}_{nl} - {}^{S}\mathscr{S}^{2}_{nl}\mathfrak{R}\left\{{}^{S}_{\xi}\mathscr{K}^{-2,0}_{nl}\right\}\right)\\
&\quad \cos 2\theta_d U_{nl}(r)^{S}\mathscr{S}^{2}_{nl}\mathfrak{R}\left\{{}^{S}_{\xi}\mathscr{K}^{-1,0}_{nl}\right\},\\
{}^{S}_{\xi}u^{R} &= \sum_{n,l} \sin 2\theta_d V_{nl}(r)\left({}^{S}\mathscr{S}^{0}_{nl}\mathfrak{R}\left\{e^{i\phi_b}{}^{S}_{\xi}\mathscr{K}^{0,+1}_{nl}\right\}\right.\\
&\quad \left. - {}^{S}\mathscr{S}^{2}_{nl}\mathfrak{R}\left\{e^{i\phi_b}\frac{{}^{S}_{\xi}\mathscr{K}^{-2,+1}_{nl} + {}^{S}_{\xi}\mathscr{K}^{+2,+1}_{nl}}{2}\right\}\right)\\
&\quad + \cos 2\theta_d V_{nl}(r)^{S}\mathscr{S}^{1}_{nl}\mathfrak{R}\left\{e^{i\phi_b}\frac{{}^{S}_{\xi}\mathscr{K}^{-1,+1}_{nl} - {}^{S}_{\xi}\mathscr{K}^{+1,+1}_{nl}}{2}\right\},\\
{}^{S}_{\xi}u^{T} &= \sum_{n,l} - \sin 2\theta_d V_{nl}(r)\left({}^{S}\mathscr{S}^{0}_{nl}\mathfrak{I}\left\{e^{i\phi_b}{}^{S}_{\xi}\mathscr{K}^{0,+1}_{nl}\right\}\right.\\
&\quad \left. - {}^{S}\mathscr{S}^{2}_{nl}\mathfrak{I}\left\{e^{i\phi_b}\frac{{}^{S}_{\xi}\mathscr{K}^{-2,+1}_{nl} + {}^{S}_{\xi}\mathscr{K}^{+2,+1}_{nl}}{2}\right\}\right)\\
&\quad - \cos 2\theta_d V_{nl}(r)^{S}\mathscr{S}^{1}_{nl}\mathfrak{I}\left\{e^{i\phi_b}\frac{{}^{S}_{\xi}\mathscr{K}^{-1,+1}_{nl} - {}^{S}_{\xi}\mathscr{K}^{+1,+1}_{nl}}{2}\right\},
\end{aligned}
\tag{21}
$$

$$
\begin{aligned}
{}^{T}_{\xi}u^{Z} &= 0,\\
{}^{T}_{\xi}u^{R} &= \sum_{n,l} \sin 2\theta_d W_{nl}(r)^{T}\mathscr{S}^{2}_{nl}\mathfrak{R}\left\{e^{i\phi_b}\frac{{}^{T}_{\xi}\mathscr{K}^{-2,+1}_{nl} - {}^{T}_{\xi}\mathscr{K}^{+2,+1}_{nl}}{2}\right\}\\
&\quad - \cos 2\theta_d W_{nl}(r)^{T}\mathscr{S}^{1}_{nl}\mathfrak{R}\left\{e^{i\phi_b}\frac{{}^{T}_{\xi}\mathscr{K}^{-1,+1}_{nl} + {}^{T}_{\xi}\mathscr{K}^{+1,+1}_{nl}}{2}\right\},\\
{}^{T}_{\xi}u^{T} &= \sum_{n,l} - \sin 2\theta_d W_{nl}(r)^{T}\mathscr{S}^{2}_{nl}\mathfrak{I}\left\{e^{i\phi_b}\frac{{}^{T}_{\xi}\mathscr{K}^{-2,+1}_{nl} - {}^{T}_{\xi}\mathscr{K}^{+2,+1}_{nl}}{2}\right\}\\
&\quad + \cos 2\theta_d W_{nl}(r)^{T}\mathscr{S}^{1}_{nl}\mathfrak{I}\left\{e^{i\phi_b}\frac{{}^{T}_{\xi}\mathscr{K}^{-1,+1}_{nl} + {}^{T}_{\xi}\mathscr{K}^{+1,+1}_{nl}}{2}\right\},
\end{aligned}
\tag{22}
$$

$$
\begin{aligned}
{}^{S}_{\varphi}u^{Z} &= \sum_{n,l} 2\sin\theta_d U_{nl}(r)^{S}\mathscr{S}^{2}_{nl}\mathfrak{I}\left\{{}^{S}_{\varphi}\mathscr{K}^{-2,0}_{nl}\right\}\\
&\quad - \cos\theta_d U_{nl}(r_R)^{S}\mathscr{S}^{2}_{nl}\mathfrak{I}\left\{{}^{S}_{\xi}\mathscr{K}^{-1,0}_{nl}\right\},\\
{}^{S}_{\varphi}u^{R} &= \sum_{n,l} 2\sin\theta_d V_{nl}(r)^{S}\mathscr{S}^{2}_{nl}\mathfrak{I}\left\{e^{i\phi_b}\frac{{}^{S}_{\varphi}\mathscr{K}^{-2,+1}_{nl} - {}^{S}_{\varphi}\mathscr{K}^{+2,+1}_{nl}}{2}\right\}\\
&\quad - \cos\theta_d V_{nl}(r)^{S}\mathscr{S}^{1}_{nl}\mathfrak{I}\left\{e^{i\phi_b}\frac{{}^{S}_{\varphi}\mathscr{K}^{-1,+1}_{nl} + {}^{S}_{\varphi}\mathscr{K}^{+1,+1}_{nl}}{2}\right\},\\
{}^{S}_{\varphi}u^{R} &= \sum_{n,l} 2\sin\theta_d V_{nl}(r)^{S}\mathscr{S}^{2}_{nl}\mathfrak{R}\left\{e^{i\phi_b}\frac{{}^{S}_{\varphi}\mathscr{K}^{-2,+1}_{nl} - {}^{S}_{\varphi}\mathscr{K}^{+2,+1}_{nl}}{2}\right\}\\
&\quad - \cos\theta_d V_{nl}(r)^{S}\mathscr{S}^{1}_{nl}\mathfrak{R}\left\{e^{i\phi_b}\frac{{}^{S}_{\varphi}\mathscr{K}^{-1,+1}_{nl} + {}^{S}_{\varphi}\mathscr{K}^{+1,+1}_{nl}}{2}\right\},
\end{aligned}
\tag{23}
$$

$$
\begin{aligned}
{}^{T}_{\varphi}u^{Z} &= 0,\\
{}^{T}_{\varphi}u^{R} &= \sum_{n,l} -2\sin\theta_d W_{nl}(r)^{T}\mathscr{S}^{2}_{nl}\mathfrak{R}\left\{e^{i\phi_b}\frac{{}^{T}_{\varphi}\mathscr{K}^{-2,+1}_{nl} + {}^{T}_{\varphi}\mathscr{K}^{+2,+1}_{nl}}{2}\right\}\\
&\quad + \cos\theta_d W_{nl}(r)^{T}\mathscr{S}^{1}_{nl}\mathfrak{R}\left\{e^{i\phi_b}\frac{{}^{T}_{\varphi}\mathscr{K}^{-1,+1}_{nl} - {}^{T}_{\varphi}\mathscr{K}^{+1,+1}_{nl}}{2}\right\},\\
{}^{T}_{\varphi}u^{T} &= \sum_{n,l} -2\sin\theta_d W_{nl}(r)^{T}\mathscr{S}^{2}_{nl}\mathfrak{I}\left\{e^{i\phi_b}\frac{{}^{T}_{\varphi}\mathscr{K}^{-2,+1}_{nl} + {}^{T}_{\varphi}\mathscr{K}^{+2,+1}_{nl}}{2}\right\}\\
&\quad + \cos 2\theta_d W_{nl}(r)^{T}\mathscr{S}^{1}_{nl}\mathfrak{I}\left\{e^{i\phi_b}\frac{{}^{T}_{\varphi}\mathscr{K}^{-1,+1}_{nl} - {}^{T}_{\varphi}\mathscr{K}^{+1,+1}_{nl}}{2}\right\},
\end{aligned}
\tag{24}
$$

where

$$
\begin{aligned}
{}^{S}\mathscr{S}^{0}_{nl} &= \dot{U}_{nl}(r_f) + \frac{1}{r_f}\left(\frac{\sqrt{l(l+1)}}{2}V_{nl}(r_f) - U_{nl}(r_f)\right),\\
{}^{S}\mathscr{S}^{1}_{nl} &= \dot{V}_{nl}(r_f) + \frac{1}{r_f}\left(\sqrt{l(l+1)}U_{nl}(r_f) - V_{nl}(r_f)\right),\\
{}^{S}\mathscr{S}^{2}_{nl} &= \frac{\sqrt{(l-1)(l+2)}}{2r_f}V_{nl}(r_f),
\end{aligned}
\tag{25}
$$

$$
{}^{T}\mathscr{S}^{1}_{nl} = \dot{W}_{nl}(r_f) - \frac{W_{nl}(r_f)}{r_f}, \quad {}^{T}\mathscr{S}^{2}_{nl} = \frac{\sqrt{(l-1)(l+2)}}{2r_f}W_{nl}(r_f),
\tag{26}
$$

and

$$
\begin{aligned}
{}^{K}_{\gamma}\mathscr{K}^{\alpha,\beta}_{nl}(\boldsymbol{x},t) &= \frac{2l+1}{4\pi}\mu w r_f \sin\theta_d \\
&\times \sum_{m=-l}^{+l}\left(P^{\alpha}_{lm}(\cos\theta_d) P^{\beta}_{lm}(\cos\theta) \mathrm{e}^{im\phi} \int_{\varphi_1}^{\varphi_2} \int_{0}^{T_f} \Delta\dot{s}_{\gamma}(\varphi,\tau) \mathrm{e}^{-im\varphi} \right. \\
&\left. \times \frac{1-\mathrm{e}^{-{}^{K}\sigma_{nl}(t-\tau)} \cos {}^{K}\omega_{nl}(t-\tau)}{{}^{K}\omega^2_{nl}} \mathrm{d}\tau \mathrm{d}\varphi \right).
\end{aligned} \tag{27}
$$

References

Aki K, Richards PG (2002) Quantitative seismology, 2nd edn. University Science Books, Sausalito, pp 491–536

Ben-Menahem A, Singh SJ (1987) Supershear accelerations and Mach-waves from a rupturing front-I. Theoretical model and implications. J Phys Earth 35:347–365

Bielak J, Loukakis K, Hisada Y, Yoshimura C (2003) Domain reduction method for three-dimensional earthquake modeling in localized regions, part I: theory. Bull Seismol Soc Am 93:817–824

Boore DM (1972) Finite-difference methods for seismic wave propagation in heterogeneous materials. In: Bolt BA (ed) Methods in computational physics. Academic Press, NY, pp 1–37

Bouchon M, Aki K (1977) Discrete wave-number representation of seismic-source wave fields. Bull Seismol Soc Am 67:259–277

Bouchon M (1980a) The motion of the ground during and earthquake 1. The case of a strike slip fault. J Geophys Res 85:356–366

Bouchon M (1980b) The motion of the ground during and earthquake 2. The case of a dip slip fault. J Geophys Res 85:367–375

Chapman CH (1978) A new method for computing synthetic seismograms. Geophys J R Astron Soc 54:481–518

Dahlen FA, Tromp J (1998) Theoretical global seismology. Princeton University Press, Princeton, pp 363–404

Gilbert F (1971) Excitation of normal modes of earth by earthquake sources. Geophys J R Astron Soc 22:223–226

Gilbert F, Helmberger DV (1972) Generalized ray theory for a layered sphere. Geophys J R Astron Soc 27:57–80

Israel M, Kovach RL (1977) Near-field motions from a propagating strike-slip fault in an elastic Half-space. Bull Seismol Soc Am 67:977–994

Kennett BLN, Kerry NJ (1979) Seismic wave in a stratified half space. Geophys J Int 57:557–583

Komatitsch D, Tromp J (1999) Introduction to the spectral element method for three-dimensional seismic wave propagation. Geophys J Int 139:806–822

Komatitsch D, Tromp J (2002) Spectral-element simulations of global seismic wave propagation-II. Three-dimensional models, oceans, rotation and self-gravitation. Geophys J Int 150:303–318

Luco JE, Apsel RJ (1983) On the Green's functions for a layered half-space. Part I. Bull Seismol Soc Am 73:909–929

Phinney RA, Burridge R (1973) Representation of elastic-gravitational excitation of a spherical earth model by generalized spherical harmonics. Geophys J R Astron Soc 34:451–487

Saikia CK, Helmberger DV (1997) Approximation of rupture directivity in regional phases using upgoing and downgoing wave fields. Bull Seismol Soc Am 87:987–998

Singh SJ, Ben-Menahem A (1969a) Eigenvibrations of the earth excited by finite dislocations-I. Toroidal oscillations. Geophys J R Astron Soc 17:151–177

Singh SJ, Ben-Menahem A (1969b) Eigenvibrations of the earth excited by finite dislocations-II. Spheroidal oscillations. Geophys J R Astron Soc 17:333–350

Song XJ, Helmberger DV (1996) Source estimation of finite faults from broadband regional networks. Bull Seismol Soc Am 86:797–804

Stump BW, Johnson LR (1982) Higher-degree moment tensors-the importance of source finiteness and rupture propagation on seismograms. Geophys J R Astron Soc 69:721–743

Takeuchi H, Saito M (1972) Seismic surface waves. In: Bruce AB (ed) Seismology: surface waves and earth oscillations, volume 11 of methods in computational physics. Academic Press, New York, pp 217–295

Tanimoto T (1984) A simple derivation of the formula to calculate synthetic long-period seismograms in a heterogeneous earth by normal mode summation. Geophys J R Astron Soc 77:275–278

Vallée M, Dunham EM (2012) Observation of far-field Mach waves generated by the 2001 Kokoxili supershear earthquake. Geophys Res Lett 39:L05311

Wald DJ, Heaton TH (1994) Spatial and temporal distribution of slip for the 1992 Landers, California, earthquake. Bull Seismol Soc Am 84:668–691

Widmer-Schnidrig R, Laske G (2009) Theory and observations-normal modes and surface wave measurements. In: Dziewonski AM, Romanowicz BZ (eds) Seismology and structure of the earth: treatise on geophysics. Elsevier, Amsterdam, pp 67–125

Woodhouse JH (1988) The calculation of the eigenfrequencies and eigenfunctions of the free oscillations of the Earth and the Sun. In: Doorbos DJ (ed) Seismological algorithms. Academic Press, San Diago, pp 321–370

Yang H-Y, Zhao L, Hung S-H (2010) Synthetic seismograms by normal-mode summation: a new derivation and numerical examples. Geophys J Int 183:1613–1632

3

Improvements on particle swarm optimization algorithm for velocity calibration in microseismic monitoring

Yue Yang · Jian Wen · Xiaofei Chen

Abstract In this paper, we apply particle swarm optimization (PSO), an artificial intelligence technique, to velocity calibration in microseismic monitoring. We ran simulations with four 1-D layered velocity models and three different initial model ranges. The results using the basic PSO algorithm were reliable and accurate for simple models, but unsuccessful for complex models. We propose the staged shrinkage strategy (SSS) for the PSO algorithm. The SSS-PSO algorithm produced robust inversion results and had a fast convergence rate. We investigated the effects of PSO's velocity clamping factor in terms of the algorithm reliability and computational efficiency. The velocity clamping factor had little impact on the reliability and efficiency of basic PSO, whereas it had a large effect on the efficiency of SSS-PSO. Reassuringly, SSS-PSO exhibits marginal reliability fluctuations, which suggests that it can be confidently implemented.

Keywords Particle swarm optimization · Staged shrinkage strategy (SSS) · Global optimization (GO) · Geophysical inversion · Microseismic velocity calibration

Y. Yang (✉) · J. Wen · X. Chen
School of Earth and Space Sciences, University of Science and Technology of China, Hefei 230026, China
e-mail: yyzf@mail.ustc.edu.cn

Y. Yang · J. Wen · X. Chen
Laboratory of Seismology and Physics of Earth's Interior, University of Science and Technology of China, Hefei 230026, Anhui, China

Y. Yang · J. Wen · X. Chen
Mengcheng National Geophysical Observatory, Hefei 230026, Anhui, China

1 Introduction

Inversion problems are basic issues in geophysics. Solutions to these problems can be classified into deterministic or probabilistic methods, or into local or global optimization methods. Most local optimization algorithms are deterministic, whereas global optimization algorithms are stochastic (Sen and Stoffa 2013).

With careful initialization, local optimizations techniques, such as the Newton, quasi-Newton (e.g., the BFGS), and conjugate-gradient (CG) methods, can obtain stable results based on gradient information after several iterations (Gill et al. 1981; Tarantola 2005). However, because the initialization may be unreliable, we cannot be certain that the iterations converge to the global minima. Compared with local optimization methods, global optimization (GO) methods, such as Simulated Annealing (SA) and genetic algorithms (GAs), typically need a huge amount of forward calculations to obtain a global optima. The crucial issues for GO methods are the convergence speed and computational efficiency (Kiranyaz et al. 2013).

Particle swarm optimization (PSO) (Russ et al. (1996)) is an efficient global optimization method. PSO has similar characteristics to evolution algorithms (EAs). These methods are stochastic (e.g., the SA) and are population-based evolutionary algorithms (e.g., the GAs) (Banks et al. 2007; Kiranyaz et al. 2013). PSO has been shown to outperform the GA methods (Angeline 1998). It has been extensively applied because of its fast convergence rate and simple implementation (Poli 2008). However, PSO is significantly slower to converge in certain circumstances, because it can converge prematurely and get trapped in local optima. Therefore, it is important to improve PSO so that it can be applied to the seismic inversion.

PSO improvements can be categorized into two sets: improvements to parameters and modifications to evolution rules. Inertia weight (Shi and Eberhart 1998) and constriction factors (Clerc 1999) have been introduced to the original PSO formulation to better control a particles' velocity. Carlisle and Dozier (2001) attempted to determine explicit and implicit off-the-shelf parameters for PSO. However, rather than offering a panacea for the parameter selection process, their work raised awareness regarding the importance of appropriate parameter values (Xu and Rahmat-Samii 2007; Zhang et al. 2005a, b). Other PSO improvements modified the particle learning rules, or were hybrids with other GO methods (Liang et al. 2006; Mendes et al. 2004; Ratnaweera et al. 2004).

In this paper, we first briefly introduce the principles of the PSO algorithm using the examples of synthetic inversion for velocity calibration in microseismic monitoring. We explore the inevitable limitations tackling with complex models. Then, we describe the staged shrinkage strategy (SSS) for improving the PSO algorithm, which avoids premature and speeds up the convergence rate. Finally, we present simulation results to demonstrate the superiority of our proposed SSS-PSO algorithm. We also investigated the influence of the built-in velocity clamping factor on the reliability and efficiency for the PSO and SSS-PSO algorithms.

2 PSO algorithm and its limitations

2.1 Fundamentals of the PSO algorithm

PSO was proposed by Eberhart and Kennedy (1995). It maintains a population of particles that represent potential solutions in the search space. These particles have two physical characteristics: location and velocity. The locations represent potential solutions. The velocities represent a kinetic property of the particles. The optimization rules follow mathematical formulas. A flow diagram for PSO is shown in Fig. 1. In a *d*-dimension model space, the particle swarm size is *N*, with x_{id} and v_{id} denoting the position and velocity of particle $i (i = 1, 2, 3, \ldots, N)$. p_{id} and p_{gd} are the personal best and global best positions for particle *i*. The rules to update a particle's velocity and position are

$$\begin{cases} v_{id}(k+1) = v_{id}(k) + c_1 r_1 [p_{id}(k) - x_{id}(k)] + c_2 r_2 [p_{gd}(k) - x_{id}(k)] \\ x_{id}(k+1) = x_{id}(k) + v_{id}(k+1) \end{cases}, \tag{1}$$

where r_1 and r_2 are random numbers in the range [0, 1], and c_1 and c_2 are acceleration constants (typically $c_1 = c_2 = 2$). The first part of the velocity update equation is called the "velocity inertia" term, the second is the "cognition" term (indicating that the particles are influenced by their own past), and the third is the "social" term (indicating that the particles learn from swarm intelligence) (Eberhart and Kennedy 1995; Shi and Eberhart 1998).

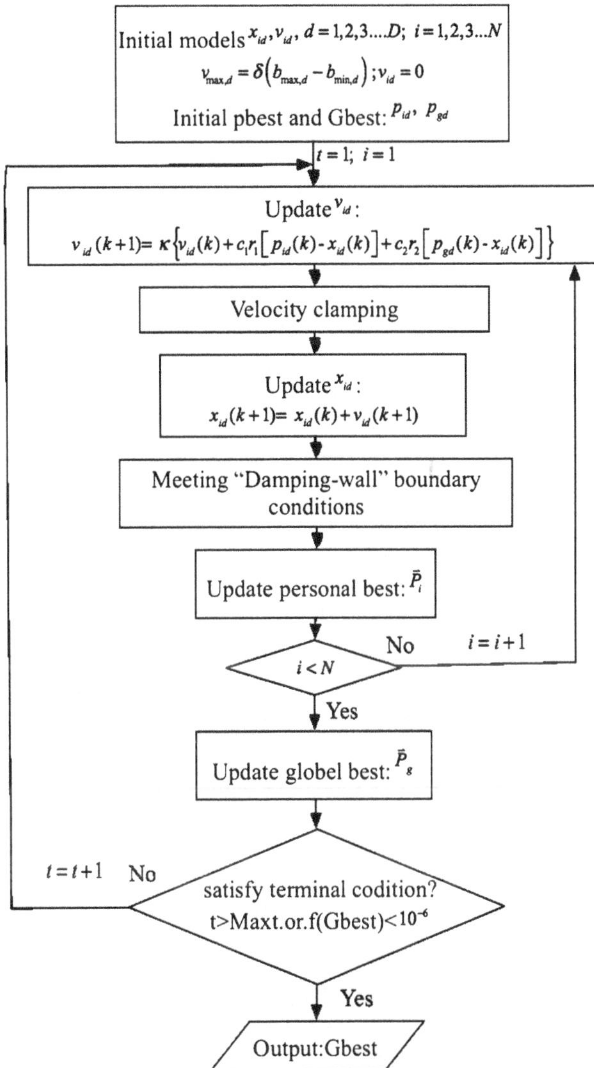

Fig. 1 Flow chart of the basic PSO implementation

The algorithm for updating the personal best is

$$\vec{P}_i(k+1) = \begin{cases} \vec{x}_i(k+1), & \text{when } F\left[\vec{x}_i(k+1)\right] < F\left[\vec{P}_i(k)\right] \\ \vec{P}_i(k), & \text{when } F\left[\vec{P}_i(k)\right] < F\left[\vec{x}_i(k+1)\right] \end{cases}, \tag{2}$$

where $\vec{P}_i$ is the personal best memory of $\vec{x}_i$. F is the objective function defined in Eq. 7.

To restrict the maximum step at each iteration and prevent overflow, the maximum velocity ($v_{\max,d}$) should be a proportion of the range of the particle search space. This proportion is called the velocity clamping factor (δ). According to Russ et al. (1996) and Clerc (1999), the mathematical relationship between these parameter is

$$v_{\max,d} = \delta\left(b_{\max,d} - b_{\min,d}\right), \quad \text{for } \delta \in (0, 1], \tag{3}$$

where $v_{\max,d}$ is the allowed maximum velocity of particles, δ is the velocity clamping factor, and $b_{\max,d}$ and $b_{\min,d}$ are the maximum and minimum location values of particles at d-th dimension.

So if the velocity update results in a step that is too large, the maximum velocity limits the velocities as follows:

$$v_{id} = \begin{cases} v_{\max,d} & \text{when} \quad v_{id} > v_{\max,d} \\ v_{id} & \text{when} \quad |v_{id}| \le v_{\max,d} \\ -v_{\max,d} & \text{when} \quad v_{id} < -v_{\max,d} \end{cases}. \tag{4}$$

Obviously, unlike acceleration constants that balance the local and global search, the velocity clamping factor controls the convergence speed. A larger velocity clamping factor allows big steps and contributes to a fast convergence rate, but increases the probability that the method will get trapped in local optima. A smaller velocity clamping factor constrains the particle step size, slows down the optimization process, and increases the particles diversity (Poli et al. 2007). Essentially, the acceleration constants and velocity clamping factor have corporative roles in the convergence of the PSO algorithm.

Inertia weights (Shi and Eberhart 1998) have been introduced to original PSO velocity update formula.

$$\begin{aligned} v_{id}(k+1) = {} & w v_{id}(k) + c_1 r_1 \left[p_{id}(k) - x_{id}(k)\right] \\ & + c_2 r_2 \left[p_{gd}(k) - x_{id}(k)\right]. \end{aligned} \tag{5}$$

Here, w is the inertia weight. The inertia weight can take various values (Rezaee Jordehi and Jasni 2013). It works with velocity clamping factor to have a better control on velocity.

Then, constriction factors (Clerc 1999) were introduced to the original PSO formula as follows:

$$\begin{aligned} v_{id}(k+1) = {} & \kappa\{v_{id}(k) + c_1 r_1 \left[p_{id}(k) - x_{id}(k)\right] \\ & + c_2 r_2 \left[p_{gd}(k) - x_{id}(k)\right]\} \end{aligned}$$

$$\kappa = \frac{2}{\left|2 - \varphi - \sqrt{\varphi^2 - 4\varphi}\right|}, \quad \varphi = c_1 + c_2, \ \varphi > 4. \tag{6}$$

where K is the constriction factor. Typically, the constriction factor is set to be $K = 4.1$, and $c_1 = c_2 = 2.05$. Although the constriction factors were designed to control particles velocity without velocity clamping process, but simulation tests have shown that constriction-based PSO performance is better using velocity clamping (Eberhart and Shi (2000)).

The boundary conditions can also affect the performance of the algorithm (Xu and Rahmat-Samii 2007; Zhang et al. 2005b). In this study, we used the "damping-wall" boundary condition as follows:

$$(x_{id}, v_{id}) = \begin{cases} (b_{\max,d}, -r v_{id}), & \text{if } x_{id} > b_{\min,d} \\ (b_{\min,d}, -r v_{id}), & \text{if } x_{id} > b_{\min,d} \end{cases}, \tag{7}$$

where $b_{\max,d}$ and $b_{\min,d}$ are the boundaries of the particles' values in the d-th dimension; and r is a random number in [0, 1].

2.2 Numerical examples of the PSO algorithm applied to velocity calibration

Velocity calibration is an indispensable data processing step in microseismic monitoring, which obtains reliable initial models for microseismic data analysis (Cipolla et al. 2012; Maxwell et al. 2010). Various methods have been applied to this problem (Pei et al. 2009; Warpinski et al. 2005; Warpinski and Du 2013). Velocity calibration is a nonlinear optimization with limited data because there is a lack of string shots and receivers. There is no accepted approach for reliably and effectively solving this problem without a priori information (log data or other resources). In this section, we present some numerical examples that applied the PSO algorithm to velocity calibration in microseismic monitoring with synthetic data.

To apply the PSO method to this problem, we chose four 1-D layered velocity models as presented in Pei et al. (2009). The models are shown in Fig. 2. We generated synthetic travel time data using the accurate two-point ray tracing method (Tian and Chen 2005). The objective function is

$$F = \sqrt{\frac{1}{n}\sum_{i=1}^{n}\left(t_{vp,i}^{\text{obs}} - t_{vp,i}^{\text{cal}}\right)^2}, \tag{8}$$

where F is the objective function, n is the number of sensor shot direct wave picks, and $t_{vp,i}^{\text{obs}}$ and $t_{vp,i}^{\text{cal}}$ are forward and inversion calculations of the direct wave travel times.

When applying the basic PSO algorithm to recover the 1-D layered velocity models with synthetic travel time data, we used a constriction factor-based formula, the velocity clamping factor $\delta = 0.1$, and the population of particles $N = 30$ (Carlisle and Dozier 2001). The maximum number of iterations for outer loop was set to $t_{\max} = 1000$, and we set the maximum for the objective function to $F \le 10^{-6}$s. The PSO algorithm terminated when either of these conditions was exceeded. Four test models are shown in Fig. 2. For each test model, we ran 300 inversions with random initial models. These 300 random initial models were generated from three different initial model ranges. There were 100 random initial models for each given range. We set three different initial model ranges to test algorithm behavior, which are shown in Fig. 3 in the cyan line with two dots at both ends at each layer.

The inversion results are shown in Fig. 3. The sub-figures in row 1 to 4 (from top to bottom) are the inversion results for models 1 to 4; the sub-figures in each row (from left to right) correspond to the inversion results of 300

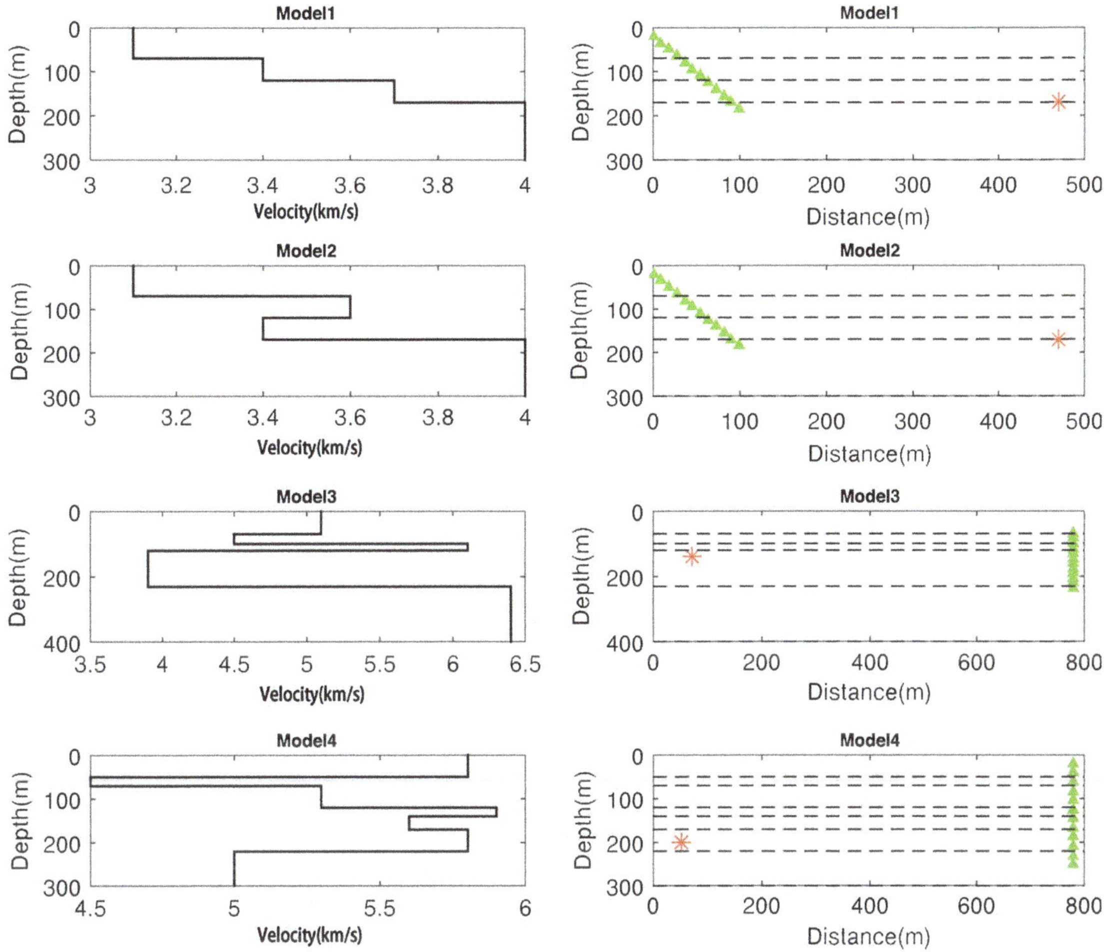

Fig. 2 Forward models and schematic geometry showing receiver arrays and perforation shot locations

simulations with three initial model ranges. For simplicity, we denote the sub-figure in row i and column j as sub-figure (i, j). Obviously, each sub-figure shows the results of 100 inversions with random initial models within a certain given model range. We can see that the inversion results for the relatively simple models (models 1 and 2) are very good, even when the initial models are quite different from the true models in sub-figs. (1, 3) and (2, 3). However, the inversion results for models 3 and 4 (relatively complicated models) are not good. Sub-figs. (3, 1) to (3, 3) show that the inversion results for model 3 are quite good in general, except for the top layer. Note that no geophones are deployed in the top layer, so there is no ray path constraint. The inversion results for model 4 are shown in sub-figs. (4, 1) to (4, 3). Although the inversion results were more accurate when the initial model ranges were narrow, they did not generally converge to the true model. In particular, the second layer (low velocity layer) was the least accurate for any of the initial model ranges. Note that we cannot guarantee that the PSO inversion convergence for this model within 1000 iterations achieves the terminal objective function $F = 10^{-6}$.

2.3 Limitations of basic PSO

In four-layer models, basic PSO iterated less than 100 times to reach the terminal objective function value. In each layer, the velocity deviations were <0.0001 km/s. The successful inversion results for models 1 and 2 indicate that basic PSO algorithm is insensitive to the initial models and stable in repeated simulations. However, the results for model 4 show that the objective function values may stop decreasing sometimes in repeated inversions. This problem could not be alleviated by increasing the maximum number of iterations from 1000 to 5000 (Fig. 4).

3 Algorithm modifications

3.1 Modifications of basic PSO

In the geophysical inversion problem, the inversion parameters have different sensitivities to the objective function. The difference is significant in some situations, as

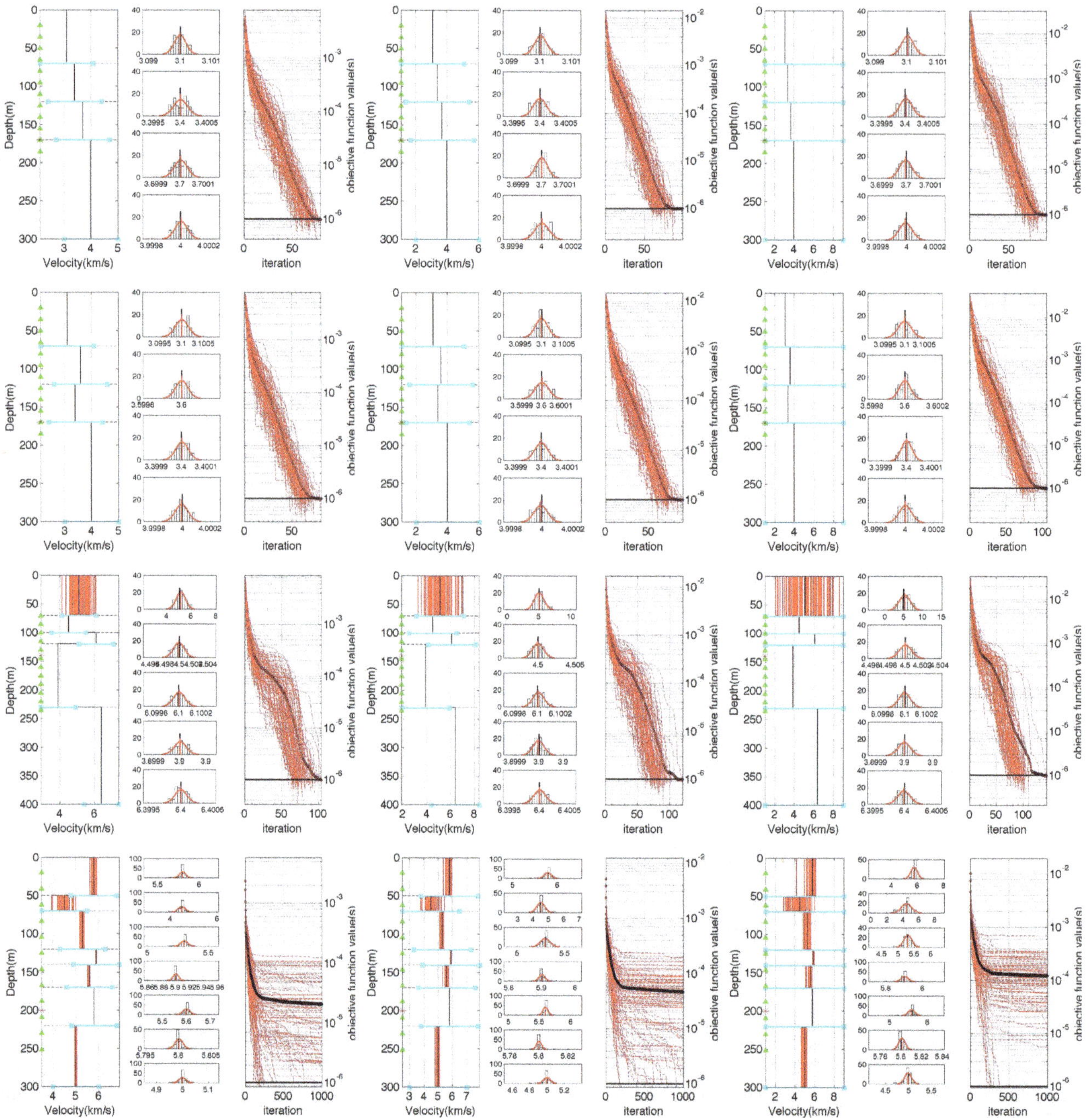

Fig. 3 PSO inversion results with 1000 maximum iterations. In each sub-plot, the plot on the left displays 100 repeated inversion results in *red*, which are occasionally overlapped by the *black lines* that represent the forward model. In each layer, the *cyan line* with two dots at both ends shows the initial model range. The plot in the *middle* gives histograms of each layers inversion results. The plot on the *right* depicts the objective function values, with the iterations represented by *red lines*. The *black line* is the average of the best values from 100 repeated inversions

revealed in model 4. The inversion results for the second layer (from the surface) in model 4 were the most scattered around the true value. The inversion parameter cannot be accurately calculated. The objective function because trapped in local optima, and stopped decreasing before satisfying the termination criteria. So modifying the search space according to the dimension and the evolutionary stage of the particles is feasible and reasonable. With this in mind, we propose SSS, which encourages the process to escape from local optima.

This modified PSO algorithm is abbreviated to SSS-PSO, and its flow chart is shown in Fig. 5. Firstly, we

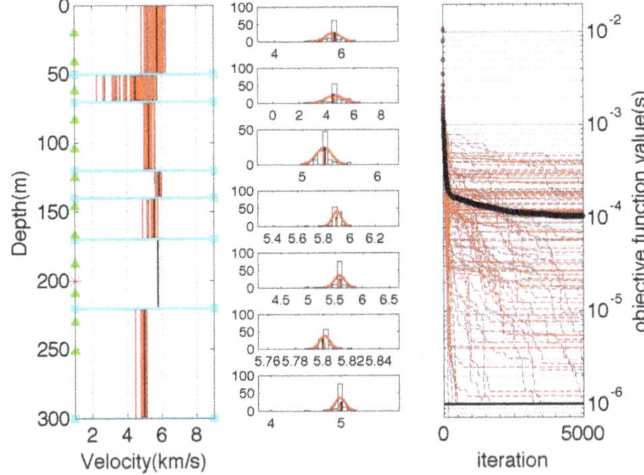

Fig. 4 PSO inversion results with 5000 maximum iterations. The plot on the *left* displays 100 repeated inversion results in *red*, which are occasionally overlapped by the *black lines* that represent the forward model. In each layer, the cyan line with two dots at both ends shows the initial model range. The plot in the *middle* gives histograms of each layers inversion results. The plot on the *right* depicts the objective function values, with the iterations represented by *red lines*. The *black line* is the average of the best values from 100 repeated inversions

initialize the global best memory ($g_{h,d}^{\mathrm{memo}}$) and update it with each iteration. h is the global best change times. Then, we calculate the maximum, minimum, and average values of the particles and their personal best positions ($x_{\mathrm{max,d}}$, $x_{\mathrm{min,d}}$, $x_{\mathrm{ave},d}$ and $y_{\mathrm{max,d}}$, $y_{\mathrm{min,d}}$, $y_{\mathrm{ave},d}$). The maximal and minimal global best positions are acquired from the global best memory ($g_{h,d}^{\mathrm{memo}}$)

$$g_{\mathrm{max},d}^{\mathrm{memo}} = \begin{cases} \max\left(g_{1:h,d}^{\mathrm{memo}}\right); & \text{when } h < h_{block} \\ \max\left(g_{h_{block}:h,d}^{\mathrm{memo}}\right); & \text{when } h \geq h_{block} \end{cases}, \quad g_{\mathrm{min},d}^{\mathrm{memo}} = \begin{cases} \min\left(g_{1:h,d}^{\mathrm{memo}}\right); & \text{when } h < h_{block} \\ \min\left(g_{h_{block}:h,d}^{\mathrm{memo}}\right); & \text{when } h \geq h_{block} \end{cases} \tag{9}$$

where h_{block} is 30 here which used to calculate $g_{h,d}^{\mathrm{memo}}$. In the early evolution stage, the particles' global best positions may not be local minimums, and should be excluded. Otherwise, the global best positions will undermine the efficiency of the boundary shrinkage. If we use a fixed length that results in $g_{\mathrm{max,\,d}}^{\mathrm{memo}}$ and $g_{\mathrm{min,\,d}}^{\mathrm{memo}}$ through blocking, we may fail to achieve the global best position. Admittedly,

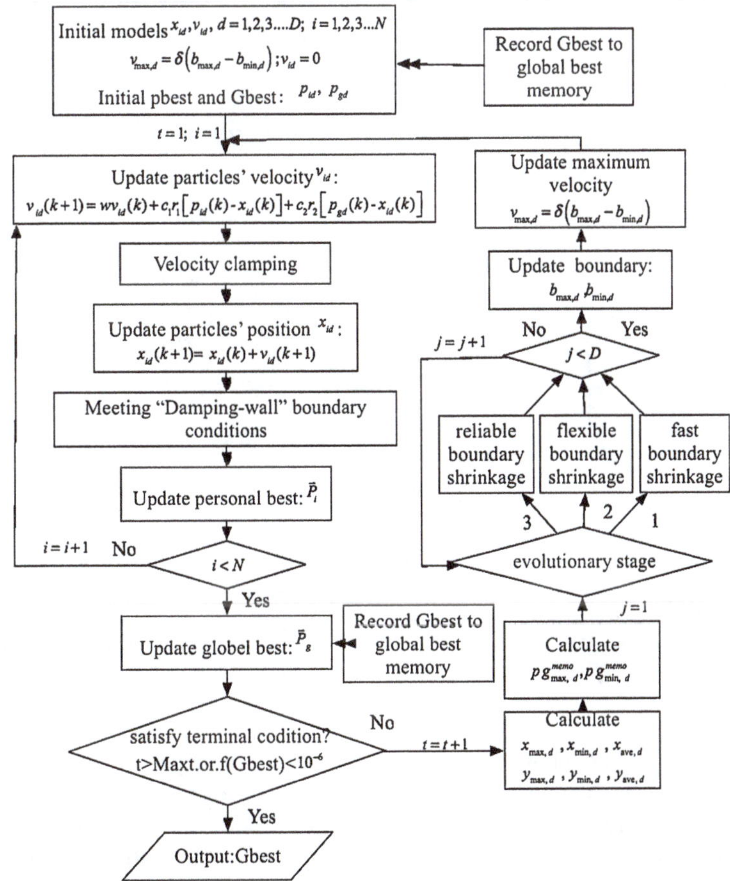

Fig. 5 Flow chart of the SSS modified PSO implementation

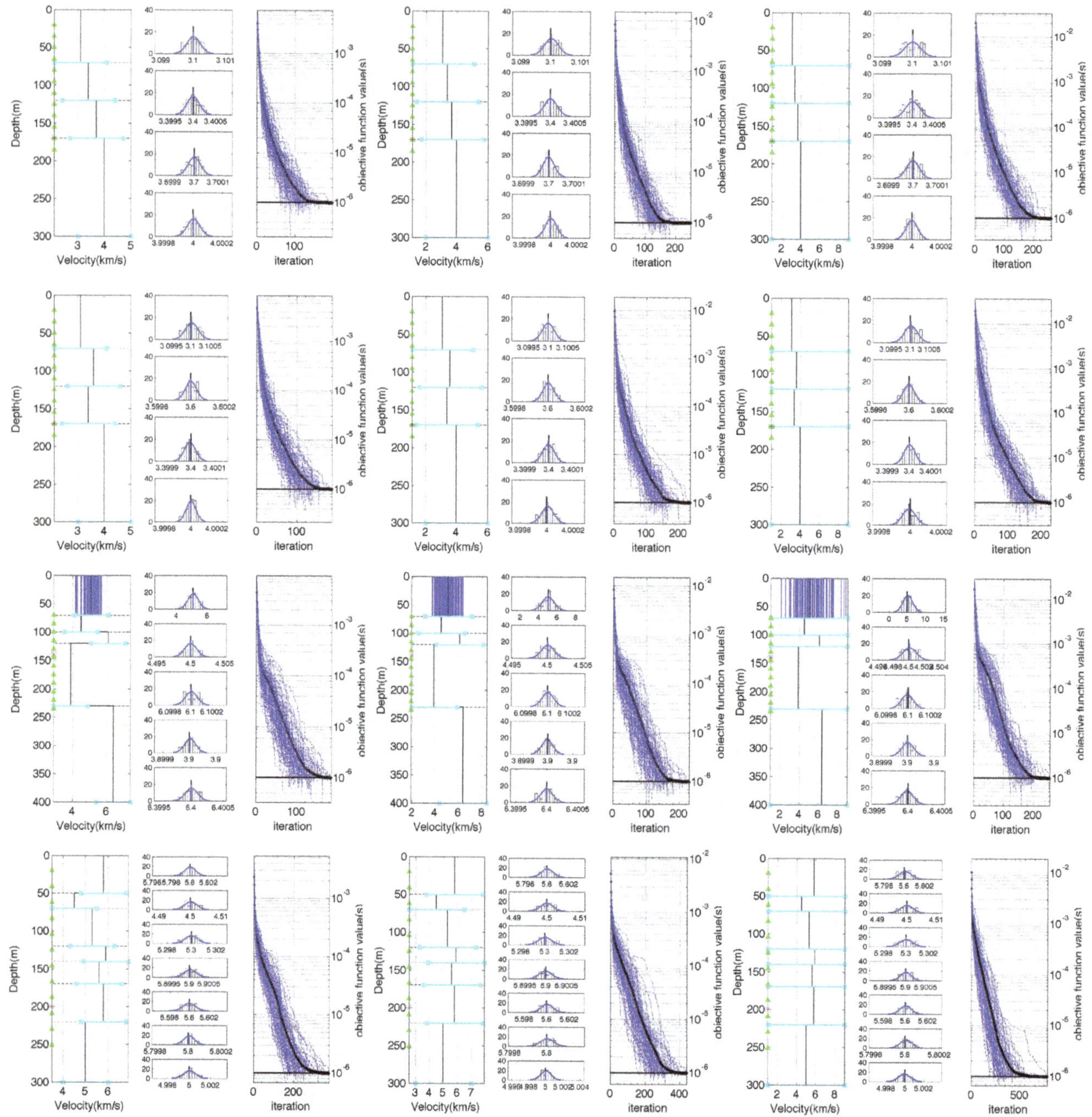

Fig. 6 SSS-PSO inversion results with 5000 maximum iterations. In each sub-plot, the plot on the left displays 100 repeated inversion results in blue, which are occasionally overlapped by the *black lines* that represent the forward model. In each layer, the *cyan line* with two dots at both ends shows the initial model range. The plot in the *middle* gives histograms of each layers inversion results. The plot on the *right* depicts the objective function values, with the iterations represented by *blue lines*. The *black line* is the average of the best values from 100 repeated inversions

more detailed work is required. However, our work suggests that $h_{\text{block}} = 30$ and the other following parameters for SSS implementations work well in velocity calibration tests.

Secondly, each inversion parameter dimension is sorted into a specified stage according to the positions of the particles. If the boundary space shrinks to 20 % of the initial space range, we defined the evolutionary situation as stage 2. If the boundary space shrinks to 4 % of the initial range, we define the situation as stage 3. Otherwise, the boundary space is 20 %–100 % of the initial range, and is in stage 1. Then, we update the boundary range according to the following rules:

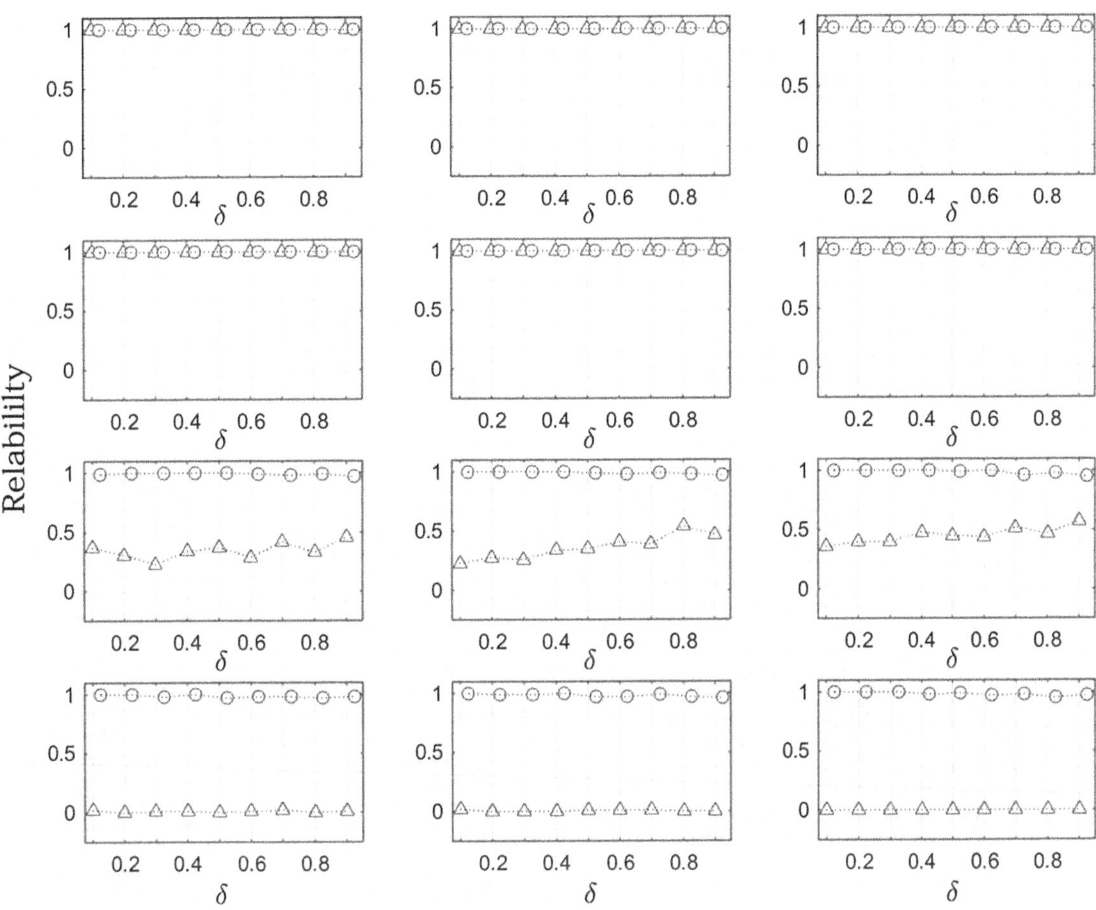

Fig. 7 Reliabilities of different models with varying velocity clamping factors. The triangle and circles represent the basic and SSS-PSO results, respectively. The plots from left to right are the results for three different initial ranges. The plots from top to bottom are the inversions for four different models. In each plot, the velocity clamping factor changed from 0.1 to 0.9

$$
\begin{aligned}
& b_d = b_{\max,d} - b_{\min,d} \\
& \begin{cases} \begin{cases} b_{\max1,d} = \max\left(g^{\text{memo}}_{\max,d} + s_4 r b_d, x_{\max,d} + s_4 r b_d, y_{\max,d} + s_4 r b_d\right) \\ b_{\max2,d} = \max\left(x_{\text{ave},d} + (s_2 + s_3 r) b_d, y_{\text{ave},d} + (s_2 + s_3 r) b_d\right) \end{cases} \\ b_{\max,d} = \min\left(b^{\text{ini}}_{\max,d}, x_{\max,d} + s_1 r b_d, \max\left(b_{\max1,d}, b_{\max2,d}\right)\right) \end{cases}, \\
& \begin{cases} \begin{cases} b_{\min1,d} = \min\left(g^{\text{memo}}_{\min,d} - s_4 r b_d, x_{\min,d} - s_4 r b_d, y_{\min,d} - s_4 r b_d\right) \\ b_{\min2,d} = \min\left(x_{\text{ave},d} - (s_2 + s_3 r) b_d, y_{\text{ave},d} - (s_2 + s_3 r) b_d\right) \end{cases} \\ b_{\min,d} = \max\left(b^{\text{ini}}_{\min,d} x_{\min,d} - s_1 r b_d, \min\left(b_{\min1,d}, b_{\min2,d}\right)\right) \end{cases}
\end{aligned}
\tag{10}
$$

where $b_{\max,d}$, $b_{\min,d}$, and b_d are the maximum, minimum, and range values of the particles' positions, $b^{\text{ini}}_{\max,d}$ and $b^{\text{ini}}_{\min,d}$ represent the initial range, r is a random number independently generated in [0,1], and s_1, s_2, s_3 and s_4 are SSS parameters. Each inversion parameter dimension exhibits a fast convergence rate in stage 1 because the parameters are set to $s_1 = 0.4$, $s_2 = 0.3$, $s_3 = 0.1$, and $s_{4=0.2}$. The convergence slows in stage 2 ($s_1 = 0.5$, $s_2 = 0.1$, $s_3 = 0.6$, and $s_4 = 0.3$) with the randomness boundary extension, and increases again in stage 3 ($s_1 = 0.3$, $s_2 = 0.3$, $s_3 = 0.3$, and $s_4 = 0.2$). Then, the boundary of the dimension shrinks to a very small range that includes the global optima.

Finally, we update the boundary conditions and maximum velocity.

3.2 Numerical experiments using the SSS-PSO algorithm

As shown in Sect. 2.2, basic PSO fails to retrieve the model parameters for model 4. We applied the SSS-PSO to this problem to determine its inversion performance. The inversion results are shown in Fig. 6. The fast and robust convergence of the objective function demonstrates the algorithm's performance. In repeated inversions, the objective function converged to the termination

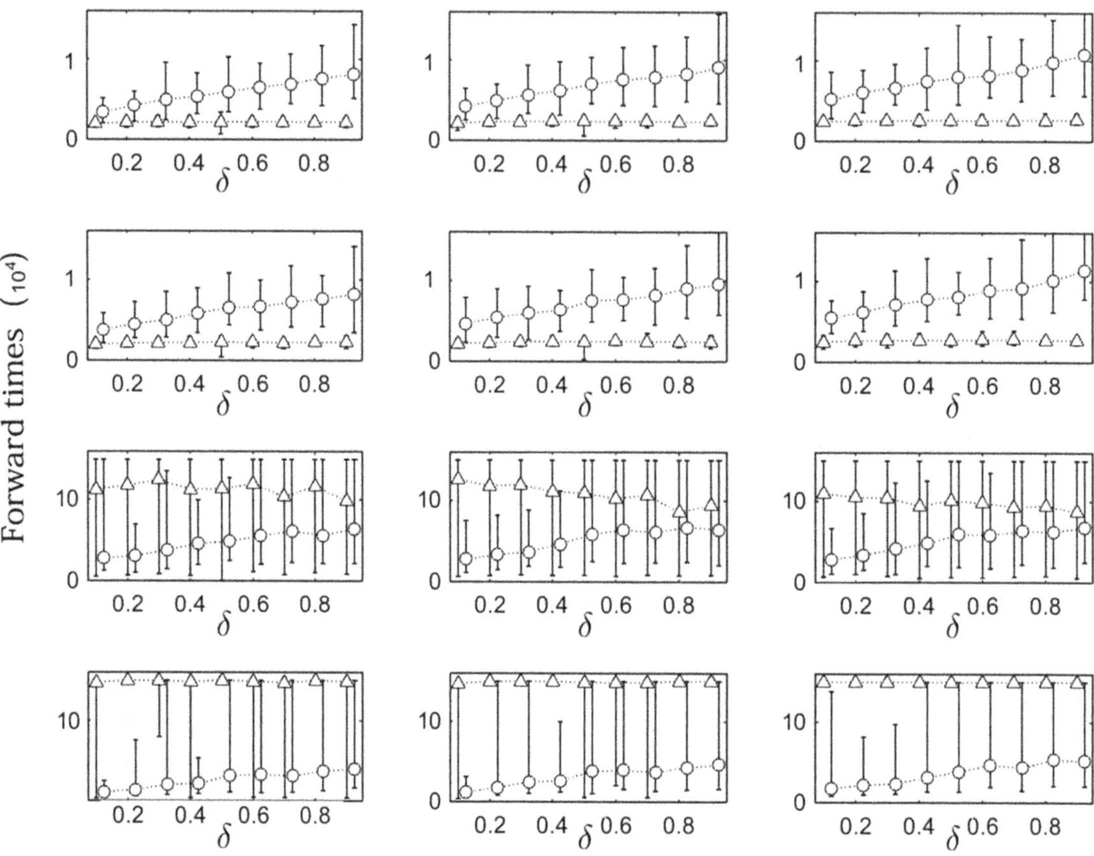

Fig. 8 *Error bars* of forward evaluations of different models with varying velocity clamping factors. The *triangle* and *circles* represent the basic and SSS-PSO results, respectively. The plots from *left to right* are the results for three different initial ranges. The plots from *top to bottom* are the inversions for four different models. In each plot, the velocity clamping factor changed from 0.1 to 0.9

criteria within the maximum number of iterations. It can be seen that although without requiring a prior information, the performance of the SSS-PSO is stable in repeated runs. These merits mean that SSS-PSO can be applied to velocity calibration application cases in complex models.

4 Sensitivity of PSO and SSS-PSO to velocity clamping factor

The velocity clamping factor (δ) is important to the performance of PSO and SSS-PSO. To investigate how δ affects the reliability and efficiency of PSO and SSS-PSO, we ran PSO and SSS-PSO inversions for nine δ values, ranging from 0.1 to 0.9. The inversion results are shown in Figs. 7 and 8.

We define the reliability of the inversion performance as follows:

$$\text{reliability}(\delta, \text{Succ}) = \frac{\text{Succ}_\delta}{\text{Total}} \times 100\,\% \tag{11}$$

where Succ_δ is the number of successful inversions in Total = 100 repeated runs. Success is defined as reaching the objective function value ($F = 10^{-6}$) within the maximum number of iterations ($t_{\max} = 5000$). Figure 7 shows the statistical reliabilities of the four models and three initial ranges. The reliabilities for all these cases were mainly insensitive to δ, although there were some small variations for model 3. Thus, the reliabilities were almost independent of δ, regardless of the high or low reliabilities. Another important feature of these results is that the SSS-PSO reliabilities were almost 100 % in all cases, whereas the basic PSO reliability was 100 % for models 1 and 2, a little above 0 % for model 4, and approximately 50 % for model 3. This indicates that the SSS-PSO inversion was systematically superior to the basic PSO inversion, and may guarantee convergence. Moreover, the reliabilities of the SSS-PSO inversion for models 3 and 4 decreased slightly as δ increased, which suggests that we should use small δ values.

Computational efficiency is another important measure for assessing an inversion algorithm. In the basic PSO and

SSS-PSO algorithms, we only need forward calculations after updating each particle position. So the time taken by forward evaluations is equal to the number of iterations multiplied by the size of the population ($N = 30$). Forward evaluations represent the most computational time, so they correlate strongly with algorithm efficiency. Therefore, it is important to determine the sensitivity of δ to the forward evaluations.

In Fig. 8, in terms of the time taken by forward simulations, the velocity clamping factor has a marginal influence on the basic PSO efficiency, but is very important to the SSS-PSO efficiency. In the basic PSO results for simple cases (Models 1 and 2), the forward simulations are stable to the changes in the velocity clamping factor and the parameter range of initial model. In models 3 and 4, on most occasions, the method used the maximum number of forward simulations, because the statistical reliability was approximately 50 % for model 3 and 0 % for model 4. For the SSS-PSO algorithm, small values of the velocity clamping factor ($\delta \approx 1$) can significantly reduce the number of forward simulations, thus the small value of δ is a preferable.

5 Conclusions

In this study, we improved the inversion performance of the basic PSO algorithm by incorporating the SSS, and applied our method to velocity calibration in miscroseismic monitoring. We demonstrated that the SSS-PSO algorithm is robust and efficient. The SSS-PSO algorithm steadily and speedily converged to the solution, regardless of the complexity of the target models and the randomness of the initial models. Furthermore, we investigated the influence of choices of the velocity clamping factor (δ) on the reliability and efficiency of the basic and SSS-PSO algorithms. We found that basic PSO is insensitive to δ, whereas SSS-PSO is not (especially the searching efficiency of). The SSS-PSO algorithm was robust to all the tested values of δ. Moreover, our simulations suggest that a small velocity clamping factor ($\delta = 0.1$) is preferred for SSS-PSO, in these microseismic velocity calibration problems.

We validated the performance of the algorithms by running 100 repeated inversions with random initial models, to differentiate the discrepancies caused by the intrinsic uncertainties of geophysical inversion problems from artifacts and errors produced by the inversion strategies. Additionally, because there is no panacea for GO problems according to the "no free lunch" theorems (Wolpert and Macready 1997), we need an inversion strategy that is aimed at specific geophysical inversion problem. Any such strategies should be tested before being applied. We cannot guarantee that the SSS-PSO will converge to global optima for any given problem, which is the same for all other GO methods. We have proposed and validated this method for microseismic velocity calibration models. The tests in this paper are limited, and further investigation is needed with noisy data, which will be proceeded in our future work.

References

Angeline PJ (1998) Evolutionary optimization versus particle swarm optimization: philosophy and performance differences, evolutionary programming VII. Springer, Berlin, pp 601–610. doi:10.1007/BFb0040811

Banks A, Vincent J, Anyakoha C (2007) A review of particle swarm optimization. Part II: hybridisation, combinatorial, multicriteria and constrained optimization, and indicative applications. Nat Comput 7(1):109–124. doi:10.1007/s11047-007-9050-z

Carlisle A, Dozier G (2001) An off-the-shelf PSO. In: Proceedings of the workshop on particle swarm optimization. Purdue School of Engineering and Technology, Indianapolis, pp 1–6

Cipolla CL, Maxwell SC, Mack MG (2012) Engineering guide to the application of microseismic interpretations. In: SPE hydraulic fracturing technology conference. Society of Petroleum Engineers, 6–8 February, The Woodlands, Texas, USA. doi:10.2118/152165-MS

Clerc M (1999) The swarm and the queen: towards a deterministic and adaptive particle swarm optimization. Evolutionary computation. doi:10.1109/CEC.1999.785513

Eberhart RC, Kennedy J (1995) A new optimizer using particle swarm theory. In: Proceedings of the sixth international symposium on micro machine and human science, pp 39–43. doi:10.1109/MHS.1995.494215

Eberhart RC, Simpson P, Dobbins R (1996) Computational intelligence PC tools, chap. 6. Academic Press Professional Inc., Diego, CA, pp 212–226

Eberhart RC, Shi Y (2000). Comparing inertia weights and constriction factors in particle swarm optimization. In: Proceedings of the 2000 congress on evolutionary computation, vol. 1. IEEE Xplore, pp 84–81. doi:10.1109/CEC.2000.870279

Gill PE, Murray W, Wright MH (1981) Practical optimization. Academic press, London, pp 1–26

Kiranyaz S, Ince T, Gabbouj M (2013) Multidimensional particle swarm optimization for machine learning and pattern recognition. Springer, Berlin, pp 101–148

Liang J, Qin K, Suganthan PN, Subramanian B (2006) Comprehensive learning particle swarm optimizer for global optimization of multimodal functions. IEEE Trans Evol Comput 10(3):281–295. doi:10.1109/TEVC.2005.857610

Maxwell SC, Underhill WB, Bennett L, Woerpel C, Martinez A (2010) Key criteria for a successful microseismic project, 19–22 September, Folrence, Italy. doi:10.2118/134695-MS

Mendes R, Kennedy J, Neves J (2004) The fully informed particle swarm: simpler, maybe better. IEEE Trans Evol Comput 8(3):204–210. doi:10.1109/TEVC.2004.826074

Pei DH, Quirein JA, Cornish BE, Quinn D, Warpinski NR (2009) Velocity calibration for microseismic monitoring: a very fast simulated annealing (VFSA) approach for joint-objective optimization. Geophysics 74(6):WCB47–WCB55. doi:10.1190/1.3238365

Poli R (2008) Analysis of the publications on the applications of particle swarm optimisation. J Artif Evol Appl. doi:10.1155/2008/685175

Poli R, Kennedy J, Blackwell T (2007) Particle swarm optimization. In: Encyclopedia of machine learning. Springer, pp 760–766

Ratnaweera A, Halgamuge SK, Watson HC (2004) Self-organizing hierarchical particle swarm optimizer with time-varying acceleration coefficients. IEEE Trans Evol Comput 8(3):240–255. doi:10.1109/TEVC.2004.826071

Rezaee Jordehi A, Jasni J (2013) Parameter selection in particle swarm optimisation: a survey. J Exp Theor Artif Intell 25(4):527–542. doi:10.1080/0952813X.2013.782348

Sen MK, Stoffa PL (2013) Global optimization methods in geophysical inversion. Cambridge University Press, Cambridge, pp 1–280

Shi Y, Eberhart R (1998) A modified particle swarm optimizer. In: The 1998 IEEE international conference on evolutionary computational proceedings. IEEE World Congress on Computational Intelligence, IEEE, pp 69–73

Tarantola A (2005) Inverse problem theory and methods for model parameter estimation. SIAM, Philadelphia, pp 1–342

Tian Y, Chen X (2005) A rapid and accurate two-point ray tracing method in horizontally layered velocity model. Acta Seismol Sin 18(2):154–161

Warpinski NR, Du J (2013) Velocity building for microseismic hydraulic fracture mapping in isotropic and anisotropic media. In: SPE hydraulic fracturing technology conference. Society of Petroleum Engineers, 4–6 February, The Woodlands, Texas, USA

Warpinski NR, Sullivan RB, Uhl J, Waltman C, Machovoie S (2005) Improved microseismic fracture mapping using perforation timing measurements for velocity calibration. SPE J 10(1):14–23. doi:10.2118/84488-PA

Wolpert DH, Macready WG (1997) No free lunch theorems for optimization. IEEE Trans Evol Comput 1:67–82. doi:10.1109/4235.585893

Xu SH, Rahmat-Samii Y (2007) Boundary conditions in particle swarm optimization revisited. IEEE Trans Antennas Propag 55(3):760–765

Zhang L-P, Yu H-J, Hu S-X (2005a) Optimal choice of parameters for particle swarm optimization. J Zhejiang University Sci-A 6(6):528–534. doi:10.1007/BF02841760

Zhang WJ, Xie XF, Bi DC (2005b) Handling boundary constraints for numerical optimization by particle swarm flying in periodic search space. In: Proceedings of the 2004 congress on evolutionary computation IEEE, vol 2, pp 2307–2311. doi:10.1109/CEC.2004.133118

4

Acoustic viscoelastic modeling by frequency-domain boundary element method

Xizhu Guan · Li-Yun Fu · Weijia Sun

Abstract Earth medium is not completely elastic, with its viscosity resulting in attenuation and dispersion of seismic waves. Most viscoelastic numerical simulations are based on the finite-difference and finite-element methods. Targeted at viscoelastic numerical modeling for multilayered media, the constant-Q acoustic wave equation is transformed into the corresponding wave integral representation with its Green's function accounting for viscoelastic coefficients. An efficient alternative for full-waveform solution to the integral equation is proposed in this article by extending conventional frequency-domain boundary element methods to viscoelastic media. The viscoelastic boundary element method enjoys a distinct characteristic of the explicit use of boundary continuity conditions of displacement and traction, leading to a semi-analytical solution with sufficient accuracy for simulating the viscoelastic effect across irregular interfaces. Numerical experiments to study the viscoelastic absorption of different Q values demonstrate the accuracy and applicability of the method.

Keywords Viscoelastic media · Viscoelastic boundary element method · Frequency-domain implementation · Viscoelastic numerical modeling

X. Guan · L.-Y. Fu (✉) · W. Sun
Institute of Geology and Geophysics, Chinese Academy of Sciences, Beijing 100029, China
e-mail: lfu@mail.iggcas.ac.cn

X. Guan
CNOOC Research Center, Beijing 100027, China

1 Introduction

The viscoelastic absorption causes energy loss during the wave propagation in real Earth media. The viscosity of the Earth shows obvious frequency dependence, especially for high-frequency components. The resulting serious dispersion reduces seismic resolution and causes difficulties for geological interpretation. The compensation of viscoelastic attenuation is an important issue in seismic data processing. Various seismic numerical methods in viscoelastic media (Dvorkin and Mavko 2006) have been widely used to understand the detailed characteristics of viscoelastic absorption. However, most viscoelastic numerical simulations are based on the finite-difference (FD) and finite-element (FE) methods. In this study, we develop an efficient viscoelastic boundary element method (BEM) for the study of viscoelastic absorption.

With the extensive publications in seismology (Aki and Richards 1980; Emmerich and Korn 1987; Liao and McMechan 1996; Carcione et al. 2002; Carcione 2010), the implementation of viscoelastic numerical modeling can be simply divided into two categories: complex-number velocity methods in the frequency domain and quality factor-based wave equation methods in the time domain. The time-domain methods use a series of viscous parameters (i.e., the standard linear body) to describe the medium viscosity. These methods are based on various approximate constant-Q models, such as Kelvin-Voigt model, Maxwell model, and standard linear solid model (SLS). Maxwell model cannot describe the elasticity creep, whereas Kelvin-Voigt model cannot describe the stress relaxation. The SLS model (Carcione 2007) can describe both the elasticity creep and the stress relaxation, more closely approximating the real law of seismic wave propagation in viscoelastic media. It is not easy to describe frequency-dependent attenuation coefficients and dispersion effects for the time-domain methods. To simplify the problem, the constant-Q model (Kjartansson 1979) assumes that the quality factor is a linear function of frequency.

Conventional viscoelastic FD modeling can accurately describe both the kinematics and dynamics characteristics of wave propagation, but requires huge computations and computer memories. As an efficient alternative, Varela (1993) proposes a one-way time-domain viscoelastic modeling algorithm with a higher computational efficiency, but involved in a single series expansion that is valid for bigger Q values. To reduce memory requirements, Lu and Hanyga (2004) use an intermediate variable to solve the fractional-order derivative, which causes huge computations. Chen and Holm (2004) apply a fractional Laplace operator to the calculation of viscous wavefields to improve computational efficiency. Treeby and Cox (2010) further explicitly decompose the dispersion and attenuation terms in the viscoelastic FD equation, greatly improving computational efficiency. Viscoelastic seismic imaging of viscosity acoustic media can compensate viscoelastic dispersion and attenuation and has been widely used to improve seismic imaging quality (Zhu 2014; Zhu et al. 2014).

Unlike the FD and FE methods that are characteristic of the implicit use of boundary conditions, the explicit use will lead to a category of semi-analytical methods. For example, frequency-domain BEMs (e.g., Dravinski 1982; Sánchez-Sesma and Campillo 1991; Fu and Mu 1994; Fu 2002), global generalized reflection/transmission matrices methods (Chen 1990, 1995, 1996), discrete wavenumber BEMs (Bouchon 1982; Bouchon et al. 1989; Fu and Bouchon 2004), global reflection/transmission BEMS (Ge and Chen 2007, 2008), dual reciprocity BEMs (Dehghan and shirzadi 2015; Dehghan and Safarpoor 2016), and some hybrid schemes (Moczo et al. 1997; Fu and Wu 2001) are more accurate in simulating reflection/transmission across irregular interfaces and have been widely used in seismology. Targeted at numerical wave propagation in layered viscoelastic media with an explicit use of boundary continuity conditions, an efficient alternative for full-waveform viscoelastic numerical modeling is proposed in this article by extending conventional frequency-domain BEMs to viscoelastic media.

We first transform the constant-Q acoustic wave equation into the corresponding wave integral representation with the Green's function accounting for viscoelastic coefficients in the frequency domain. The frequency-domain BE method is used to solve the integral equation for each subregion in multilayered media, and then the boundary conditions between subregions are used to assemble the BE submatrix from each subregion into a global system of matrices. In general, the resultant global coefficient matrix is sparse and narrow banded and can be solved by an improved block Gaussian elimination method. To show the applicability of the method, we present numerical examples with viscoelastic media to study the viscoelastic effect of different Q values on wave propagation.

2 Viscoelastic integral equations for multilayered media

Seismic response $u(\boldsymbol{r})$ for steady-state scalar wave propagation with a constant velocity v satisfies the following scalar Helmholtz equation

$$\nabla^2 u(\boldsymbol{r}) + k^2 u(\boldsymbol{r}) = -s(\boldsymbol{r}, \omega), \tag{1}$$

where the wavenumber $k = \omega/v$ and $s(\boldsymbol{r},\omega)$ is the body force. Assuming the source point is located at $\boldsymbol{r}_0$, the source term can be expressed as

$$s(\boldsymbol{r}, \omega) = S(\omega)\delta(\boldsymbol{r} - \boldsymbol{r}_0), \tag{2}$$

where $S(\omega)$ is the source spectrum and $\delta(\boldsymbol{r} - \boldsymbol{r}_0)$ is the delta function.

As is well known, wave propagation simulation in the frequency domain is easy to incorporate the viscoelastic coefficient into wave equation by expressing the acoustic velocity as a plural form. The resultant viscoelastic wave equation with a complex velocity plays an attenuation role in simulating wave propagation in viscoelastic media. We use the following complex-velocity expression (Ravaut et al. 2004) for the viscoelastic wave equation,

$$\frac{1}{\bar{v}} = \frac{1}{v}\left[1 + \frac{\mathrm{i}}{2Q}\mathrm{sign}(\omega)\right], \tag{3}$$

where $\mathrm{i} = \sqrt{-1}$, $\bar{v}$ is the complex velocity corresponding to the real velocity v, and Q is the quality factor.

Replacing the real velocity in Eq. (1) with the complex velocity by Eq. (3), seismic response $u(\boldsymbol{r})$ for viscoelastic wave propagation satisfies the following equation

$$\nabla^2 u(\boldsymbol{r}) + (1 - \alpha\mathrm{i})k^2 u(\boldsymbol{r}) = -s(\boldsymbol{r}, \omega), \tag{4}$$

where the attenuation coefficient α can be computed by the quality factor Q. Defining the complex wavenumber as $k_\alpha = \sqrt{(1 - \alpha\mathrm{i})k}$, the equation can be further compacted as

$$\nabla^2 u(\boldsymbol{r}) + k_\alpha^2 u(\boldsymbol{r}) = -s(\boldsymbol{r}, \omega), \tag{5}$$

Consider 2D steady-state scalar wave propagation in a homogenous region Ω bounded by an irregular boundary Γ. The seismic response $u(\boldsymbol{r})$ at location $\boldsymbol{r} \in \Omega$ can be

composed of the incident wavefield $f(\boldsymbol{r})$ and the boundary wavefield $u^{s}(\boldsymbol{r})$ scattered by the irregular boundary Γ

$$u(\boldsymbol{r}) = f(\boldsymbol{r}) + u^{s}(\boldsymbol{r}). \tag{6}$$

The incident wavefield in the background medium can be expressed as

$$f(\boldsymbol{r}) = \int_{\Omega} s(\boldsymbol{r}', \omega) G(\boldsymbol{r}, \boldsymbol{r}') \mathrm{d}\boldsymbol{r}' = s(\omega) G(\boldsymbol{r}, \boldsymbol{r}_0), \tag{7}$$

which can be given directly by the characteristics of the source in the numerical implementation. Based on the Helmholtz integral representation formulas for Eq. (1), the boundary wavefield satisfies the boundary integral equation

$$u^{s}(\boldsymbol{r}) = \int_{\Gamma} \left[G(\boldsymbol{r}, \boldsymbol{r}') \frac{\partial u(\boldsymbol{r}')}{\partial n} - u(\boldsymbol{r}') \frac{\partial G(\boldsymbol{r}, \boldsymbol{r}')}{\partial n} \right] \mathrm{d}\boldsymbol{r}', \tag{8}$$

where $G(\boldsymbol{r}, \boldsymbol{r}')$ and $\frac{\partial G(\boldsymbol{r}, \boldsymbol{r}')}{\partial n}$ are the free-space Green's functions, as the basic solution of the integral equation of displacement and stress in the background of the homogeneous viscoelastic medium, respectively. $\partial/\partial n$ denotes differentiation with respect to the outward normal of the boundary Γ. The viscoelastic Green's function in the complex domain has the similar form as the elastic Green's function in the real domain, that is, $G(\boldsymbol{r}, \boldsymbol{r}') = \mathrm{i}\mathrm{H}_0^{(1)}(k_\alpha |\boldsymbol{r}' - \boldsymbol{r}|)/4$ for 2D problems and $G(\boldsymbol{r}, \boldsymbol{r}') = \mathrm{e}^{ik_\alpha |\mathbf{r}' - \mathbf{r}|}/(4\pi |\boldsymbol{r}' - \boldsymbol{r}|)$ for 3D problems, where $\mathrm{H}_0^{(1)}$ denotes the complex Hanker function.

Substituting Eqs. (7), (8) in (6) and considering the "boundary naturalization" of the integral equation (that is, a limit analysis when the "observation point" $\boldsymbol{r}$ approaches the boundary Γ and tends to coincide with the "scattering point" $\boldsymbol{r}' \in \Gamma$), we obtain the following boundary integral equation

$$\int_{\Gamma} \left[G(\boldsymbol{r}, \boldsymbol{r}') \frac{\partial u(\boldsymbol{r}')}{\partial n} - u(\boldsymbol{r}') \frac{\partial G(\boldsymbol{r}, \boldsymbol{r}')}{\partial n} \right] \mathrm{d}\boldsymbol{r}' + S(\omega) G(\boldsymbol{r}, \boldsymbol{r}_0) = \begin{cases} u(\boldsymbol{r}) & \boldsymbol{r} \in \Omega \\ C(\boldsymbol{r}) u(\boldsymbol{r}) & \boldsymbol{r} \in \Gamma, \\ 0 & \boldsymbol{r} \notin \bar{\Omega} \end{cases} \tag{9}$$

where $\bar{\Omega} = \Omega + \Gamma$ and the coefficient $C(\boldsymbol{r})$ depends on the local geometry at $\boldsymbol{r}$ on Γ. $C(r) = \theta/2\pi$ with θ the opening angle at $\boldsymbol{r}$ in the direction of Ω. The boundary integral Eq. (9) for viscoelastic wave propagation is a Fredholm integral equation of the second kind. According to the Fredholm theorems, we can prove that the solution of Eq. (9) exists and is unique for an internal boundary value problem (all $\boldsymbol{r}$, $\boldsymbol{r}' \in \bar{\Omega}$) with both Neumann and Dirichlet boundary conditions. The solution is also stable because the singularity that arises when $\boldsymbol{r} \to \boldsymbol{r}'$ in the Green's function is only apparent and can be removable (Fu and Mu 1994). For numerical calculations, however, particular techniques are required for the evaluation of the weakly singular integrals, for instance, the Bouchon's discrete wavenumber expansion of the Green's functions (Bouchon 1982) and the analytical treatment (Fu and Mu 1994) that is based on the fact that the asymptotic behavior of the integral kernels can be exactly represented by their static counterparts.

3 Numerical discretization of the viscoelastic integral equation

In this section, the frequency-domain boundary element method is used to solve the viscoelastic boundary integral equation for a full-wave solution in multilayered media. The problem to be studied is illustrated by a multilayered viscoelastic model as shown in Fig. 1. In this model, there are $M + 1$ homogeneous layers over a free space, with each layer bounded by two irregular interfaces and a source embedded in arbitrary layer. For simplicity, we restrict the present study to the 2D acoustic problem (or SH problem). For instance, the elastic properties of the mth layer are described by the velocity v_m, density ρ_m, and attenuation coefficient a_m. The seismic response $u(\boldsymbol{r})$ satisfies the following boundary conditions: the continuities of displacement and traction across interfaces and the radiation boundary conditions imposed on the far-field behavior at infinity.

The collocation method has been extensively used for numerical solutions of all types of integral equations. The numerical solution of Eq. (9) by the collocation method involves several steps. First, the discretization of Eq. (9) can be done in each layer by numerical methods such as the collocation method or weighted residual method. Then, all equations are assembled into a set of simultaneous matrix equations by using the boundary conditions of continuity

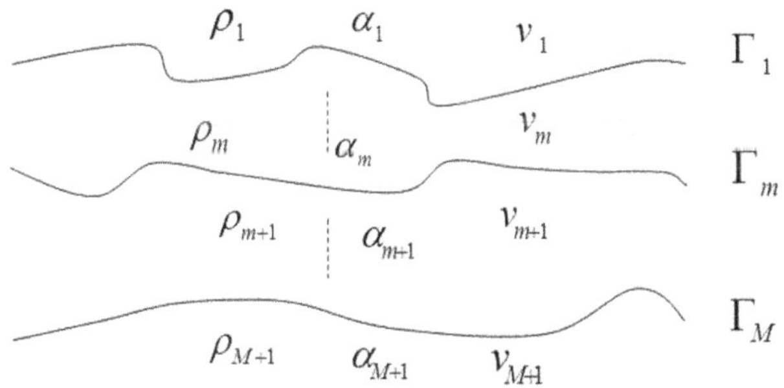

Fig. 1 Configuration of a multilayered viscoelastic model

for displacement and traction across all interfaces. This global matrix is sparse or narrow banded, depending on the structure of the model.

We discretize all interfaces in the mth layer into L boundary elements denoted by Γ_e ($e = 1, 2, \ldots, L$) resulting in a total of N nodes. In the collocation method, interpolation shape functions Φ are used so that all the variables ($\boldsymbol{r}$, u, and $\partial u/\partial n$) are approximated by the linear combination of their nodal values over an element Γ_e defined geometrically between the nodes I_1 and I_2, for example,

$$\begin{cases} u(\xi) = \sum_{l=I_1}^{I_2} u(\boldsymbol{r}_l)\Phi_l(\xi) \\ \dfrac{\partial u(\xi)}{\partial n} = \sum_{l=I_1}^{I_2} t(\boldsymbol{r}_l)\Phi_l(\xi) \end{cases}, \tag{10}$$

where ξ denotes the local coordinate of an element and t is the normal gradient of u with respect to the outward normal to the boundary. Then, Eq. (9) for $i = 1$ to N is transformed into

$$\sum_{j=1}^{N} \left[H_{ij}u(\boldsymbol{r}_j) - G_{ij}t(\boldsymbol{r}_j)\right] = f(\boldsymbol{r}_i), \tag{11}$$

where the coefficients H_{ij} and G_{ij} denote a concentrated force generated at the jth scattering point and applied at the ith observation point, which can be calculated by numerically integrating the scalar product of the Green's function with interpolation shape functions over elements,

$$H_{ij} = \sum_{e=1}^{L} \int_{\Gamma_e} \frac{\partial}{\partial n} G(\boldsymbol{r}_i, \boldsymbol{r}'(\xi))\Phi_j(\xi)\mathrm{d}\boldsymbol{r}'(\xi) + C(\boldsymbol{r}_i)\delta_{ij}, \tag{12}$$

$$G_{ij} = \sum_{e=1}^{L} \int_{\Gamma_e} G[\boldsymbol{r}_i, \boldsymbol{r}'(\xi)]\Phi_j(\xi)\mathrm{d}\boldsymbol{r}'(\xi), \tag{13}$$

where δ_{ij} is the Kronecker delta function. These integrals can be evaluated by the Gaussian integration algorithm.

After the discretization of Eq. (9) is done for all the layers, the resulting numerical equations are assembled into a global matrix equation by the boundary conditions of continuity for displacement and traction across all interfaces. For instance, the continuities of displacement and traction across the interface Γ_m are given by

$$\begin{cases} u_m^-(\boldsymbol{r}) = u_m^+(\boldsymbol{r}) \\ \dfrac{\rho_{m+1}v_{m+1}^2 t_m^-(\boldsymbol{r})}{(1-\alpha_{m+1}i)} = -\dfrac{\rho_{m+1}v_m^2 t_m^+(\boldsymbol{r})}{(1-\alpha_m i)}, \end{cases} \tag{14}$$

where "−" denotes the top side of Γ_m toward the mth layer and "+" denotes the underside of Γ_m toward the $(m+1)$th layer. The boundary continuity condition of traction can be further compacted as

$$t_m^+(\boldsymbol{r}) = -\eta_m t_m^-(\boldsymbol{r}),$$

where

$$\eta_m = \frac{(1-\alpha_{m+1}i)\rho_m v_m^2}{(1-\alpha_m i)\rho_{m+1}v_{m+1}^2}$$

After applying boundary continuity conditions to boundary integral equations, Eq. (11) can be expressed as a matrix form:

$$\begin{bmatrix} H_{1^-1^-} & -G_{1^-1^-} & & & & & & \\ \cdots & \cdots & \cdots & \cdots & & & & \\ \cdots & \cdots & \cdots & \cdots & & & & \\ & & H_{m^+m^+} & \eta_m G_{m^+m^+} & H_{m^+(m+1)^-} & G_{m^+(m+1)^-} & & \\ & & H_{(m+1)^-k^+} & \eta_m G_{(m+1)^-m^+} & H_{(m+1)^-(m+1)^-} & G_{(m+1)^-(m+1)^-} & & \\ & & & & \cdots & \cdots & \cdots & \cdots \\ & & & & \cdots & \cdots & \cdots & \cdots \\ & & & & & & H_{M^+M^+} & \eta_M G_{M^+M^+} \end{bmatrix} \begin{bmatrix} U_1 \\ \vdots \\ \vdots \\ U_m \\ T_m \\ \vdots \\ \vdots \\ T_M \end{bmatrix} = \begin{bmatrix} F \\ \vdots \\ \vdots \\ 0 \\ 0 \\ \vdots \\ \vdots \\ 0 \end{bmatrix}, \tag{15}$$

Equation (15) expresses wave propagation through the entire model, where the boundary coefficient submatrices are calculated by Eqs. (12), (13) and the source vector on the right side can be computed by Eq. (7). After solving the linear system of Eq. (15) for u and t at all the nodes, we can compute the seismic response at any receiving point in the medium through back substitution of Eq. (9).

In the numerical implementation, we use an improved block Gaussian elimination method to solve the matrix equation for the improvement in implementation efficiency. In seismic exploration, the source is generally located in the near-surface and receivers are deployed at the free surface. In such a case, we calculate and assemble the matrices $\boldsymbol{H}$ and $\boldsymbol{G}$ from deep to shallow to eliminate coupling data on the interface between two adjacent layers until the calculation reaches the surface. The resulting final matrix equation is then solved for u and t at all the nodes inside the surface layer. Note that the maximum memory amount required for the global coefficient matrix is limited to the total node number of the biggest layer, which saves memory and reduces computational costs. Fewer elements per wavelength will reduce the size of the resultant coefficient matrices. To improve computation speed, we adopt a variable element dimension technique in the program implementation. Since a discretization rate of three points per wavelength (Campillo 1987) is sufficient to make the numerical noise level negligible for general applications, the element dimension for each computational frequency is computed according to the medium velocity and the frequency, and then the model is automatically discretized. This improves the efficiency of the BE method.

4 Numerical examples

Figure 2 shows a viscoelastic homogeneous model with a velocity of 2500 m/s and three sources at different depths 1000, 2000, and 3000 m. Figure 3 shows synthetic seismograms with surface survey for $Q = 10$ and $Q = 100$, respectively. We see that the seismic resolution with $Q = 100$ is much better than that with $Q = 10$. Particularly, the synthetic amplitudes with $Q = 10$ decrease fast with increasing source depths, with the synthetic seismogram for the source at depth 3000 m almost invisible, whereas the synthetic amplitudes with $Q = 100$ decrease slowly with increasing propagation distances. The

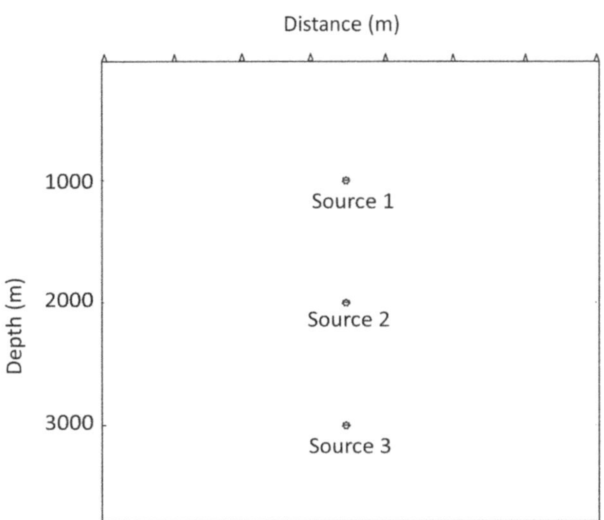

Fig. 2 Geometry of a viscoelastic homogeneous model with three sources at different depths

synthetic amplitudes from three sources at different depths are almost the same.

To access the effect of different Q values on wave propagation, we calculate zero-offset synthetic seismograms with $Q = 10, 20, 50, 100, 200, 500$, and 1000, respectively. The resulting synthetic records and their frequency spectra are shown in Figs. 4, 5 and 6 for different Q values and different source depths. We see that the amplitudes decrease obviously with decreasing Q from 1000 to 10. For the same propagation distance (depth), the decrease of Q values lowers the dominant frequency and amplitude of seismograms. For the same Q value, the resulting seismograms become weaker with increasing depths. Particularly, the seismograms with $Q = 10$ and $Q = 100$ are almost invisible as shown in Fig. 6. For the seismograms with Q over 100, variations in amplitude are not significant, that is, the

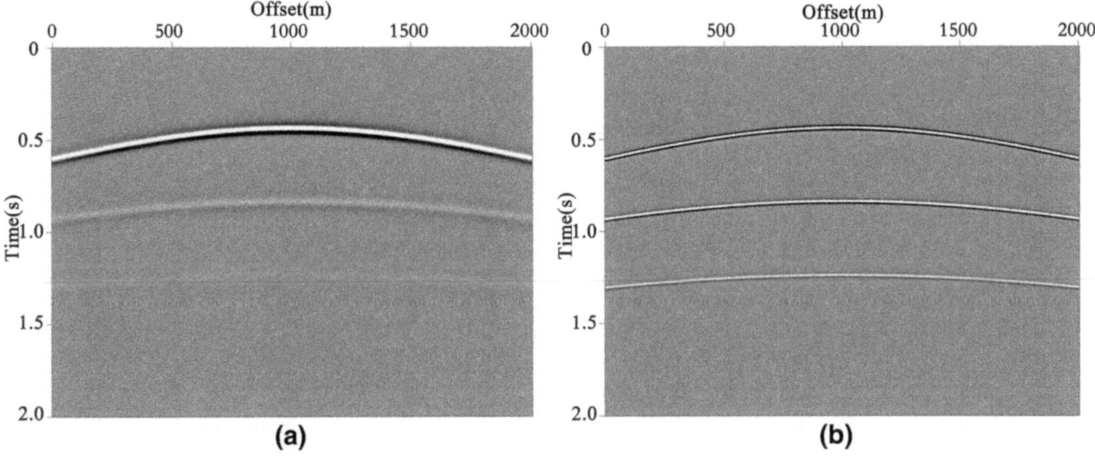

Fig. 3 Synthetic seismograms with surface survey for $Q = 10$ (**a**) and $Q = 100$ (**b**), respectively

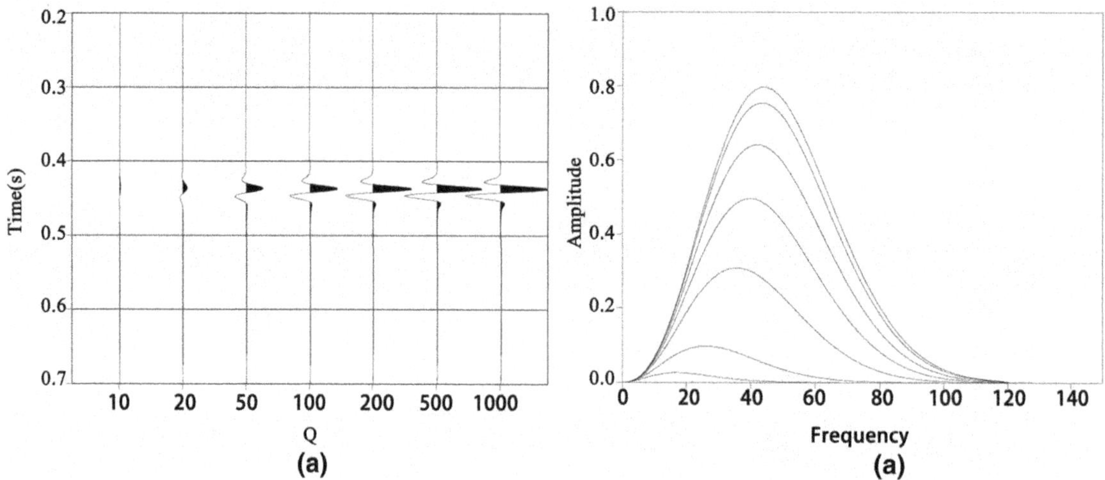

Fig. 4 Zero-offset synthetic seismograms (**a**) with the source at depth 1000 m and their frequency spectra (**b**) for different Q values

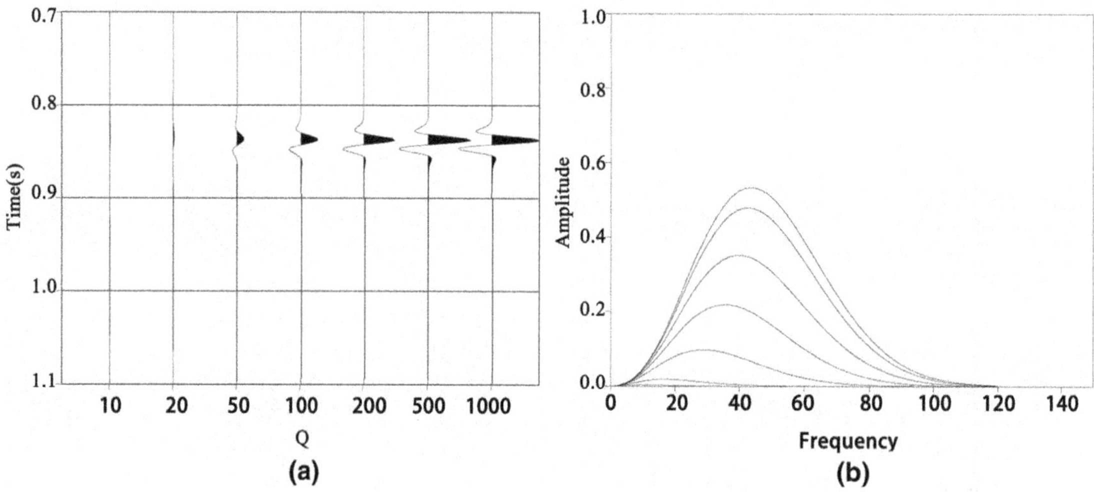

Fig. 5 Zero-offset synthetic seismograms (**a**) with the source at depth 2000 m and their frequency spectra (**b**) for different Q values

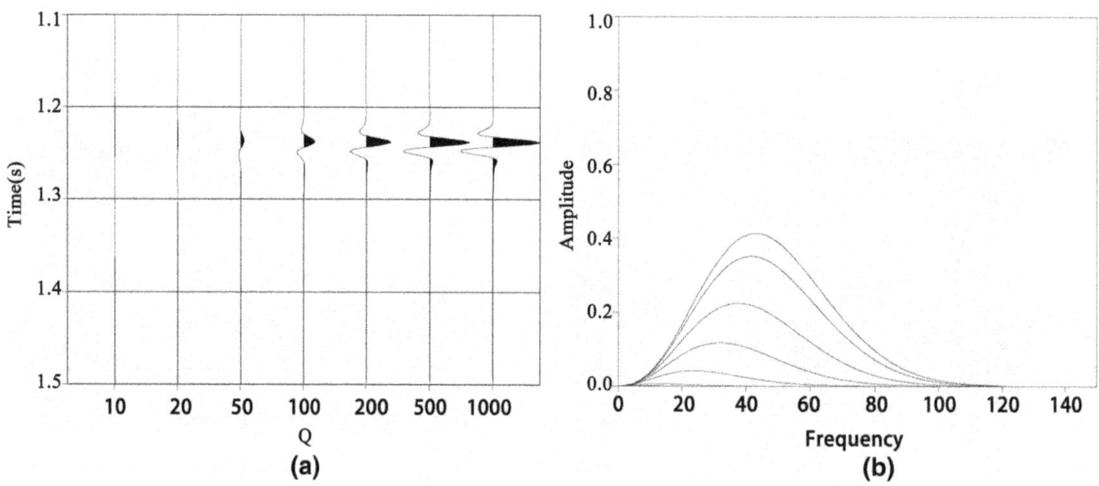

Fig. 6 Zero-offset synthetic seismograms (**a**) with the source at depth 3000 m and their frequency spectra (**b**) for different Q values

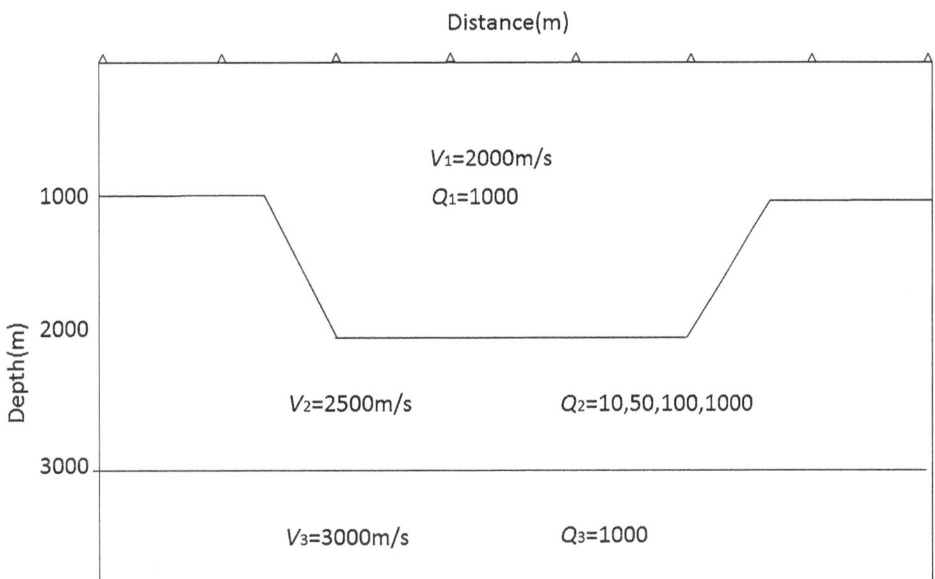

Fig. 7 Simple multilayered viscoelastic model

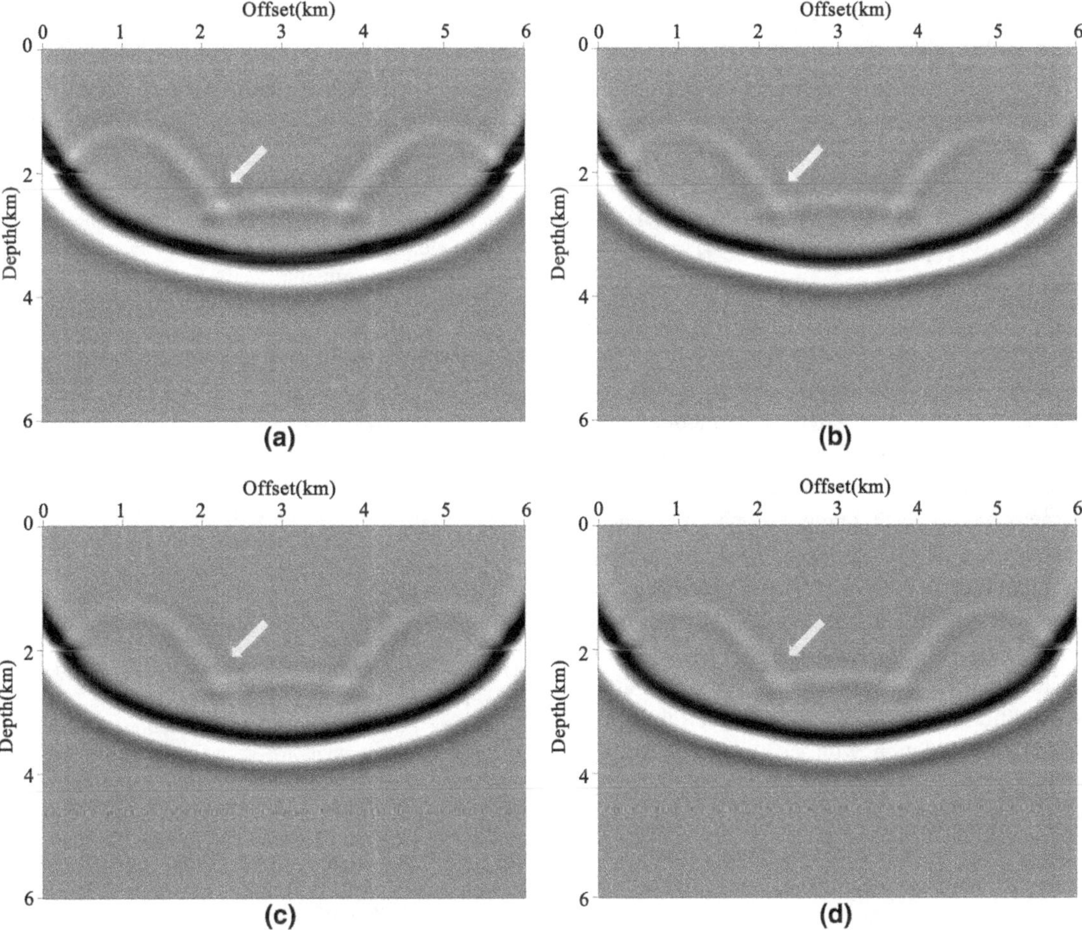

Fig. 8 Snapshots at 1000 ms of waves propagating a simple multilayered viscoelastic model with $Q_2 = 10$ (**a**), 50 (**b**), 100 (**c**), and 1000 (**d**)

influence on amplitudes attenuation is small for high-enough Q values. In conclusion, these figures show a prominent influence of attenuation on wave amplitudes

A simple sedimentary model is shown in Fig. 7 with parameters for each layer displayed in the figure. Numerical simulations are conducted with different Q_2 values of 10, 50, 100, and 1000. Figure 8 shows the snapshots at 1000 ms for $Q_2 = 10$, 50, 100, and 1000, respectively. We see that the variation of Q_2 values has less effect on wave propagation for small difference in Q between layers. Otherwise, the reflection and transmission of waves are possible to reflect the variation of Q values.

5 Conclusions

Most viscoelastic numerical simulations are based on the finite-difference and finite-element methods. In this article, we present an efficient alternative for full-waveform method to simulate wave propagation in multilayered viscoelastic media. First, the constant-Q acoustic wave equation is transformed into a corresponding viscoelastic integral equation in terms of viscoelastic Green functions. We then apply the conventional frequency-domain BEM to the viscoelastic integral equation for accurate viscoelastic numerical modeling. We extend the frequency-domain viscoelastic BEM to multilayered media using the boundary conditions of continuity for displacement and traction across all interfaces. The resultant global coefficient matrix is sparse and narrow banded, depending on the complexity of the model. We use an improved block Gaussian elimination method to solve the sparse matrix equation for the improvement in implementation efficiency. To improve computation speed in the program implementation, we adopt a variable element dimension technique based on the discretization rate of three points per wavelength. The element dimension for each computational frequency is computed according to the medium velocity and the frequency, and then, the model is automatically discretized. Compared to the finite-difference and finite-element methods, the viscoelastic boundary element method enjoys a distinct characteristic of the explicit use of boundary continuity conditions of displacement and traction, leading to a semi-analytical solution with sufficient accuracy for simulating the viscoelastic effect across irregular interfaces. Numerical experiments to study the viscoelastic effect of different Q values on the attenuation and dispersion of seismic waves demonstrate the accuracy and applicability of the method.

Acknowledgements This research was supported by the National Natural Science Foundation of China (No. 41130418) and the Strategic Leading Science and Technology Programme (Class B) of the Chinese Academy of Sciences (No. XDB10010400).

References

Aki K, Richards PG (1980) Quantitative seismology: theory and methods, vol 1. W. H. Freeman, San Francisco

Bouchon M (1982) The complete synthesis of seismic crustal phases at regional distances. J Geophys Res 82:1735–1741

Bouchon M, Campillo M, Gaffet S (1989) A boundary integral equation-discrete wavenumber representation method to study wave propagation in multilayered media having irregular interfaces. Geophysics 54:1134–1140

Campillo M (1987) Modeling of *SH*-wave propagation in an irregularly layered medium—application to seismic profiles near a dome. Geophys Prospect 35:236–249

Carcione JM (2007) Wave fields in real media: wave propagation in anisotropic, anelastic, porous and electromagnetic media, 2nd edn. Elsevier, Amsterdam

Carcione JM (2010) A generalization of the Fourier pseudospectral method. Geophysics 75:A53–A56

Carcione JM, Cavallini F, Mainardi F, Hanyga A (2002) Time domain seismic modeling of constant-Q wave propagation using fractional derivatives. Pure Appl Geophys 159:1719–1736

Chen XF (1990) Seismogram synthesis for multilayered media with irregular interfaces by global generalized reflection/transmission matrices method. I. Theory of two-dimensional SH case. Bull Seismol Soc Am 80:1696–1724

Chen XF (1995) Seismogram synthesis for multi-layered media with irregular interfaces by the global generalized reflection/transmission matrices method. II. Applications of 2-D SH case. Bull Seismol Soc Am 85:1094–1106

Chen XF (1996) Seismogram synthesis for multi-layered media with irregular interfaces by the global generalized reflection/transmission matrices method. III. Theory of 2D P-SV case. Bull Seismol Soc Am 86:389–405

Chen W, Holm S (2004) Fractional Laplacian time-space models for linear and nonlinear lossy media exhibiting arbitrary frequency power-law dependency. J Acoust Soc Am 115: 1424–1430

Dehghan M, Safarpoor M (2016) The dual reciprocity boundary elements method for the linear and nonlinear two-dimensional time-fractional partial differential equations. Math Methods Appl Sci 39:3979–3995

Dehghan M, Shirzadi M (2015) The modified dual reciprocity boundary elements method and its application for solving stochastic partial differential equations. Eng Anal Bound Elem 58:99–111

Dravinski M (1982) Influence of interface depth upon strong ground motion. Bull Seismol Soc Am 72:597–614

Dvorkin JP, Mavko G (2006) Modeling attenuation in reservoir and nonreservoir rock. Lead Edge 25:194–197

Emmerich H, Korn M (1987) Incorporation of attenuation into time domain computations of seismic wave fields. Geophysics 52:1252–1264

Fu LY (2002) Seismogram synthesis for piecewise heterogeneous media. Geophys J Int 150:800–808

Fu LY, Bouchon M (2004) Discrete wavenumber solutions to numerical wave propagation in piecewise media heterogeneous media. I. Theory of two-dimensional SH case. Geophys J Int 157:481–498

Fu LY, Mu YG (1994) Boundary element method for elastic wave forward modeling. Acta Geophys Sinica 37:521–529

Fu LY, Wu RS (2001) A hybrid BE-GS method for modelling regional wave propagation. Pure Appl Geophys 158:1251–1277

Ge ZX, Chen XF (2007) Wave propagation in irregularly layered elastic models: a boundary element approach with a global reflection/transmission matrix propagator. Bull Seismol Soc Am 97:1025–1031

Ge ZX, Chen XF (2008) An efficient approach for simulating wave propagation with the boundary element method in multilayered media with irregular interfaces. Bull Seismol Soc Am 98:3007–3016

Kjartansson E (1979) Constant-Q wave propagation and attenuation. J Geophys Res 84:4737–4748

Liao Q, McMechan GA (1996) Multifrequency viscoacoustic modeling and inversion. Geophysics 61:1371–1378

Lu JF, Hanyga A (2004) Numerical modeling method for wave propagation in a linear viscoelastic medium with singular memory. Geophys J Int 159:688–702

Moczo P, Bystrický E, Kristek J, Carcione JM, Bouchon M (1997) Hybrid modeling of P-SV seismic motion at inhomogeneous viscoelastic topographic structures. Bull Seismol Soc Am 87:1305–1323

Ravaut C, Operto S, Improta L, Virieux J, Herrero A, Dell'Aversana P (2004) Multiscale imaging of complex structures from multifold wide-aperture seismic data by frequency-domain full-waveform tomography: application to a thrust belt. Geophys J Int 159:1032–1056

Sánchez-Sesma FJ, Campillo M (1991) Diffraction of P, SV and Rayleigh waves by topographic features: a boundary integral formulation. Bull Seismol Soc Am 81:2234–2253

Treeby BE, Cox BT (2010) Modeling power law absorption and dispersion for acoustic propagation using the fractional Laplacian. J Acoust Soc Am 127:2741–2748

Varela CL (1993) Modeling of attenuation and dispersion. Geophysics 58:1167–1173

Zhu T (2014) Time-reverse modelling of acoustic wave propagation in attenuating media. Geophys J Int 197:483–494

Zhu T, Harris JM, Biondi B (2014) Q-compensated reverse time migration. Geophysics 79:S77–S87

5

Inversion of ocean-bottom seismometer (OBS) waveforms for oceanic crust structure: a synthetic study

Xueyan Li · Yanbin Wang · Yongshun John Chen

Abstract The waveform inversion method is applied—using synthetic ocean-bottom seismometer (OBS) data—to study oceanic crust structure. A niching genetic algorithm (NGA) is used to implement the inversion for the thickness and P-wave velocity of each layer, and to update the model by minimizing the objective function, which consists of the misfit and cross-correlation of observed and synthetic waveforms. The influence of specific NGA method parameters is discussed, and suitable values are presented. The NGA method works well for various observation systems, such as those with irregular and sparse distribution of receivers as well as single receiver systems. A strategy is proposed to accelerate the convergence rate by a factor of five with no increase in computational complexity; this is achieved using a first inversion with several generations to impose a restriction on the preset range of each parameter and then conducting a second inversion with the new range. Despite the successes of this method, its usage is limited. A shallow water layer is not favored because the direct wave in water will suppress the useful reflection signals from the crust. A more precise calculation of the air-gun source signal should be considered in order to better simulate waveforms generated in realistic situations; further studies are required to investigate this issue.

Keywords Waveform inversion · OBS · Oceanic crustal structure · Niching genetic algorithm

Submitted to: Earthquake Science.

X. Li · Y. Wang (✉) · Y. J. Chen
Department of Geophysics, School of Earth and Space Sciences, Peking University, Beijing 100871, China
e-mail: ybwang@pku.edu.cn

1 Introduction

Seismic waveform inversion is a technique to extract quantitative information on subsurface structure by fitting the synthetic seismograms with that of observation. Through comparison with a travel-time-based method, full-waveform inversion can improve the resolution of the model as it uses both the amplitude and phase information of various seismic phases contained in seismograms. As the wave equation is directly solved in this method, higher-order effects such as diffractions and multiple scattering are accounted for automatically (Pratt 1999). Many studies on the regional continental crust and upper-mantle structure have been carried out using waveform inversion (Das and Nolet 1998; Sherrington et al. 2004; Shibutani et al. 1996; Zheng et al. 2006; Zhu et al. 2006). Full-waveform inversion is a strong nonlinear optimization process. For gradient-based local searching methods, the final model is strongly related to the initial model and may be trapped into local minima. It is computationally time-consuming to perform large-scale forward modeling of seismic wave field propagation (Virieux and Operto 2009).

For regions where subsurface structure can be approximated by a layered model, global optimization methods, such as genetic algorithm and simulated annealing, can be applied to full-waveform inversion to improve its efficiency and convergence (Sen and Stoffa 2013). A layered 1D velocity structure is the most simplified, but fundamental, approximation of the Earth. It plays a significant role in accurate seismic location, calculation of Green's function, and can serve as the initial model for inversion of fine 2D or 3D velocity structure. A more accurate layered crust model could also contribute to improved study of the deeper Earth structure. Recently, waveform inversion has been conducted at several regions to extract layered continental crustal and upper-mantle structure using the data recorded from local seismic networks with global optimization methods (Abdelwahed and Zhao 2014; Chang and Baag 2006; Li and Lei 2014a, b; Li et al. 2007, 2012).

The oceanic crustal and upper-mantle structure is less well investigated than continental cases because of lack of seismic observation data. Researchers attempted to invert waveform data obtained from ocean reflection experiment for elastic parameters of oceanic crust (Igel et al. 1996; Mendes et al. 1990) and crust and upper-mantle velocity structure from surface waves (Cara and Lévêque 1987; Debayle and Lévêque 1997). In recent years, ocean-bottom seismometer (OBS) techniques have provided increasing amount of seismic data to study the oceanic crust of localized areas, such as north-eastern Japan (Takahashi et al. 2004), Lucky Strike segment (Seher et al. 2010), Southwest Indian Ridge (Li et al. 2015) and north-eastern South China Sea (Zhao et al. 2010). Full-waveform inversion has been applied to OBS data to improve the resolution of oceanic crust structure (Jian et al. 2014; Operto et al. 2006; Borisov and Singh 2015).

This paper attempts to extend a full-waveform inversion scheme used for continental crust and upper-mantle structure studies to the investigation of the oceanic crust using OBS data. Synthetic OBS data generated from an air-gun source in water is used. One significant difference between land and OBS observation is the presence of the water layer, which produces multiples and strong direct waves that cannot be used in waveform inversion. The validity and efficiency of OBS data methods are studied with the use of synthetic numerical models. Considering a limited number and sparse distribution of OBS stations, compared with the land seismic network, an attempt is made to perform inversion for a layered oceanic crust model. The parameters of interest include thickness and P-wave velocity of the sedimentary layer and crust. A niching genetic algorithm (NGA) is employed to perform the global optimization for the waveform inversion. The influence and performance under different observation systems of some key parameters in the NGA method are investigated, and a strategy is proposed to accelerate the convergence rate without increasing the computational complexity. This strategy is verified through comparative tests.

2 Waveform inversion method

Waveform inversion is a highly nonlinear problem. There are many traditional methods that can be used to determine seismic velocity structure through the use of waveform inversion, such as the conjugate gradient method, the grid search method, genetic algorithms (GA), and simulated annealing. Their common point is that they can only give one minimum of the objective function as the final model, and it is possible that the solution may converge to a local minimum if the initial model is quite far away from the real model (Maurice et al. 2003). However, NGA can search different minima by simulating the evolutionary processes in biology, such as crossover, mutation, selection, and competition (Mahfoud 1995). Hence, NGA is quite suitable for a multimodal optimization problem in geophysics. Koper et al. (1999) developed the NGA and applied it in a teleseismic waveform inversion for the source parameters of the M_w 7.2 Kuril Islands earthquake in 1996. Maurice et al. (2003) inferred the crustal and upper-mantle structure under southernmost South America through the use of NGA. Lawrence and Shearer (2006) gave a constrained seismic velocity and density for the mantle transition zone, and Li et al. (2012) inferred the 1D crustal structure under south-eastern Gansu, China, by applying NGA to regional waveform inversion.

Several researches have investigated the effects of different objective functions and parameters in NGA, such as the number of models in each subpopulation, and the critical separation radius (Koper et al. 1999; Li and Lei 2014a). In this paper, further discussion of NGA implementation is presented based on these previous studies, and an attempt is made to illustrate some other factors that could influence the resultant final model.

The NGA works as follows: first of all, initial models are created and then divided into n subpopulations, or demes. A deme represents a group of models that are close to each other in the whole model space, and there is a distance between different demes. We set each deme to contain m models. Traditional genetic algorithm approach is carried out in each deme to work out the second deme, but in NGA the similarity of each member of this deme is calculated with respect to the best model from the former ones. Similarity is an important parameter. The simplest approach is to define the distance, D, of two models, $\mathbf{x}$ and $\mathbf{y}$, as the arithmetic average of the normalized separation of the model parameters (Koper et al. 1999), as shown in Eq. (1),

$$D(x,y) = \frac{1}{m}\sum_{i=1}^{m}\frac{|x_i - y_i|}{b_i - a_i}, \tag{1}$$

where m represents the number of parameters we need to search, x_i and y_i are the ith parameters of models $\mathbf{x}$ and $\mathbf{y}$, and b_i and a_i are the upper and lower bounds of the ith parameters. The similarity varies from 0, for two identical models, to 1 for two models at opposite ends of the search boundary. If one's similarity exceeds a specified criterion (called R_c), then that model is given a high penalty and is eliminated in the next generation. This process continues until it reaches the final generation.

3 NGA implementation

3.1 Model parameterization

The waveform inversion assumes a layered, laterally homogeneous model and seeks to invert for crustal and upper-mantle thickness and P-wave velocity. The S-wave velocity for each layer is derived from the relationship $v_S = kv_P$, where k can be determined by the Poisson ratio of each layer. For the crustal layer, k is set to 0.577, which means that the Poisson ratio is equal to 0.25 for this case. For the sediment layer, k is equal to 0.489. The synthetic model is designed by considering the oceanic crust model given by Rao et al. (2012) and Arnulf et al. (2014). The model parameters are listed in Table 1.

3.2 Objective function and synthetic seismograms

The primary component of the objective function is the average value of root-mean-square residual between the synthetic and observed seismograms and their cross-correlation in the time domain. This is shown in Eq. (2):

$$\mathrm{Cost} = \frac{1}{2} \times \frac{\sqrt{\sum_i \sum_j (O_{ij} - S_{ij})^2}}{N_w \times \sqrt{\sum_i \sum_j O_{ij}{}^2}} + \frac{1}{2} \times \left[1 - \frac{1}{N_w} \times \sum_i \frac{\max(O_i \times S_i)}{\sqrt{O_i \times O_i}\sqrt{S_i \times S_i}} \right], \tag{2}$$

where $O(t) \times S(t) = \int O(\tau)S(t-\tau)\mathrm{d}\tau$, and N_w is the number of waveforms used. The first term of this formula is the root-mean-square error, O_{ij} and S_{ij} are the amplitudes of the observed and synthetic seismograms of the ith component at the jth sampling point. The second term contains the cross-correlation of the synthetic and observed waveform, which shows their similarity. O_i and S_i are the ith waveform of the observed and synthetic seismograms. This objective function takes into account both the amplitude and phase information of each seismic phase (Li et al. 2012).

A seismic wave field generated from an air-gun source in the water is studied. Synthetic waveforms are calculated using a reflectivity method (Fuchs and Müller 1971). In comparison with land-based observation, a major difference to OBS data is the existence of the water layer, which produces multiples in the water and strong direct waves that cannot be used in the inversion for subsurface structure. In the forward modeling, the water multiples are removed by omitting them from the calculation since they are useless for the crustal structure, though they are of a high amplitude. The direct wave propagating in the water cannot be ignored, and poses the biggest challenge to this study. The very strong direct wave and relatively weak series of reflection waves from the crust are visible in Fig. 1. With the increase of the offset, the direct wave tends to become mixed up with the first reflection phase. This prevents the study of a rather large area, restricting the study region in which the direct wave and reflection wave are separated. The useful reflection signals for inversion are selected, and the direct wave is ignored.

Table 1 Theoretical model used in this study

Layer	Thickness (km)	v_P (km/s)	v_s (km/s)	ρ (g/cm^3)
Water	3.2	1.5	0	1.05
Sediment	1.8	2.3	1.125	2.00
1	3.0	5.5	3.176	2.55
2	2.5	6.7	3.868	2.85
3	4.0	7.1	4.099	3.10
Upper mantle		8.1	4.677	3.25

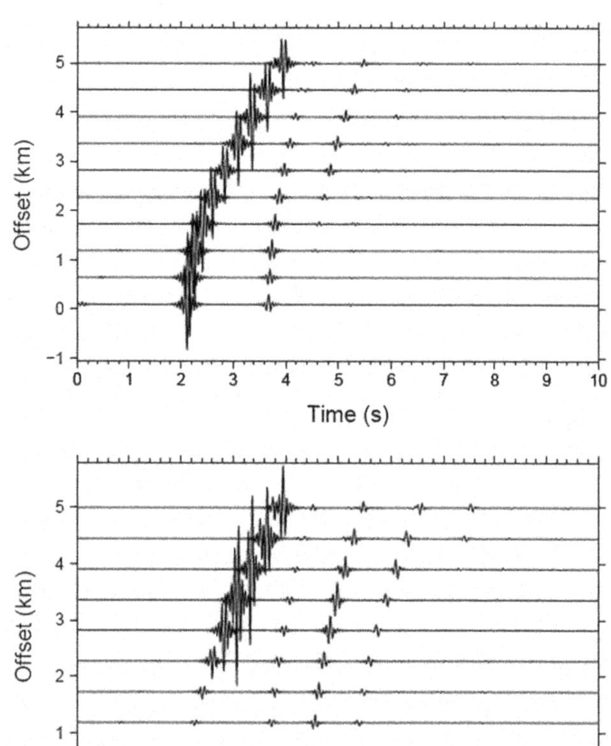

Fig. 1 Synthetic seismograms calculated by the reflectivity method. The *upper* and *lower* panels represent the vertical component and the radial component, respectively

The seismic source is an explosive type, and a Ricker wavelet is used to imitate the air-gun signal. The source is located at a depth of 20 m in the water layer. Considering that the real air-gun signal contains more high-frequency components, a Butterworth band-pass filter is applied to the source with corner frequencies of 5 and 15 Hz, and a central frequency of 10 Hz.

The first phase with a large amplitude shown in Fig. 1 is the direct wave, which adds a large noise signal to the reflected waves. Since they are not overlapped, a time window can be used to pick up the subsequent reflections for waveform inversion.

4 Numerical test of parameters in NGA method

4.1 General test

As shown in Table 1, a total of five P-wave velocities and four thicknesses of layers are searched. The searching ranges are within ±0.5 km/s for v_P and ±30 % for the thickness around the preset values described in Table 1. In the first test, ten receivers are set, in line, at the surface of the sediment layer, each between 100 meters and 5 km in horizontal distance apart from the source, as shown in Fig. 2. The NGA inversion is calculated for 500 generations with ten demes, with each deme containing ten models. Seven different NGA runs are performed, using different random model generators (by changing the random number value in the NGA). The average of seven results is taken, with the minimum value of the objective function in each run being used for the final model, with their standard deviation as the error. The probabilities of crossover and mutation are set to 0.90 and 0.10, respectively.

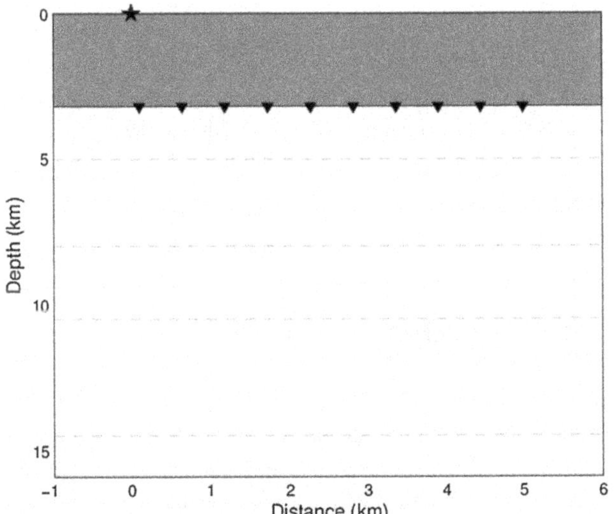

Fig. 2 Descriptive sketch for the distribution of air-gun source (★) and OBS receivers (*triangles*)

The results shown in Table 2 and Fig. 3 show the comparison of the inversion result with the true model. Figure 4 shows the cost-generation curve. It is clear that the convergence rate slows down gradually, and that the change becomes negligible after hundreds of generations. It is evident from Fig. 5 that the waveform is a good fit.

4.2 Number of models in each deme

The number of models (N_m) in each deme has a great effect on the result of the NGA inversion. If N_m is too small, then the rate of convergence will be low if there is poor diversity in each subpopulation. This could easily be solved by increasing the number of models, which would also lead to more computing time. So there is a trade-off between improving the convergence rate and cutting down the computing cost. A test is performed with N_m equal to 6, 8, 10, 12, and 16. It is evident from Fig. 6 that the results of the test with the smallest value of N_m have the lowest quality, and that quality improves as N_m increases. It was found that once N_m has reached a value greater than 10, then there is no further apparent change in the cost value. For an appropriate balance of accuracy and time requirement, N_m equal to 10 was selected.

4.3 Critical distance R_c

It is always the hope that the inversion method can search the whole solution space and avoid converging on the local minima as much as possible. In NGA, this is carried out by choosing a suitable R_c value, which controls how different subpopulations migrate into different niches of the solution space. If R_c equals 0, then all demes will inhabit global minimum as no artificial distinction exists between each deme, like performing several GA at the same time. However, if R_c is too large, then only the best model evolves with time, while the others are artificially ruled out and become "frustrate". According to Koper et al. (1999), the cost value of a model that is "frustrate" always remains high and shows little improvement over time, being reinitialized randomly after every generation. This is undesirable behavior, because competition and selection between useful demes is required in order to efficiently search for the optimum model.

Four different R_c values are selected for this test: 0.01, 0.1, 0.2, and 0.3. Values greater than 0.3 resulted in poor performance in pretest and are not discussed. It is evident that the cost value is not so favorable when R_c is equal to 0.3, and from Fig. 7 there is no obvious distinction between the other tests. It is only by examining the results of the 200th generation that some subpopulations are separated when R_c equals 0.1 or 0.2. With R_c equal to 0.01, most

Table 2 Comparison of results from waveform inversion with the theoretical model, for the general test described in Sect. 4.1

Theoretical model thickness H (km)	Inversion result thickness $H \pm \sigma$ (km)	Theoretical model P-wave velocity v_P (km/s)	Inversion result P-wave velocity $v_P \pm \sigma$ (km/s)
1.8	1.798 ± 0.002	2.3	2.298 ± 0.002
3.0	3.003 ± 0.010	5.5	5.504 ± 0.016
2.5	2.515 ± 0.258	6.7	6.539 ± 0.199
4.0	3.702 ± 0.521	7.1	6.870 ± 0.230
		8.1	7.668 ± 0.098

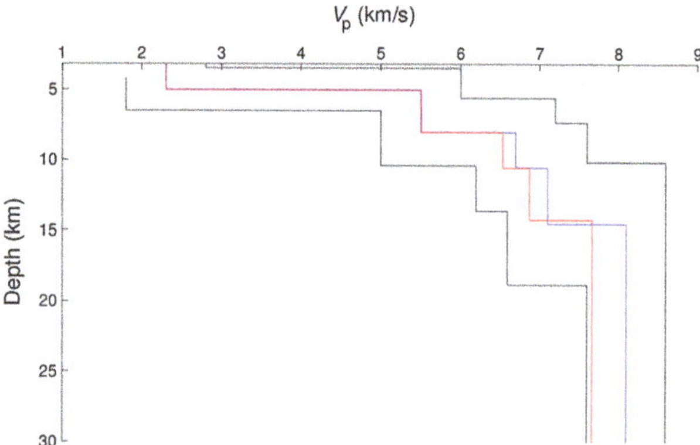

Fig. 3 Comparison of the average model (*red line*) from seven NGA inversions with the true models (*blue line*). The preset ranges for velocity and depth, over which the model parameters are allowed to change in the inversion, are shown by *black lines*

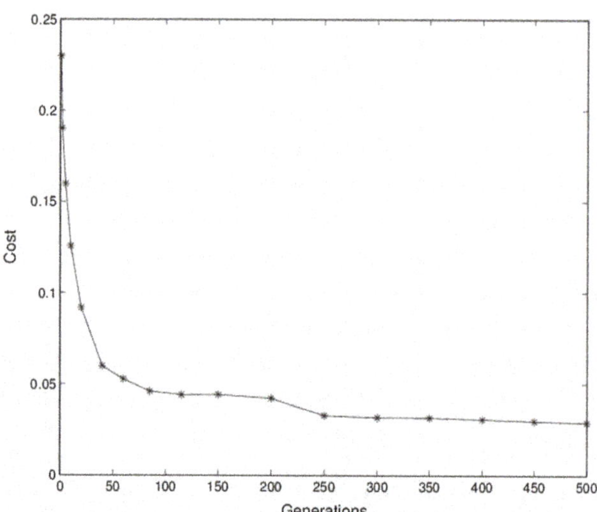

Fig. 4 Cost versus generations. Under the general test in which the total number of generations is 500, R_c is 0.1, and the number of models is 10

subpopulations converge to the global minimum. Considering the competition between subpopulations, it is desirable to set R_c between 0.1 and 0.2 in order to make our NGA different from the traditional GA, while at the same time ensuring that none, or very few, subpopulations become "frustrate"—ensuring the search efficiency of the NGA.

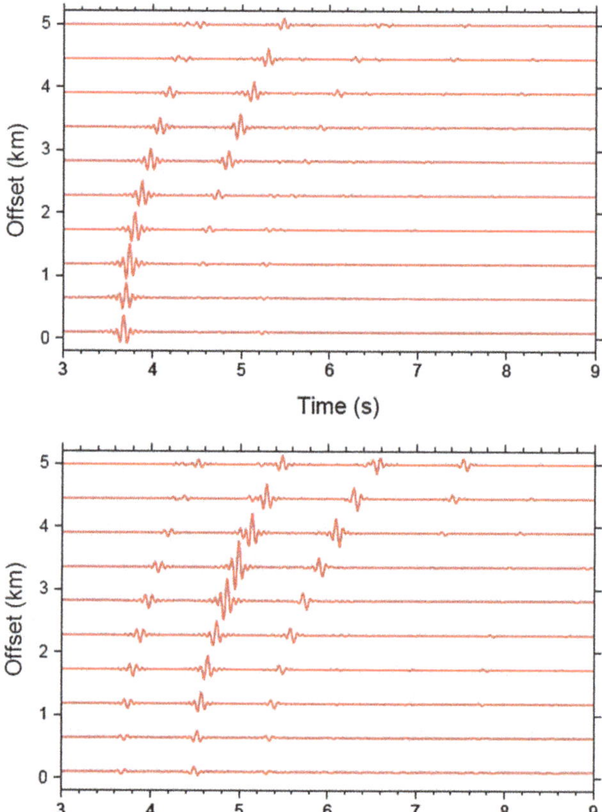

Fig. 5 Comparison of vertical (*upper panel*) and radial (*lower panel*) component waveforms (*black*) used for the inversion with those (*red*) calculated from average final models in the seven subpopulations

5 Effect of different observation systems

In the previous discussion, the receivers were always arranged in a line, with the goal of achieving an along-axis profile. However, there are several ways to set the observation system based on different research purposes. This section discusses the feasibility of NGA inversion methods using a variety of observation systems.

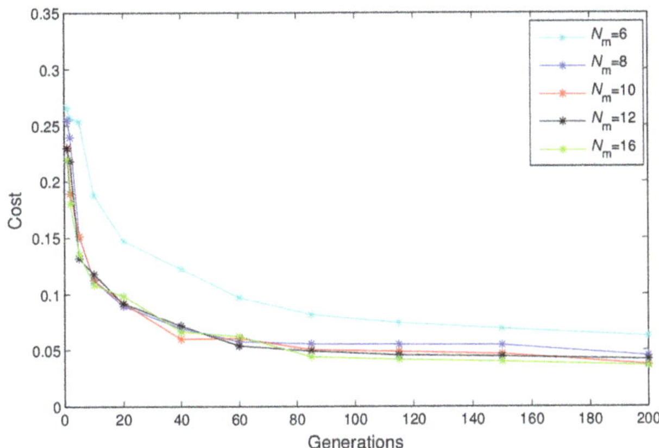

Fig. 6 Cost versus generations, for the tests described in Sect. 4.2. The results are obtained for five different N_m values

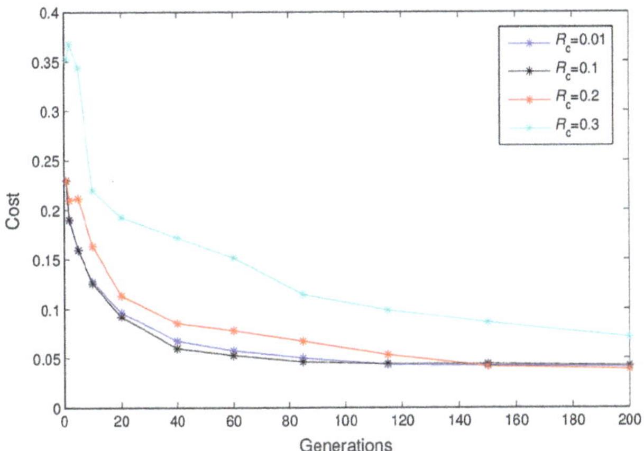

Fig. 7 Cost versus generations, for the tests described in Sect. 4.3. The results are obtained for four different R_c values

5.1 Irregular and sparse distribution of receivers

Four receivers are set at different distances from the source and azimuths with respect to the source as shown in Fig. 8. In some cases, it was not possible to set a sufficient number of OBS in line, due to the constraints of the situation. Furthermore, data from the irregular distribution of OBS give the average 1D structure of a wider area, while linear arrangement just focuses on the profile. Therefore, it is necessary to perform tests for such an OBS network. The results of these tests are presented in Table 3 and Figs. 9, 10, and 11. A good fitting of the waveform is still observed.

5.2 Single receiver

Even though the observation system could include many accurately timed, high-gain seismometers, a single-station method still has advantages, of which economic cost is an important one. Tests were performed in order to examine whether single OBS data are sufficient to explore the local

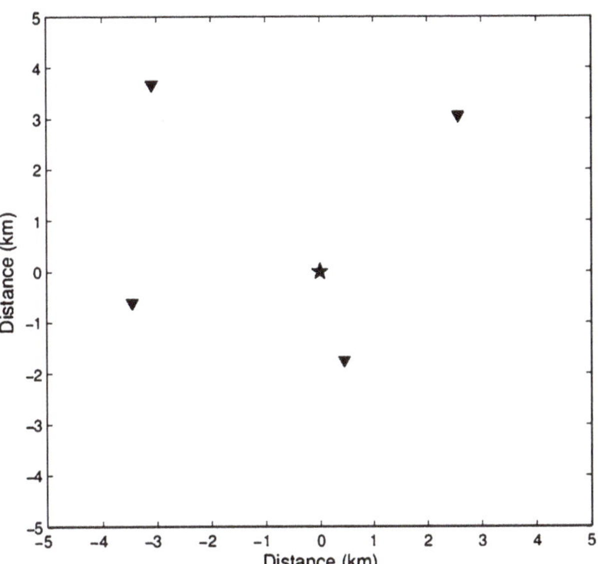

Fig. 8 The distribution of source (*black star*) and receivers (*black triangles*) in the horizontal plane

crust structure. The distance of the OBS to the source is 3.5 km. The results of the tests are presented in Table 4 and Figs. 12, 13, and 14. The waveform fitting is sufficiently good to validate the single receiver method.

6 Strategy to speed up convergence

Through observation of the evolution of cost values in the previous numerical tests, it is apparent that calculating for more generations gives a final model with a lower cost value, but that this comes at the expense of significantly increased time consumption. Because significant changes rarely take place after about 100 generations, as shown by the cost-generation curves, it is proposed to use the result of the first few generations to further narrow the range of each parameter to conduct a second inversion. Generally speaking, a smaller search range leads to a better inversion result, and this can be judged by comparison of the value of the object function.

In this section, the two experiments described in Sect. 5 are revisited, with the results of the 15th generation used. The mean value and standard deviation σ of each parameter are calculated, and used to set the upper and lower bound of a new search range equal to mean $\pm 3\sigma$. A coefficient smaller than 3 would be more favorable, but would increase the risk that the parameter value of the real model would not be included in the search range, thus making this inversion a failure. For this reason, 3 is chosen as the coefficient, which gives a limited restriction on the range as shown in Tables 5 and 6.

Table 3 Comparison of results from waveform inversion with the theoretical model, for the tests described in Sect. 5.1

Theoretical model thickness H (km)	Inversion result thickness $H \pm \sigma$ (km)	Theoretical model P-wave velocity v_P (km/s)	Inversion result P-wave velocity $v_P \pm \sigma$ (km/s)
1.8	1.805 ± 0.015	2.3	2.306 ± 0.018
3.0	3.189 ± 0.418	5.5	5.518 ± 0.076
2.5	2.505 ± 0.386	6.7	6.570 ± 0.220
4.0	4.034 ± 0.772	7.1	7.065 ± 0.216
		8.1	7.967 ± 0.227

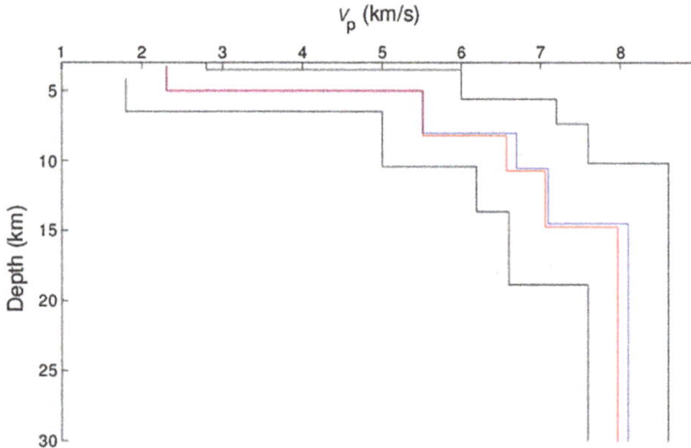

Fig. 9 Comparison of the average model (*red line*) from seven NGA inversions with the true models (*blue line*). The preset ranges for velocity and depth in which the model parameters are allowed to change in the inversion are shown by *black lines*, for the tests described in Sect. 5.1

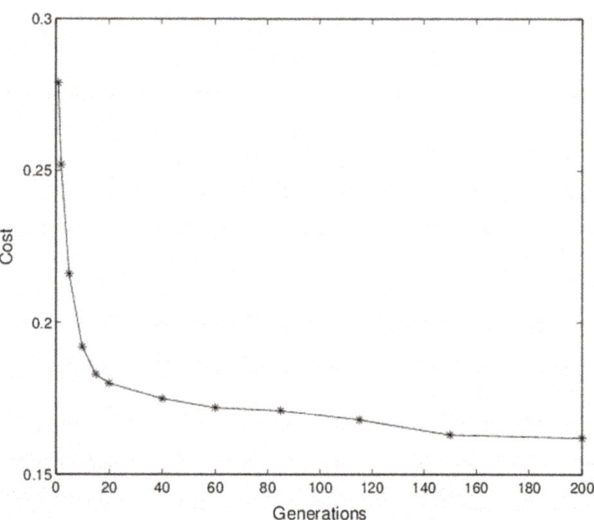

Fig. 10 Cost versus generations. This result is from the situation in which OBS are located irregularly and sparsely, for the tests described in Sect. 5.1

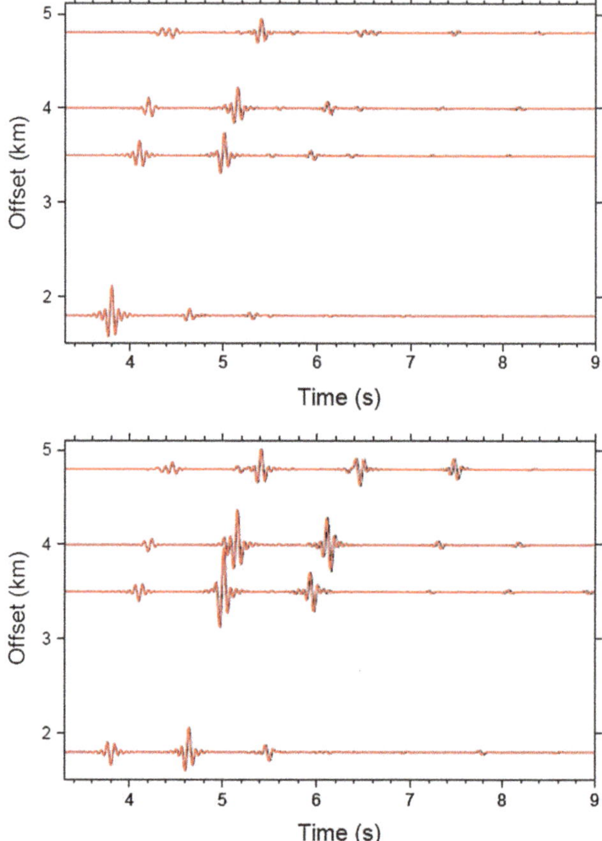

Fig. 11 Comparison of vertical (*upper panel*) and radial (*lower panel*) component waveforms (*black*) used for the inversion with those (*red*) calculated from average final models in the seven inversions, for the tests described in Sect. 5.1

From Tables 5 and 6, it can be seen that the range restriction greatly limits the range of the top two layers, especially the first layer, and that improvements on the upper layers have significant consequences. Because each reflection wave phase travels through upper layers, if the velocity and thickness values deviate away from the actual ones too much, then it will be difficult for all of the seismic phases to arrive at the correct time, which increases the difficulty of waveform fitting. Therefore, better estimation of the upper layers is crucial, and the range restriction strategy contributes to a reduction of its uncertainty.

The essence of this strategy is to reduce the searching range of model parameters from the first inversion within a

Table 4 Comparison of results for the tests described in Sect. 5.2, with a source receiver offset of 3.5 km

Theoretical model thickness H (km)	Inversion result thickness $H \pm \sigma$ (km)	Theoretical model P-wave velocity v_P (km/s)	Inversion result P-wave velocity $v_P \pm \sigma$ (km/s)
1.8	1.796 ± 0.023	2.3	2.296 ± 0.024
3.0	2.924 ± 0.277	5.5	5.488 ± 0.026
2.5	2.499 ± 0.555	6.7	6.575 ± 0.216
4.0	4.299 ± 0.612	7.1	7.006 ± 0.293
		8.1	8.164 ± 0.229

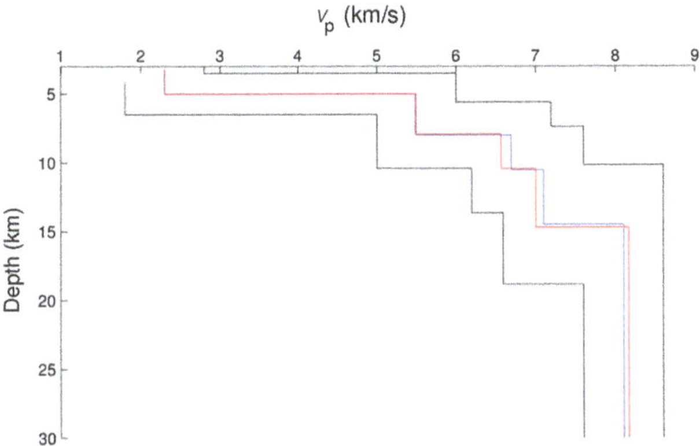

Fig. 12 Comparison of the average model (*red line*) from seven NGA inversions with the true models (*blue line*). The preset ranges for velocity and depth, in which the model parameters are allowed to change in the inversion, are shown by *black lines*, for the tests described in Sect. 5.2

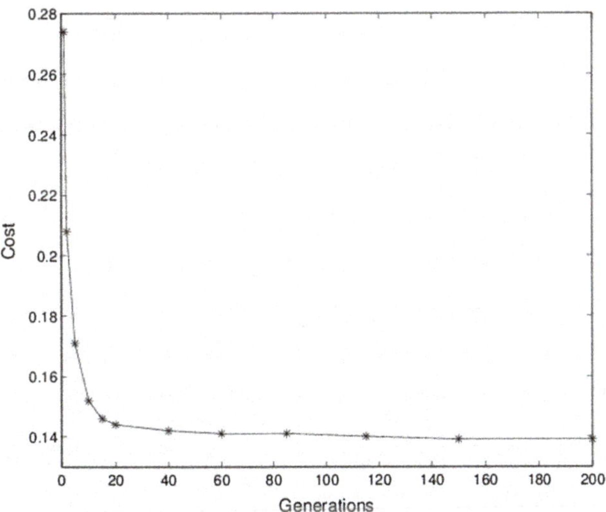

Fig. 13 Cost versus generations. This result is from the situation in which the OBS is located 3.5 km away, for the tests described in Sect. 5.2

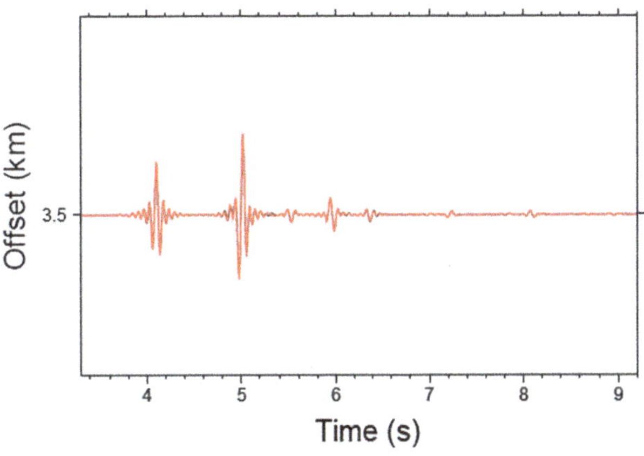

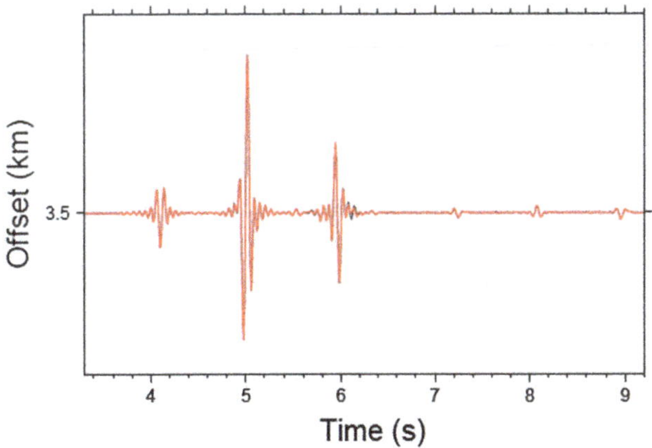

Fig. 14 Comparison of vertical (*upper panel*) and radial (*lower panel*) component waveforms (*black*) used for the inversion with those (*red*) calculated from average final models in the seven subpopulations, for the tests described in Sect. 5.2

few generations, and then to carry out a new, second, inversion with the narrowed range. It is seen that convergence on the global minimum is more likely after several generations in the second inversion. By applying this strategy, it takes less than 50 generations to obtain a result that has a similar cost value as was found from the ordinary inversion as shown in Fig. 15. A time saving of almost 80 % is made in this way, demonstrating that the computational efficiency is greatly enhanced.

This strategy utilizes a simple "mean $\pm k\sigma$" approach, but there remains a question of optimal selection of the constant k. This work set k equal to 3, and the results show

Table 5 Comparison of the search range, before and after restriction, for the tests described in Sect. 5.1

Model thickness H_0 (km)	Search range before restriction	Search range after restriction
1.8	1.26–2.34	1.72–1.91
3.0	2.10–3.90	2.35–3.50
2.5	1.75–3.25	1.91–3.25
4.0	2.80–5.20	2.80–5.20
Model V_P v_{P0} (km/s)	Search range before restriction	Search range after restriction
2.3	1.80–2.80	2.21–2.43
5.5	5.00–6.00	5.07–5.91
6.7	6.20–7.20	6.20–7.20
7.1	6.60–7.60	6.60–7.60
8.1	7.60–8.60	7.60–8.60

Table 6 Comparison of the search ranges before and after restriction, for the tests described in Sect. 5.2

Model thickness H_0 (km)	Search range before restriction	Search range after restriction
1.8	1.26–2.34	1.66–1.89
3.0	2.10–3.90	2.10–3.90
2.5	1.75–3.25	1.75–3.25
4.0	2.80–5.20	2.80–5.20
Model V_P v_{P0} (km/s)	Search range before restriction	Search range after restriction
2.3	1.80–2.80	2.16–2.39
5.5	5.00–6.00	5.02–5.96
6.7	6.20–7.20	6.23–7.14
7.1	6.60–7.60	6.60–7.60
8.1	7.60–8.60	7.60–8.60

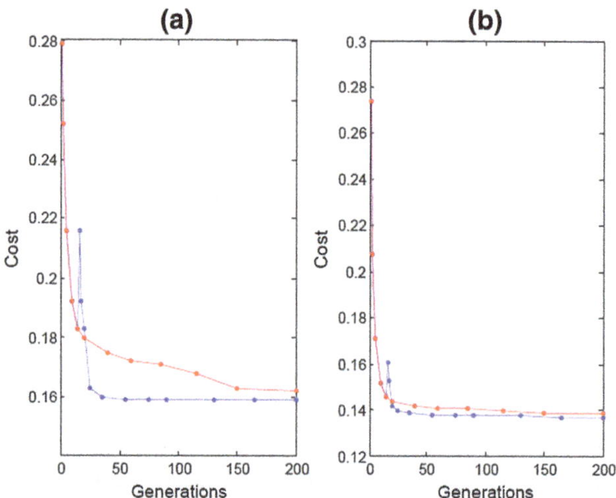

Fig. 15 Comparison of cost-generation curves of prerange-restriction test (*red*) with those with the range restriction strategy applied (*blue*): **a** the irregular and sparse OBS distribution described in Sect. 5.1 and **b** the single OBS test described in Sect. 5.2. The sharp rise is caused by the reshuffling of all the parameter values from the new, restricted ranges to determine the initial model for the second inversion

that 3 is a suitable value, but it is not proven to be the ideal value. It may be difficult to prove the existence of an optimal k value, and so the value should be based on practical experience of real cases.

7 Discussion and conclusions

This work attempted to invert the waveform recorded by OBS using a synthetic model, for the purpose of studying the structure of the oceanic crust. Tests were carried out on the parameters of a niching genetic algorithm in order to determine appropriate values. Parameter values of N_m equal to 10 and of R_c between 0.1 and 0.2 were found to be favorable. Different observation systems, irregular distribution and single station, were studied, and results showed that both offered quite good waveform fitting. A strategy was proposed to accelerate the convergence rate and its effectiveness was verified. In conclusion, using OBS data for waveform inversion is effective in the study of a regional oceanic crust structure that can be approximated by a layered model.

Limitations as well as advantages were found. Most significantly, this method is confined to a circular range of several kilometers, where the direct wave in water does not overlap with the useful reflection signals. By setting the thickness of the water layer to 3.2 km, a circular area of 5 km radius is available. This radius will be smaller if the depth of the sea level is shallower, as the direct wave will suppress the reflected wave at a close distance. Measures should be taken in order to study larger areas, and in particular, an effective method is required to ignore the direct wave while not adding artificial noise to the reflection signals. In addition, this study used a Ricker wavelet to approximate the air-gun source signal, but the real seismic source is much more complicated. Many researchers have discussed improved approximation of the real bubble behavior (e.g., Johnson 1994; Schulze-Gattermann 1972), but this was not included in this study. This study did not consider the influence of different source time functions, but if an accurate expression of air-gun source time function is calculated, then this will offer a better simulation of the real situation.

Acknowledgments This research was supported by the National Natural Science Foundation grant No. 41174034 and the Major State Basic Research Development Program of China (973 Program).

References

Abdelwahed MF, Zhao D (2014) Genetic waveform modeling for the crustal structure in Northeast Japan. J Asian Earth Sci 89:66–75

Arnulf AF, Harding AJ, Singh SC, Kent GM, Crawford WC (2014) Nature of upper crust beneath the Lucky Strike volcano using elastic full inversion of streamer data. Geophys J Int 196:1471–1491

Borisov D, Singh SC (2015) Three-dimensional elastic full waveform inversion in a marine environment using multicomponent ocean-bottom cables: a synthetic study. Geophys J Int 201:1215–1234. doi:10.1093/gji/ggv048

Cara M, Lévêque JJ (1987) Waveform inversion using secondary observables. Geophys Res Lett 14(10):1046–1049

Chang SJ, Baag CE (2006) Crustal structure in southern Korea from joint analysis of regional broadband waveforms and travel times. Bull Seismol Soc Am 96(3):856–870

Das T, Nolet G (1998) Crustal thickness map of the western United States by partitioned waveform inversion. J Geophys Res 103(B12):30021–30038

Debayle E, Lévêque JJ (1997) Upper mantle heterogeneities in the Indian Ocean from waveform inversion. Geophys Res Lett 24(3):245–248

Fuchs K, Müller G (1971) Computation of synthetic seismograms with the reflectivity method and comparison with observations. Geophys J Int 23(4):417–433

Igel H, Djikpéssé H, Tarantola A (1996) Waveform inversion of marine reflection seismograms for P impedance and Poisson's ratio. Geophys J Int 124(2):363–371

Jian H, Singh SC, Chen YJ, Li J (2014) Imaging of lower-crustal magma chambers at an ultraslow spreading ridge segment using elastic waveform inversion of a sparse OBS dataset. In: AGU Fall Meeting Abstracts, vol 1. p 4752

Johnson DT (1994) Understanding air-gun bubble behavior. Geophysics 59(11):1729–1734

Koper KD, Wysession ME, Wiens DA (1999) Multimodal function optimization with a niching genetic algorithm: A seismological example. Bull Seismol Soc Am 89(4):978–988

JF Lawrence, Shearer PM (2006) Constraining seismic velocity and density for the mantle transition zone with reflected and transmitted waveforms. Geochemistry, Geophysics, Geosystems. doi: 10.1029/2006GC001339

Li C, Lei J (2014a) Numerical tests for effects of various parameters in niching genetic algorithm applied to regional waveform inversion. Earthq Sci 27(5):541–551. doi:10.1007/s11589-014-0095-7

Li C, Lei J (2014b) Crustal velocity structure under southwestern Yunnan from regional waveform inversion. Chin Sci Bull 59:3398–3415. doi: 10.1360/N972014-00407Li (**in Chinese with English abstract**)

Li H, Michelini A, Zhu L (2007) Crustal velocity structure in Italy from analysis of regional seismic waveforms. Bull Seismol Soc Am 97(6):2024–2039

Li SH, Wang YB, Liang ZB, He SL, Zeng WH (2012) Crustal structure in southeastern Gansu from regional seismic waveform inversion. Chin J Geophys 55(4):1186–1197. (**in Chinese with English abstract**)

Li J, Jian H, Chen YJ, Singh SC, Ruan A, Qiu X, Zhao M, Wang X, Niu X, Ni J, Zhang J (2015) Seismic observation of an extremely magmatic accretion at the ultraslow spreading Southwest Indian Ridge. Geophys Res Lett 42(8):2656–2663

Mahfoud SW (1995) Niching methods for genetic algorithms. Ph.D. dissertation, University of Illinois at Urbana-Champagne

Maurice R, Stacey D, Wiens DA, Koper KD, Vera E (2003) Crustal and upper mantle structure of southernmost South America inferred from regional waveform inversion. J Geophys Res. doi:10.1029/2002JB001828

Mendes M, Beydoun W, Planchon JM, Tarantola A (1990) Seismic identification of a karstified reservoir in the Adriatic Sea. Geophysics 55(12):1639–1644

Operto S, Virieux J, Dessa JX, Pascal G (2006) Crustal seismic imaging from multifold ocean bottom seismometer data by frequency domain full waveform tomography: application to the eastern Nankai trough. J Geophys Res 111:B09306. doi:10.1029/2005JB003835

Pratt RG (1999) Seismic waveform inversion in the frequency domain, Part 1: theory and verification in a physical scale model. Geophysics 64(3):888–901

Rao PP, Rajput S, Ashalatha B, Shankar U, Sain K, Naidu MS, Rriveni V, Thakur NK (2012) Lithospheric structure model of Central Indian Ocean Basin using ocean bottom seismometer data. J Earth Sci Eng 2:344–359

Schulze-Gattermann R (1972) Physical aspects of the "airpulser" as a seismic energy source. Geophys Prospect 20(1):155–192

Seher T, Crawford WC, Singh SC, Cannat M, Combier V, Dusunur D (2010) Crustal velocity structure of the Lucky Strike segment of the Mid-Atlantic Ridge at 37 N from seismic refraction measurements. J Geophys Res. doi:10.1029/2009JB006650

Sen MK, Stoffa PL (2013) Global optimization methods in geophysical inversion. Cambridge University Press, Cambridge

Sherrington HF, Zandt G, Frederiksen A (2004) Crustal fabric in the Tibetan Plateau based on waveform inversions for seismic anisotropy parameters. J Geophys Res. doi:10.1029/2002JB002345

Shibutani T, Sambridge M, Kennett B (1996) Genetic algorithm inversion for receiver functions with application to crust and uppermost mantle structure beneath eastern Australia. Geophys Res Lett 23(14):1829–1832

Takahashi N, Kodaira S, Tsuru T, Park JO, Kaneda Y, Suyehiro K, Kinoshita H, Abe S, Nishino M, Hino R (2004) Seismic structure and seismogenesis off Sanriku region, northeastern Japan. Geophys J Int 159(1):129–145

Virieux J, Operto S (2009) An overview of full-waveform inversion in exploration geophysics. Geophysics 74(6):WCC1–WCC26

Zhao M, Qiu X, Xia S, Xu H, Wang P, Wang TK, Lee CS, Xia K (2010) Seismic structure in the northeastern South China Sea: S-wave velocity and v_P/v_S ratios derived from three-component OBS data. Tectonophysics 480(1):183–197

Zheng T, Chen L, Zhao L, Xu W, Zhu R (2006) Crust-mantle structure difference across the gravity gradient zone in North China Craton: seismic image of the thinned continental crust. Phys Earth Planet Inter 159(1):43–58

Zhu L, Mitchell BJ, Akyol N, Cemen I, Kekovali K (2006) Crustal thickness variations in the Aegean region and implications for the extension of continental crust. J Geophys Res 111:B01301. doi:10.1029/2005JB003770

6

P-wave tomography and relation between shallow and deep structures beneath the Songliao basin

Qiyan Yang · Qingju Wu · Xiaojun Ma · Fengxue Zhang · Yanrui Sheng

Abstract We selected relative travel-time residuals from teleseismic waveform data using the waveform correction method and imaged the P wave velocity structure beneath Northeast China. In combination with other geophysical data, we discussed the relation between the shallow and deep structures of the area. The results show that there is a primary high-velocity zone with some high- and low-velocity distribution characters beneath the Songliao basin. The low-velocity anomalies may extend down to the upper mantle, and may be the result of material upwelling. The low-velocity anomaly beneath the southern part of the Songliao basin is connected to those beneath the Changbaishan and A'ershan volcanic areas. It may be an upwelling channel from the mantle beneath the Songliao basin and adjacent area. This finding indicates the Songliao basin was a result of asthenospheric upwelling caused by subduction of the Pacific plate under the Eurasian plate.

Keywords Tomography · Northeast China · Ray · Relation between shallow-deep structures

Q. Yang · Y. Sheng
Earthquake Administration of Hebei Province,
Shijiazhuang 050021, China

Q. Yang · Q. Wu (✉) · X. Ma · F. Zhang
Institute of Geophysics, China Earthquake Administration,
Beijing 100081, China
e-mail: wuqj@cea-igp.ac.cn

X. Ma
Earthquake Administration of Ningxia Autonomous Region,
Yinchuan 751000, China

1 Introduction

The Songliao basin is in the center of the Northeast China region, with the Central Asian fold belt (the Great Xing'an Range) to the west, the uplifted Changbaishan volcanic region to the east, and the Sino-Korean Craton to the south. The Songliao basin is commonly believed to be formed due to rifting in the Mesozoic Era (Hu et al. 1998; Ren et al. 2001; Liu and Niu 2011). In recent years, some studies have used tomography (P wave, Pn wave, S wave, and ambient noise), receiver functions, and S wave attenuation to reveal deep dynamic characteristics in the area (Zhao 2004; Lei and Zhao 2005; Sun 2012; Zhang et al. 2013, 2014; Pan et al. 2014). The previous research has shown that the Songliao basin was probably formed as a result of rifting during the Mesozoic. The eastern margin of the Songliao basin, where Cenozoic volcanism is active, has a relatively thin crust, with a upwelling channel of material from the mantle. The lithospheric mantle possibly underwent destruction and transformation, and lithospheric delamination may have occurred and greatly affected the Songliao basin (Song et al. 2010; Yu et al. 2012). Other studies have used the Bouguer gravity anomaly (Wei et al. 2014), magnetic (Shu et al. 2003), geothermal (Shi 2004), seismic and geological profiles (Wan 2012) to study the tectonic evolution of the Songliao basin. The dynamic studies paid more attention to the deep mantle structure beneath the basin; while tectonic evolution studies focused on shallow structure change. However the relation between shallow and deep structures in Songliao basin still remains poorly understood. In this study, we investigated lateral and vertical variations of the mantle using P wave tomographic data and other geophysical results to determine the relation between the shallow and deep structures beneath the Songliao basin.

2 Seismic data and method

The recent release of broadband waveform data of permanent regional seismic networks in Northeast China from the China Earthquake Administration (CEA) (Zheng et al. 2009), from the NECESSA Array, and from our temporary networks provided a rare opportunity to investigate the crustal structure of this area. There were a total of 137 CEA broadband stations, 129 NECESSA Array stations, and 116 temporary stations in the study area (115°E–135°E and 38°N–52°N) (Fig. 1). We collected waveform data from 606 earthquakes of magnitude $M_S \geq 5.0$ occurred between May 2009 and August 2011 recorded at an epicentral distance of 30°–90° (Fig. 2).

The criteria used to select teleseismic events were as follows: (1) events should have an epicentral distance of 30°–90° and try to avoid effects of mantle and complex structure of the core-mantle boundary on seismic wave travel time; (2) events should have magnitude greater than M_S 5.0 to ensure that the seismic waves arrived at stations with a high signal-to-noise ratio; (3) there should be more than 10 records for each event. After selection, we performed the following pretreatments on the data: RMEAN, RTREND, and band-pass filter with frequencies range from 0.02～0.1 Hz. We also used the wave cross-correlation method (VanDecar and Crosson 1990; Rawlinson and Kennett 2004) to pick up the residuals. Finally, we visually checked all the seismograms and selected a total of 90,724 usable relative travel-time residuals.

This study uses seismic ray travel-time tomography to study the velocity structure of Northeast China. This method involves the following steps: firstly subdivide the research region and mesh the grid; secondly pick up ray travel time in seismic data, seismic ray tracking; thirdly build the inversion equation and solve the matrix equation; and finally evaluate the results of resolution and reliability. The results provide some reliable velocity anomaly characteristics and provide good constraints for the solution of these velocity anomalies to elucidate basin formation and the geodynamic process. In this article, our main focus is P wave velocity abnormal structures.

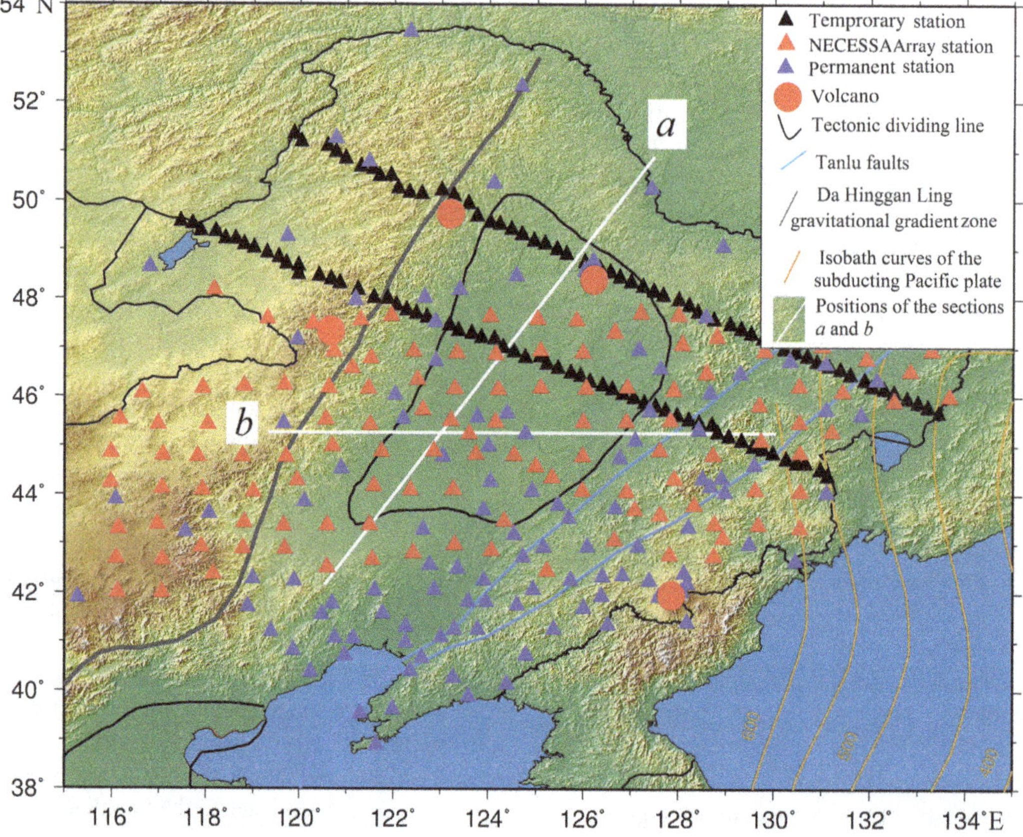

Fig. 1 Tectonic setting and station distribution in the studied area

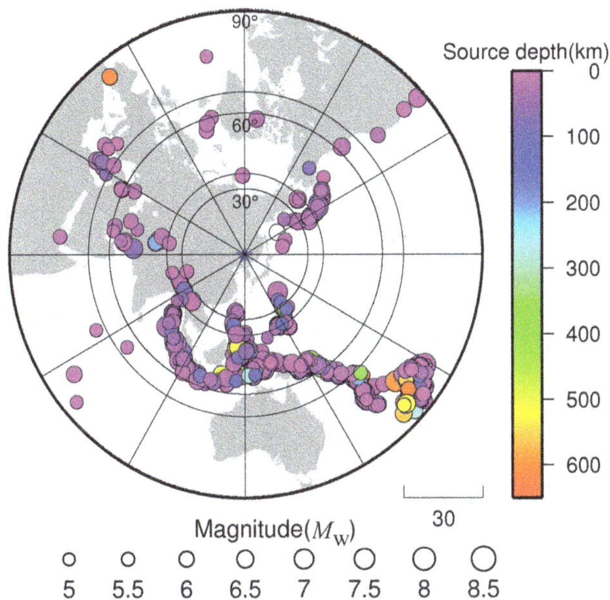

Fig. 2 Distribution of teleseismic events in the study

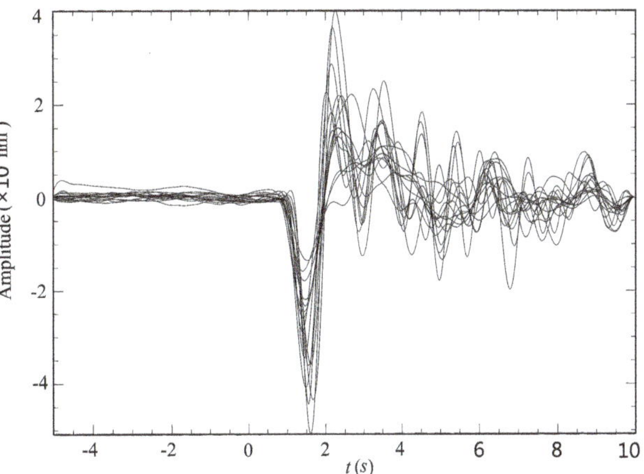

Fig. 3 Stack results obtained using multi-channel cross-correlation

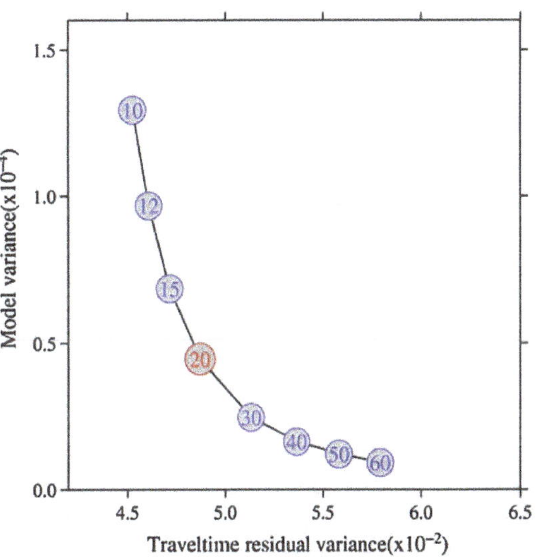

Fig. 4 Tradeoff curves of model variance and travel-time residual variance for different values of damping coefficient (values shown in *circles*)

3 Reliability analyses of inversion results

3.1 Multi-channel cross-correlation, damping coefficient, and residual analyses

To validate the multi-channel cross-correlation results, we obtained the travel-time residuals from the original data. The stack results obtained using multi-channel cross-correlation is shown in Fig. 3. It shows that the cross-correlation results are reliable. We adopted the LSQR method (Paige 1982) with damping inversion. The damping coefficient in the algorithm controls the convergence speed and constrains the smoothness of the model. From the trade-off curve (Fig. 4) of model variance and travel-time residual variance for different values of damping coefficient, it can be found apparently when the damping coefficient is 20, the model is relatively smooth and the mean square is relatively small. Thus, in the inversion, we adopted a damping coefficient of 20.

Figure 5 shows the distributions of the relative travel-time residuals before and after inversion. The relative travel-time residuals before the inversion are most focused on the time period −1.0 s to 1.0 s. After inversion, the residual distribution has a smaller range, with most values lying between −0.4 s and 0.4 s, and has the characteristics of a normal distribution.

3.2 Resolution tests

Several inversion results could be obtained for multiplicity. Thus, we performed a checkerboard test to verify inversion results to show the reasonable P wave structure beneath the studied area. The checkerboard test process is as follows: we firstly built an initial model I and created a disturbance model II in which positive and negative velocity anomalies are increased or reduced respectively on the basis of the initial velocity model I. Secondly using the relative residual mentioned above as the observation, we take the model II as the initial model for imaging inversion, and take the inversion result as the model III. Finally, we compared abnormal scale between models II and III, and the minimum abnormal body should be the resolution. In this study, we firstly created checkerboard anomalies alternating between 3 % faster and slower velocity anomalies. The studied area was separated into several blocks and the size of each block is approximately 50 km × 60 km. We then

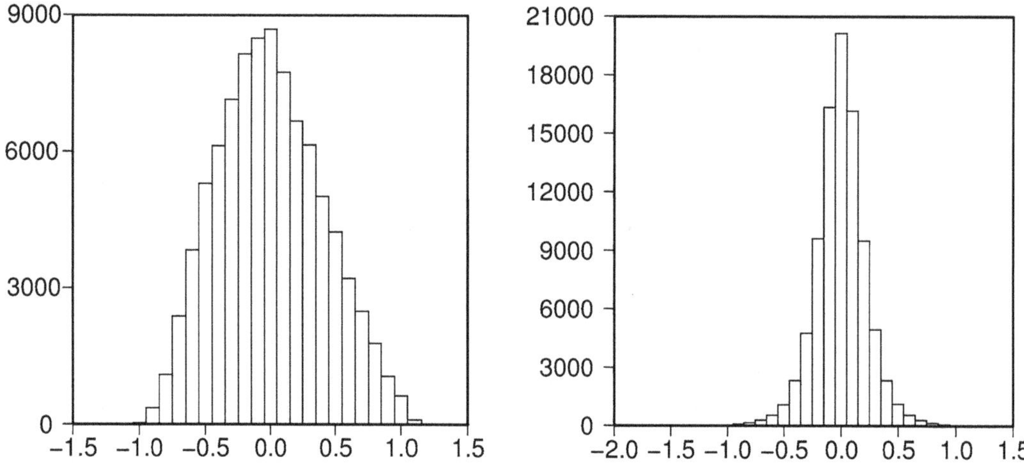

Fig. 5 Distributions of the relative travel-time residuals before (*left*) and after (*right*) inversion

generated a synthetic travel-time dataset using earthquakes and stations identical to that of the observed differential travel-time data. Random noise ranging from −0.15 s to 0.15 s was added to the synthetic data. Figure 6 shows the results of the recovered checkerboards at eight different depths. The checkerboard model fit well in the depth less than 800 km. The largest and the smallest blocks are 16.0° × 20.0° and 0.500° × 0.625° laterally and 800 km and 50 km vertically. The study region was parameterized with the smallest blocks. The test indicates that the P wave travel-time data have good resolution at length scales of approximately 200 km in the study region. Note that the amplitudes of the velocity anomalies in the recovered model decay steadily with depth. For structures below 400 km depth, significant smearing occurs at the edge of the study area, and the recovered velocity perturbations are less than half of the input values.

4 Results and discussion

There have been many studies on seismic structure of the Songliao basin. Liu and Niu (2011) found that the crustal thickness varies from 27.9 km beneath the eastern flank of the Songliao basin to 40.7 km beneath the Da Hinggan Ling region and suggested there would be a mantle upwelling using tele-seismic receiver function method. Zhang et al. (2013, 2014) found there is a high-velocity zone beneath the Songliao basin with some high- and low-velocity anomalies using tele-seismic tomographic data. The low-velocity anomaly was found at the depth of 400 km beneath the southern part of the Songliao basin. Tele-seismic receiver function and seismic images studies by Tang et al. (2014) also verified this characteristics of high- and low-velocities anomalies distributed jointly in the depths of 100 km and 260 km and low velocities zone at the depth of 420 km beneath the Songliao basin.

We selected the one-dimensional Earth velocity model IASP91 (Kennett and Engdahl 1991) as the initial test model, and adopted travel-time tomographic inversion of P wave to image the velocity structure. The imaging results for each depth are shown in Fig. 7. The Tanlu faults and the gravitational gradient zone are roughly the east and west boundaries of the Songliao basin, which is basically consistent with the main boundaries of the velocity anomalies. The region of Northeast China can be divided into three parts: east, middle, and west zones by velocity anomaly. The east and west zones are low-velocity anomalies: Changbaishan and A'ershan, respectively. The middle zone is the Songliao basin; in which there is some high- and low-velocities anomalies in the background of high-velocity zone. The high-velocity zone mainly distributes from depth of 50 km to 600 km, while the low-velocity anomalies zones mainly concentrate from depth of 50 km to 400 km. We also determined one low-velocity anomaly zone at the center of the Songliao basin in depth of 50–100 km and some low-velocity anomalies zones in the north and south of the Songliao basin with depths range from 150 km to 400 km. We deduced that the low-velocity anomalies extend down to the upper mantle, which may be the result of mantle material upwelling. Our results are similar to, but not completely consistent with, the other studied results (Li et al. 2006; Li and van der Hilst 2010; Wei et al. 2012; Zhang et al. 2013, 2014). The mixture of high and low-velocity anomalies suggests that the lithospheric mantle may have been destroyed and transfered, and hence the formation of the Songliao basin may have been greatly affected by lithospheric delamination. The low-velocity anomaly beneath the southern part of the Songliao basin area is connected to the those beneath the Changbaishan and A'ershan volcanic areas. This result

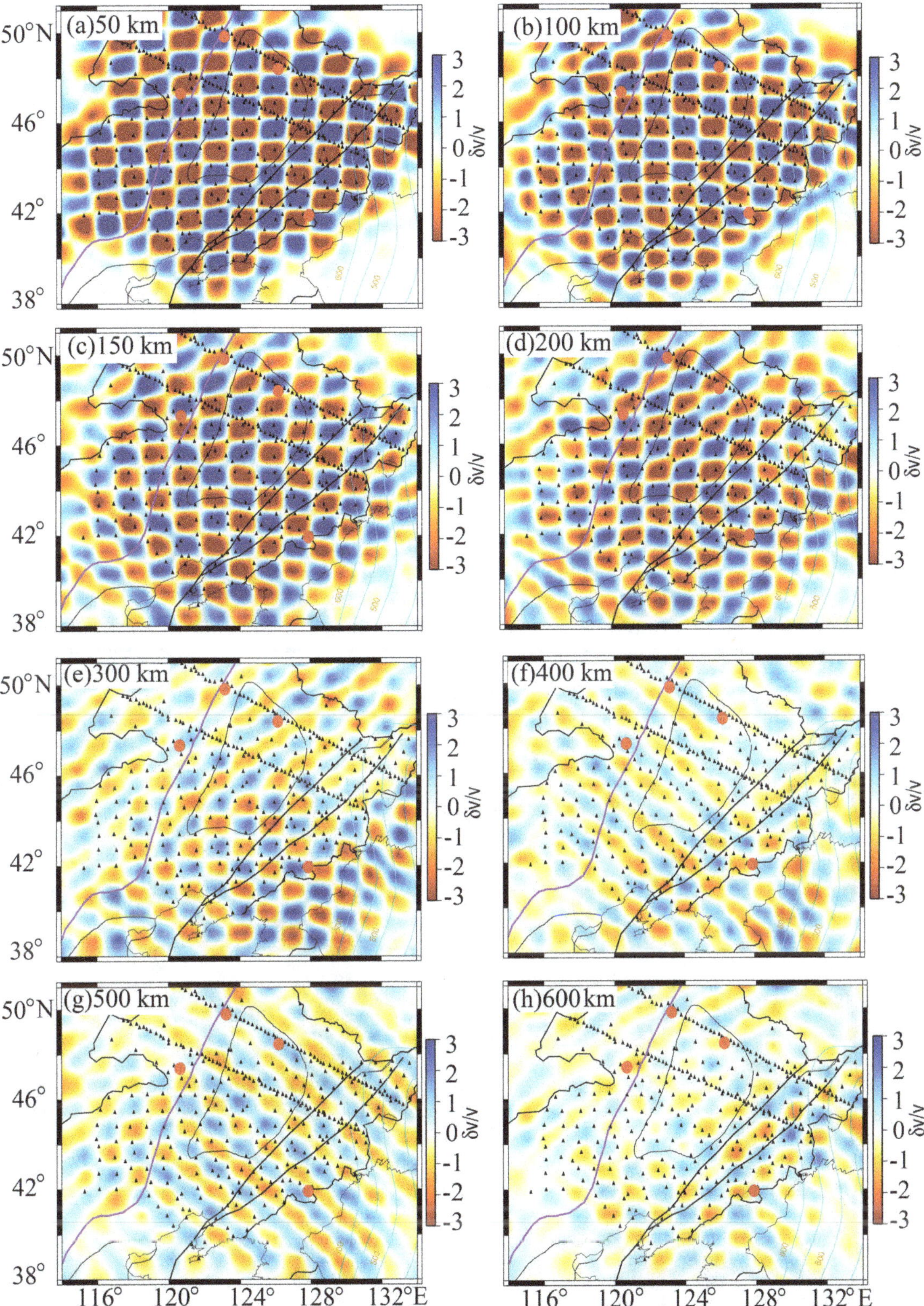

Fig. 6 Results of the checkboard resolution test of depth of 50 km (**a**), 100 km (**b**), 150 km (**c**), 200 km (**d**), 300 km (**e**), 400 km (**f**), 500 km (**g**) and 600 km (**h**) . The *blue* and *red* stand for high- and low-velocities, respectively. The *purple lines* are gravitational gradient zone. The *black lines* represent the Tanlu faults; The *gray lines* represent the the tectonic dividing line; The *blue lines* represent the isobath curves of the subducting Pacific plate, the *red dots* stand for the volcano and the *black triangles* denote stations

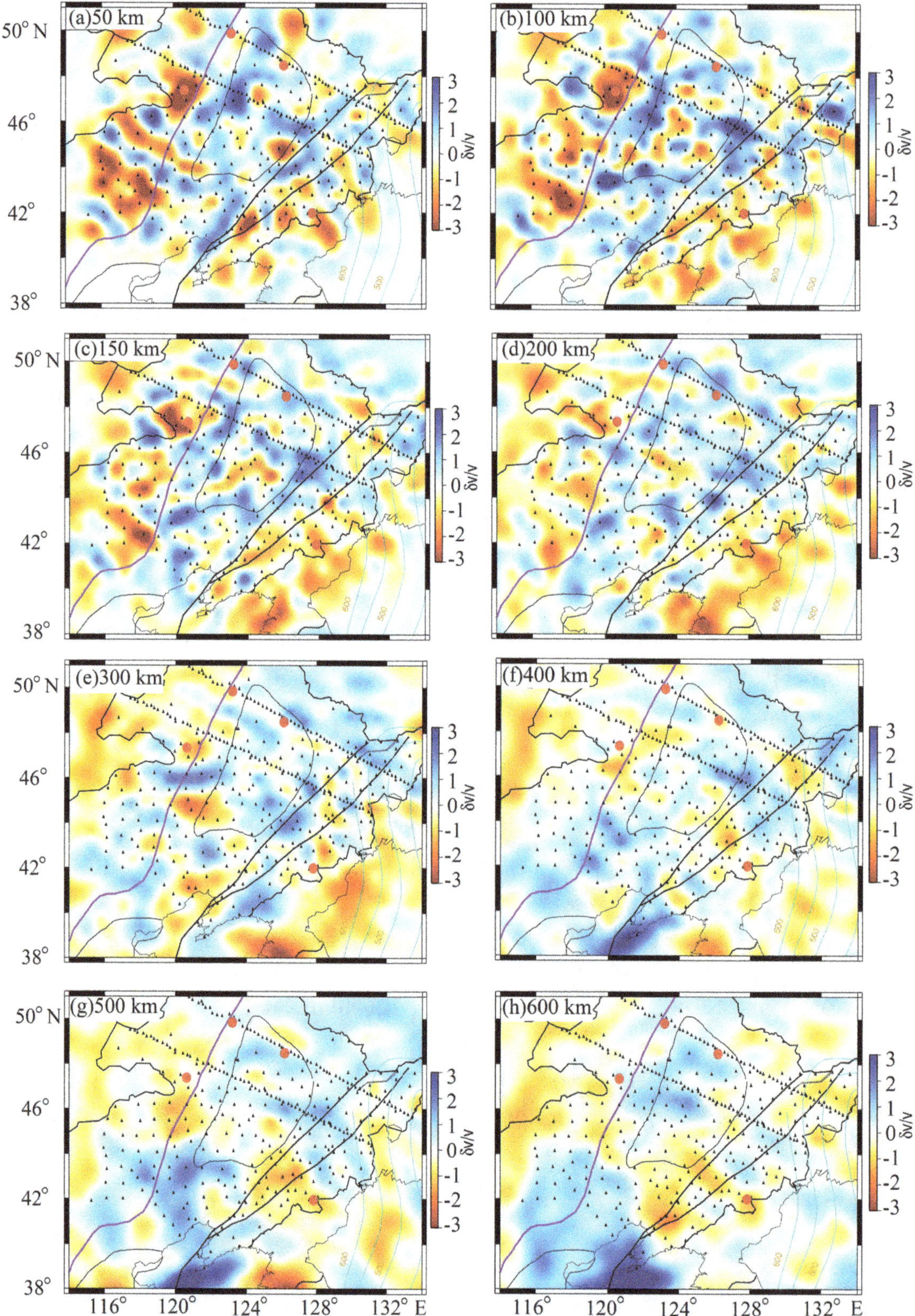

Fig. 7 Velocity structure of depth of 50 km (**a**), 100 km (**b**), 150 km (**c**), 200 km (**d**), 300 km (**e**), 400 km (**f**), 500 km (**g**) and 600 km (**h**) inverted using P wave travle-time tomography. The *blue* and *red* stand for high- and low-velocities, respectively. The *purple lines* are gravitational gradient zone. The *black lines* represent the Tanlu faults; The *gray lines* represent the the tectonic dividing line; The *blue lines* represent the isobath curves of the subducting Pacific plate, the *red dots* stand for the volcano and the *black triangles* denote stations

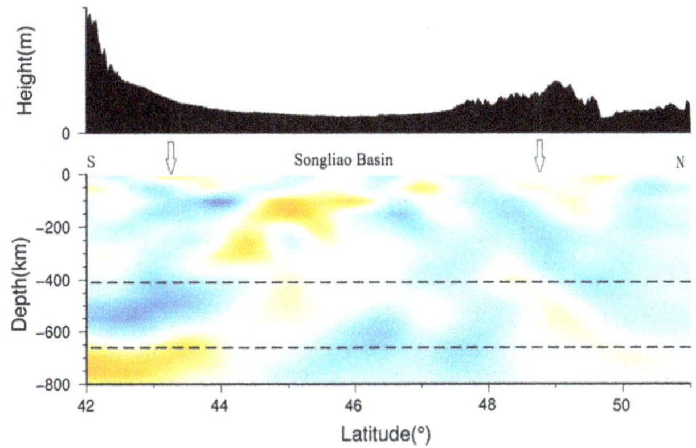

Fig. 8 Vertical section across the Songliao basin from south to north (line *a* in Fig. 1) (42°N, 120°E) to (51°N, 129°E). The *upper* figure shows the topographical relief in this vertical section; the *lower* is the velocity anomaly imaging inverted by tomography. *Dashed lines* are the 410 km and 660 km discontinuities, the *arrow* in the figure give the boundary of the basin

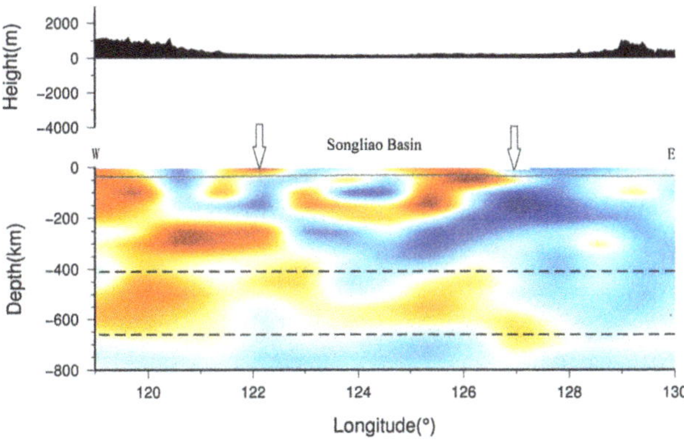

Fig. 9 Vertical section across the Songliao basin from west to east (line *b* in Fig. 1) (45°N, 119°E) to (45°N, 130°E). The *upper* figure shows the topographical relief in this vertical section; the *lower* figure is the velocity anomaly imaging inverted by tomography. *Dashed lines* are 410 and 660 km discontinuities, the *arrow* in the figure give the boundary of the basin

reveals that there is an upwelling material channel from the mantle beneath the Songliao basin and its surrounding area.

As mentioned above, although many studies have been performed in the Songliao basin using various geological and geophysical methods (Shu et al. 2003; Zhao 2004; Lei and Zhao 2005; Sun 2012; Wan 2012; Zhang et al. 2013, 2014; Pan et al. 2014; Wei et al. 2014), the relation between shallow and deep structures in the Songliao basin still remains poorly understood. Figures 8 and 9 show the topographical relief and imaging inverted by tomography of two vertical sections across Songliao basin from (42°N, 120°E) to (51°N, 129°E) and (45°N, 119°E) to (45°N, 130°E), respectively. The figures suggest the Changbaishan maybe a result of mantle material upwelling originated from depth near 700 km. The receiver function results shows obvious lateral variation in the upper mantle transition zone, which maybe result from mantle material upwelling (Liu and Niu 2011; Tang et al. 2014). The geothermal gradient of the Songliao basin which has a high heat value, is slightly higher than that of other basins elsewhere in the world (Nie 2004). The strain fields of Northeast and North China show NW-SE extension is weaken from east to west. The maximum strain occurs in the middle of the Tanlu fault zone, which is the boundary of the Songliao basin (Li 2013). The subducting Pacific slab and the mantle upwelling may have a remote effect.

The suggestion of mantle material upwelling is consistent with fact that in the eastern margin of Songliao basin, Cenozoic volcanism is active and crust is thin. Despite the relatively high altitude in the region, the Songliao basin has a relatively thin crust, which is probably related to Mesozoic rifting (Liu and Niu. 2011). The previous studies (Liu and Niu 2011; Gao and Li 2014; Pan et al. 2014; Wei et al. 2014) suggested that variations in crustal thickness apparently play an important role in controlling tectonic activity in the study region. Studies of the regional variations in crustal attenuation in Northeast China and the adjacent regions shows that the highest attenuation (low Q_0 values) is observed beneath the Songliao plain, which has thick sedimentary deposits and strong tectonic activity (Sun 2012). The structure of the crust in the North Songliao basin, inferred from the Bouguer gravity anomaly, is divided into three layers: upper, middle, and lower crusts (Wei et al. 2014). The seismic profiles through the two uplifts of the Songliao basin show that the crustal structure has strong vertical and lateral heterogeneities; obvious differences can be observed between the upper and lower crustal reflections (Yang et al. 2003). This result reveals that the Songliao basin experienced multistage tectonic activities, while the distribution of basalt magma in the basin possibly confirms the existence of an upwelling material channel from the mantle that affected both the crust and the tectonic activity.

Our results indicate that the formation and evolution of the Songliao basin were related to the subduction of the Pacific plate under the Eurasian plate as well as asthenosphere upwelling, both of which produced jointly crustal extension. The subduction induced one-way circulation of the asthenosphere and diapirism of the mantle magma, leading to stretching and thinning of the lower crust under the shearing action and brittle stretching of the upper crust to form a rift basin. As the subduction of the west Pacific plate steepened, the subduction zone retreated eastward and the melted mantle material extended continuously eastward. As the mantle cooled, the Songliao basin subsided and developed into a depression basin.

5 Conclusions

We selected relative travel-time residuals from teleseismic waveform data using the waveform correction method and obtained 90,724 usable relative travel-time residuals. The teleseismic waveform data were recorded by a total of 382 stations in one permanent and two temporary networks in Northeast China. We used the data to image the P wave velocity structure beneath Northeast China.

The results show that mostly a high-velocity zone exists with many high and low velocity anomalies areas beneath the Songliao basin. The low-velocity anomalies may extend down to the mantle and are possibly induced by material upwelling. The distribution of high and low velocity anomalies suggests that the lithospheric mantle possibly experienced destruction and transformation, and the formation of Songliao basin was possibly greatly affected by lithospheric delamination. The low-velocity anomaly beneath the southern part of the Songliao basin is connected to the low-velocity anomalies beneath the Changbaishan and A'ershan volcanic areas.

Formation and evolution of the Songliao basin were caused by subduction of the Pacific plate under the Eurasian plate, which induced asthenospheric upwelling and hence crustal extension. This induced one-way circulation of the asthenosphere, and the diapirism of the mantle magma led to stretching and thinning of the lower crust as a result of shearing. Brittle stretching of the upper crust formed a rift basin. When the subduction of the west Pacific plate became steeper, the subduction zone retreated to the east, and the melted material in the mantle extended continuously eastward. As the mantle cooled, the Songliao basin underwent subsidence and developed as a depression basin.

Acknowledgments Waveform data for this study are provided by Data Management Center of China National Seismic Networks at Institute of Geophysics, China Earthquake Administration (SEISDMC, doi:10.7914/SN/CB). And we also thank NECESSA Array group for sharing their data. We are also grateful anonymous reviewers for their valuable comments and suggestions. This work was supported by the National Natural Science Foundation of China (Grant No. 41274088), International Science and Technology Cooperation Program of China (ISCTP) (Grant No.2011DFB20210), and the Earthquake Science and Technology Spark Plan Project of Hebei province, China (Grant NO.DZ20150420030).

References

Gao YG, Li YH (2014) Crustal thickness and *Vp*/*Vs* in the Northeast Chviina-North China region and its geological implication. Chin J Geophys 57(3):847–857. doi:10.6038/cig20140314 (in Chinese with English abstract)

Hu WS, Cai CF, Wu ZY, Li JM (1998) Structural style and its relation to hydrocarbon exploration in the Songliao Basin, Northeast China. Mar Petroleum Geol 15:41–55 (in Chinese with English abstract)

Kennett BLN, Engdahl ER (1991) Travel times for global earthquake location and phase identification. Geophys J Int 105(2):429–465

Lei JS, Zhao DP (2005) P-wave tomography and origin of the Changbai intraplate volcano in Northeast Asia. Tectonophysics 397(3–4):281–295

Li Y (2013) The impact on the Northeast and North China caused by the Tohoku-Oki Earthquake in Japan in 2011. Master Dissertation. Beijing, Institute of Geology, China Earthquake Administration (in Chinese with English abstract)

Li C, van der Hilst RD (2010) Structure of the upper mantle and transition zone beneath Southeast Asia from travel time tomography. J Geophys Res 115:B07308. doi:10.1029/2009JB006882

Li C, van der Hilst RD, Toksoz MN (2006) Constraining P-wave velocity variations in the upper mantle beneath Southeast Asia. Phys Earth Planet Inter 154(2):180–195

Liu H, Niu F (2011) Receiver function study of the crustal structure of Northeast China: Seismic evidence for a mantle upwelling beneath the eastern flank of the Songliao Basin and the Changbaishan region. Earthq Sci 24:27–33. doi:10.1007/s11589-011-0766-6

Nie FJ (2004) Study on tectonic evolution and sequences stratigraphy and ridden traps for petroleum in Dougtumo District of the Songiiao Basin. Beijing, Institute of Geology, China Earthquake Administration (in Chinese with English abstract)

Paige CC. 1982. Saunders MA. Algorithm 583 LSQR: sparse linear equations and least squares problems. ACM Trans, 8(2):195–209

Pan JT, Li YH, Wu QJ, Yu DX (2014) Ambient noise tomography in northeast China. Chin J Geophys 57(3):812–821. doi:10.6038/cjg20140311 (in Chinese with English abstract)

Rawlinson N, Kennett BLN (2004) Rapid estimation of relative and absolute delay times across a network by adaptive stacking. Geophys J Int 157(1):332–340

Ren JY, Tanaki K, Li ST, Zhang JX (2001) Late Mesozoic and Cenozoic rifting and its dynamic setting in Eastern China and adjacent areas. Tectonophysics 344:175–205

Shi L (2004) Study of the tectonic-thermal evolution of the Songliao basin and quantitative assessment of its geothermal resource: a case study on the Dumeng district. Ph.D Dissertation. GuangZhou, Guangzhou Institute of Geochemistry Chinese Academy of Sciences (in Chinese with English abstract)

Shu LS, MU YF, Wang B (2003) The oil gas bearing strata and the structure features in the Songliao Basin, NE China. J Stratigr 27(4):340–347

Song LZ, Zhao ZH, Jiao GH, Sun P, Luo X, Jiang XH, Wang ZH, Zeng FY, Liao WD (2010) Geochemical characteristics of Early Cretaceous volcanic rocks from Songliao basin, Northeast China, and its tectonic impications. Acta Petrol Sin 26(4):1182–1194

Sun L (2012) Velocity structure of uppermost mantle from Pn tomography and tomography of S-wave attenuation in Northeast China. Ph. D Dissertation. Beijing, Institute of Geophysics, China Earthquake Administration (in Chinese with English abstract)

Tang YC, Obayashi M, Niu FL, Grand SP, Chen John YS, Kawakatsu H, Tanaka S, Ning Y, Ni James F (2014) Changbaishan volcanism in northeast China linked to subduction-induced mantle upwelling. Nat Geosci 7:470–475. doi:10.1038/NEGO2166

VanDecar JC, Crosson RS (1990) Determination of teleseismic relative phase arrival times using multi-channel cross-correlation and least squares. Bull Seismol Soc Am 80(1):150–169

Wan SS (2012) The study on tectonic evolution in North Songliao Basin. Master Dissertation. Chengdu University of Technology, ChengDu (in Chinese with English abstract)

Wei D-Y, WU Yan-G, Huan H-F (2014) The Bouguer anomaly field source in North of Songliao Basin. North China Earthquake Sciences. 32(1):1–4. doi:10.3969/j.iassn.1003-1375.2014.01.001 (in Chinese with English abstract)

Wei W, Xu JD, Zhao DP, Shi YL (2012) East Asia mantle tomography: new insight into plate subduction and intraplate volacanism. J Asian Earth Sci 60:88–103. doi:10.1016/j.jseaes.2012.08.001

Yang BJ, Tang JR, Li QX, Wang JM, Faisal SA, Li RL, Wang HZ, Li ZL, Zhang H (2003) The reflection structure of the crust in Songliao basin uplifts and “disconnect” Moho interface. Sci China (Series D) 33(2):170–176

Yu WX, LU JL,Zhang QL, Wang BH, Zhang YX (2012) A study on petrological geochemistry of the early cretaceous volcanic rocks from Changling Rift,Songliao Basin,China. ACTA Mineral Sin, 32(1):83-92

Zhang FX, Wu QJ, Li YH (2013) The traveltime tomography study by teleseismic P wave data in the Northeast China area. Chin J Geophys 56(8):2690–2700. doi:10.6038/cjg20130818 (in Chinese with English abstract)

Zhang FX, Wu QJ, Li YH (2014) A travel time tomography study by telescismic S wave data in the Northeast China area. Chin J Geophys 57(1):88–101. doi:10.6038/cjg201410109 (in Chinese with English abstract)

Zhao DP (2004) Global tomographic images of mantle plumes and subducting slabs: insight into deep Earth dynamics. Phys Earth Planet Inter 146(1–2):3–34

Zheng XF, Ouyang B, Zhang DN, Yao ZX, Liang JH, Zheng J (2009) Technical system construction of Data Backup Center for China Seismograph Network and the data support to researches on the Wenchuan earthquake. Chin J Geophys 52(5):1412–1417. doi:10.3969/j.issn.0001-5733.2009.05.031 (in Chinese with English abstract)

7

Crustal velocity structure of central Gansu Province from regional seismic waveform inversion using firework algorithm

Yanyang Chen · Yanbin Wang · Yuansheng Zhang

Abstract The firework algorithm (FWA) is a novel swarm intelligence-based method recently proposed for the optimization of multi-parameter, nonlinear functions. Numerical waveform inversion experiments using a synthetic model show that the FWA performs well in both solution quality and efficiency. We apply the FWA in this study to crustal velocity structure inversion using regional seismic waveform data of central Gansu on the northeastern margin of the Qinghai-Tibet plateau. Seismograms recorded from the moment magnitude (M_W) 5.4 Minxian earthquake enable obtaining an average crustal velocity model for this region. We initially carried out a series of FWA robustness tests in regional waveform inversion at the same earthquake and station positions across the study region, inverting two velocity structure models, with and without a low-velocity crustal layer; the accuracy of our average inversion results and their standard deviations reveal the advantages of the FWA for the inversion of regional seismic waveforms. We applied the FWA across our study area using three component waveform data recorded by nine broadband permanent seismic stations with epicentral distances ranging between 146 and 437 km. These inversion results show that the average thickness of the crust in this region is 46.75 km, while thicknesses of the sedimentary layer, and the upper, middle, and lower crust are 3.15, 15.69, 13.08, and 14.83 km, respectively. Results also show that the P-wave velocities of these layers and the upper mantle are 4.47, 6.07, 6.12, 6.87, and 8.18 km/s, respectively.

Keywords Seismic waveform inversion · Crustal velocity structure · Central Gansu Province · Firework algorithm

Y. Chen · Y. Wang (✉)
Department of Geophysics, School of Earth and Space Sciences, Peking University, Beijing 100871, China
e-mail: ybwang@pku.edu.cn

Y. Zhang
Lanzhou Institute of Seismology, China Earthquake Administration, Lanzhou 730000, China

1 Introduction

Determining the structure of the Earth is one of the main aims of geophysical research. In this context, the crustal velocity structure is very important for accurately locating earthquakes, inverting seismic source moment tensors, and rapidly evaluating earthquake intensity distributions. To achieve this, a layered one-dimensional (1D) velocity structure can be used as an initial model for two-dimensional (2D) and three-dimensional (3D) inversions, while regional seismic waveform inversion is an effective tool that can be used to determine the former. However, as seismic waveform inversion problems are usually highly nonlinear and have non-unique solutions (Maurice et al. 2003; Li and Lei 2014a), linearized and gradient-based methods have traditionally been applied because of their computational efficiency. One shortcoming of these methods is that they often rely on the initial models and can converge on local minima; alternative approaches that do not depend on gradient-based searching within the model space can be used instead to improve the probability of convergence on the global minimum.

Among currently available search methods, one class of optimization algorithms inspired by biological evolution has been widely applied to geophysical inversions. Li et al. (2007), for example, used a genetic algorithm (GA) for the inversion of regional seismic waveforms to obtain the crustal structural velocity in Italy, while Abdelwahed and Zhao (2014) applied a micro-GA waveform inversion method in their study of crustal structure in northeastern Japan. Similarly, Maurice et al. (2003), Li et al. (2012), and Li and Lei (2014b) all applied a niching GA (NGA) in their work on crustal velocity inversions to get teleseismic and regional velocity structures, while Li et al. (2016) applied an NGA to the inversion of ocean-bottom waveforms from seismometers in order to determine the structure of the oceanic crust.

The FWA is a recently proposed novel swarm intelligence-based method that is used for nonlinear global optimization. This algorithm was inspired by firework explosions and is applied in global optimization problems that relate to complex objective functions. Numerical tests of the FWA applied to the inversion of crustal velocity structures have been carried out using synthetic models; this approach has been validated and is known to have higher accuracy and level of convergence compared to NGA (Ding et al. 2015). In this study, we apply the FWA to the inversion of regional seismic waveforms to determine the 1D crustal velocity structure in central Gansu on the northeastern margin of the Qinghai-Tibet plateau. An M_W 5.4 earthquake took place near Minxian in Gansu Province on July 22, 2013, and was recorded by densely distributed permanent recording stations across the province. These recordings include three seismic waveform components that provide valuable data for investigations of regional crustal velocity structure in central Gansu. Thus, several geophysical studies have subsequently been carried out in this region, including the work of Shen et al. (2013) and He et al. (2014), who applied receiver function techniques to measure crustal thickness. In this study, we utilize regional seismic waveform inversion to obtain velocity structures within the crust, important data that augments our geophysical understanding of this region.

2 The FWA

Tan and Zhu (2010) proposed the use of FWA for the global optimization of complex objective functions. The FWA is a kind of searching method similar to GA, the search process used by this algorithm mimics the explosion of fireworks. In other words, when a firework is detonated, a shower of sparks is generated into the local space; in the FWA, these sparks search the neighboring model space in the vicinity of a specific location (Tan et al. 2013).

We applied a FWA variant called the Attract-Repulse FWA (AR-FWA) in this study (Ding et al. 2013). The core components of this variant are FWA search and attract-repulse mutations, as illustrated by the flowchart in Fig. 1. The main steps of the FWA are as follows:

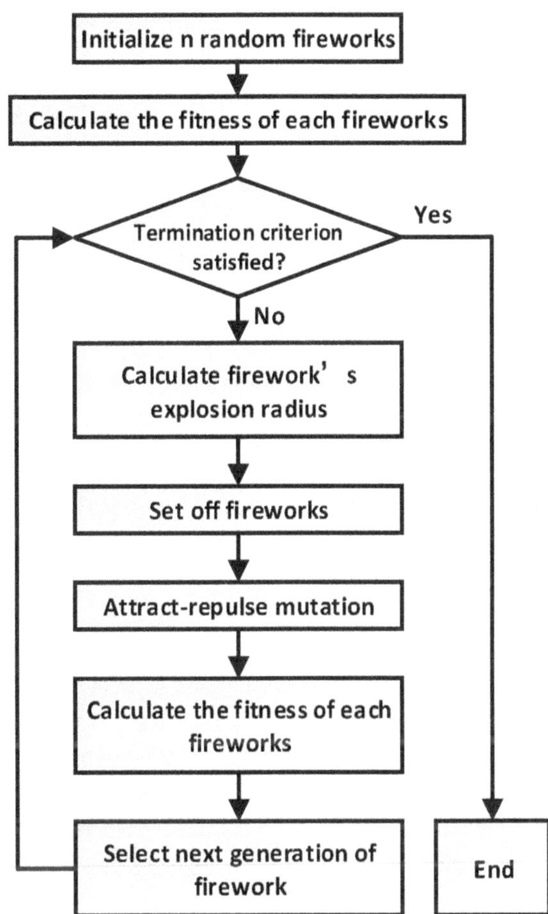

Fig. 1 Flowchart of FWA (After Tan and Zhu 2010)

1. In the initialization step, n fireworks are randomly generated within the search range of a parameter.
2. A misfit value for each firework is then calculated using the objective function in Eq. (3).
3. The radius of explosion of each firework, A_i, is then calculated using the misfit value in Eq. (1).
4. Fireworks are detonated, with each generating m sparks randomly distributed within an explosion radius A_i. Misfit values are then calculated for each spark, with each firework replaced by its current best-fit (i.e., minimum misfit value) spark.
5. Attract-repulse mutation is then randomly applied to non-best-fit fireworks to increase their swarm diversity.
6. Mutated and best-fit fireworks are then chosen for use in the next generation, and this work flow is repeated from (2) until a predetermined termination criterion is satisfied.

The radius of a given firework explosion, A_i, is as follows:

$$A_i = \hat{A} \cdot \left(\frac{f(\boldsymbol{x}_i) - y_{\min} + \xi}{\sum_{i=1}^{n} (f(\boldsymbol{x}_i) - y_{\min}) + n \cdot \xi} \right) + \delta. \quad (1)$$

In this expression, $\hat{A}$ denotes the maximum explosion radius, n is the number of fireworks, x_i indicates the ith firework, and $f(\boldsymbol{x}_i)$ is the objective function value for the ith firework among n fireworks. Thus, $y_{\min} = \min(f(\boldsymbol{x}_i), i = 1, 2, \ldots n)$, the minimum (best fit) objective function value of n fireworks, and ξ denote machine precision to avoid zero division error. A small number, δ, is also introduced to this expression to guarantee a nonzero radius and to avoid stalls in the search process.

Thus, according to Eq. (1), a firework with a better misfit value will generate sparks within a smaller radius, and vice versa. Applying this strategy enhances exploitation of a greater number of potential positions, while searches will continue in the case of worse fit spaces.

As fireworks are set off and m sparks are randomly generated within neighboring spaces, the value of the kth dimension of the jth spark generated by the ith firework, $\text{spark}_{i,j,k}$, will have a random displacement with $\text{firework}_{i,k}$, $\text{spark}_{i,j,k} = \text{firework}_{i,k} + \text{rand}(-1, 1) \times A_{i,k}$, and $(i = 1, 2, \ldots, n;\ j = 1, 2, \ldots, m;\ k = 1, 2, \ldots, D)$. As sparks will exploit neighbor-solution spaces, this process is referred to as a FWA search. Thus, as the misfit value of each spark is calculated, each firework will be replaced by its current best-fit spark (Ding et al. 2015).

Attract-repulse mutation maintains FWA search swarm diversity while a further parameter, c, is introduced to control the mutation probability of each non-best-fit firework. Thus, during mutation, each dimension (d) of an ith firework has a 50% probability of being moved to a new position according to a random number, s, and the distance between non-best-fit and best fit fireworks. This random number, s, has a uniform distribution within the range $(1 - \delta s,\ 1 + \delta s)$; thus, if s is less than 1, non-best-fit fireworks will be 'attracted' by the best fit to occupy the currently optimal position, while if s is greater than 1, these non-best-fit fireworks will be 'repulsed' to explore further spaces. This choice between 'attract' and 'repulse' reflects the balance between exploitation and exploration, while δs was set to 0.8 to enable faster convergence, as suggested by Ding et al. (2013). In order to ensure that all fireworks and sparks remain within the defined space, a mapping strategy was applied if mutated fireworks or generated sparks fell outside this range. Thus, if $\boldsymbol{x}_{i,d}$ is outlying searching boundary of dimension d, while $\boldsymbol{x}_{i,d}$ is mapped onto the search range using Eq. (2), as follows:

$$\boldsymbol{x}_{i,d} = \boldsymbol{b}_d^l + \left| \boldsymbol{x}_{i,d} - \boldsymbol{b}_d^l \right| \% \left(\boldsymbol{b}_d^u - \boldsymbol{b}_d^l \right). \quad (2)$$

In this expression, b_d^l and b_d^u denote the lower and upper boundaries of a search range of dimension d, while % refers to the modular arithmetic operation.

We chose mutated and best-fit fireworks for use in the next generation, calculating the misfit value and explosion radius, A_i, of each new firework using Eqs. (3) and (1), respectively.

3 Robustness

In order to validate the performance of the FWA, Tan and Zhu (2010) carried out a series of experimental comparisons using nine benchmark test functions including standard and clonal particle swarm optimization (PSO). These experimental results show that the FWA outperforms the two others in terms of both optimization accuracy and speed of convergence. In addition, Ding et al. (2015) applied both PSO and AR-FWA approaches to the inversion of regional waveforms using synthetic models and data; the results of this study show that both PSO and AR-FWA are more efficient at avoiding premature in numerical tests, compare to the widely used evolutionary algorithms, including GA, NGA, and differential evolution. These experimental results also show that PSO and AR-FWA approaches generate models that converge very closely with synthetic examples.

In this study, we numerically test the robustness of AR-FWA in determining crustal velocity structure. To do this, we used a series of synthetic seismograms as observational data that were calculated using the reflectivity method (Fuchs and Müller 1971) and a given velocity model. The focal depth of these seismograms is 18.8 km, while nine stations range between 146 and 437 km away from the epicenter. We constructed a synthetic four-layered isotropic medium model on top of a half-space for this study, which corresponds to the sedimentary layer, the upper, middle, and lower crust, as well as the uppermost mantle. The P-wave velocity, v_P, and the thickness, h, of each layer are presented in Table 1, while the density, ρ, of each layer was obtained using PREM (Dziewonski and Anderson 1981) and the crust 2.0 model (Bassin et al. 2000); these are 2.65, 2.75, 2.8, 3.1, and 3.37 g/cm^3, respectively.

We determined nine 1D velocity model parameters for FWA inversion, including v_P and h for the four crustal layers and v_P for the upper mantle. The search range considered in this study is delimited by the lower and upper dashed line boundaries in Fig. 2. We inverted two velocity structure models in this test, one in which the velocity of layers roughly increases with depth (model A), and another that contains a low-velocity middle crust (model B). These two models were used to determine the inversion robustness for the model that contains a low-velocity layer.

We set c to 0.9 in each numerical robustness test, while the numbers of fireworks, n, and sparks, m, were fixed at ten in each generation, depending on the preceding FWA

Table 1 True and FWA inversion results for models A and B

Given v_P (km/s)	Inverted v_P A (km/s)	Inverted v_P B (km/s)	Given h (km)	Inverted h A (km)	Inverted h B (km)
5.5	5.490 ± 0.012	5.532 ± 0.008	3	3.067 ± 0.034	3.075 ± 0.139
6.1	6.100 ± 0.005	6.083 ± 0.009	14	14.038 ± 0.139	14.153 ± 0.593
6.4 (5.6 for B)	6.389 ± 0.003	5.587 ± 0.015	19	18.350 ± 0.288	18.656 ± 0.624
6.92	6.882 ± 0.012	6.897 ± 0.019	14	14.273 ± 0.090	13.971 ± 0.113
8.1	8.085 ± 0.003	8.091 ± 0.003			

(Ding et al. 2015) and NGA inversion (Li et al. 2012, 2016), as well as computing efficiency. Because there are $n \times m$ forward computing and objective function evaluations at each generation, larger values of n or m will require a longer computation time. Therefore, since a total of 100 (10 × 10) sparks will be generated in each generation and there are 500 generations in each FWA inversion, an optimizer can conduct up to 5×10^4 function evaluations in each run. We ran our FWA inversions five times separately using different random seed values; because each inversion will find one most optimal model, our approach generates five models after five inversions (runs). A final inverted model was then obtained by averaging the five most optimal models and calculating their standard deviations as the final model error.

The mean values and standard deviations for each parameter in our final model inversion are listed in Table 1 and illustrated in Fig. 2. The results show that mean values for inverted parameters are close to true preset values, while standard errors are relatively small, between 0.03% and 3.3%. The convergence of misfit values (Fig. 2) between the two models is also very small; our accurate and stable numerical test results show that the FWA is suitable for the inversion of waveforms.

4 Application of the FWA in central Gansu Province

The region of China investigated in this study lies to the northeast of the Qinghai-Tibet plateau (34°–39°N, 100°–106°E). The M_W 5.4 earthquake took place near Min County in Gansu Province, China, on July 22, 2013. The epicenter of this event was adjacent to the Lintan-Tanchang fault, on the boundary between the Qilian and Qaidam blocks (Fig. 3). Shen et al. (2013) obtained crustal thickness of this region by using receiver function method, and the average crustal thickness beneath the five same stations we used from their results is 46.76 km. According to the seismic topography inversion results published by Zhou et al. (2006), the crustal structure of this region comprises four layers. The first of these corresponds to sediments at depths down to 3 km, while the second to fourth layers comprise upper (between 3 and 17 km), middle (between 17 and 36 km), and lower crust (between 36 km and the depth of Moho).

The parameters of the M_W 5.4 Minxian earthquake are shown in Table 2. The focal depth of this event is 18.8 km. We used the reflectivity method developed by Fuchs and Müller (1971) for forward modeling. This widely applied method is known for its high degree of accuracy and speed (Jia and Zhang 2008; Li et al. 2012). Our inversion model includes four layers on top of a half-space, including a sedimentary layer, upper, middle, and lower crust as well as the uppermost mantle. Each layer is characterized by a P-wave velocity, v_P, a S-wave velocity, v_S, a density, ρ, and a thickness, h. We maintained the velocity ratio between the P- and S-waves, $v_P/v_S = 1.73$, while layer densities were calculated using PREM and the crust 2.0 model as 2.65, 2.75, 2.8, 3.1, and 3.37 g/cm^3, respectively. We determined a total of nine parameters via waveform inversion, including four layer thicknesses and five P-wave velocities (Fig. 4).

The v_P search ranges we applied to the sedimentary layer and to the upper, middle, and lower crust were between 4.0 and 6.4 km/s, 5.5 and 6.6 km/s, 5.75 and 7.0 km/s, and 6.0 and 7.45 km/s, respectively. At the same time, the v_P search range we applied to the upper mantle was between 7.3 and 8.4 km/s, while the range of search thicknesses for each layer was between 1.0 and 15.0 km, 5.0 and 20.0 km, 5.0 and 25.0 km, and 5.0 and 25.0 km, respectively. We randomly generated an initial inversion model within each search range.

Instrument responses and observed trends in seismic waveform data were removed in each case, and data were filtered through a 0.02 Hz to 0.1 Hz Butterworth band-pass filter to remove high frequencies influenced by small-scale lateral heterogeneous variation. The objective function we used for waveform inversion consists of a weighted combination of root mean square residuals (RMR) and the correlation of observed and synthetic waveforms in the time domain, as follows:

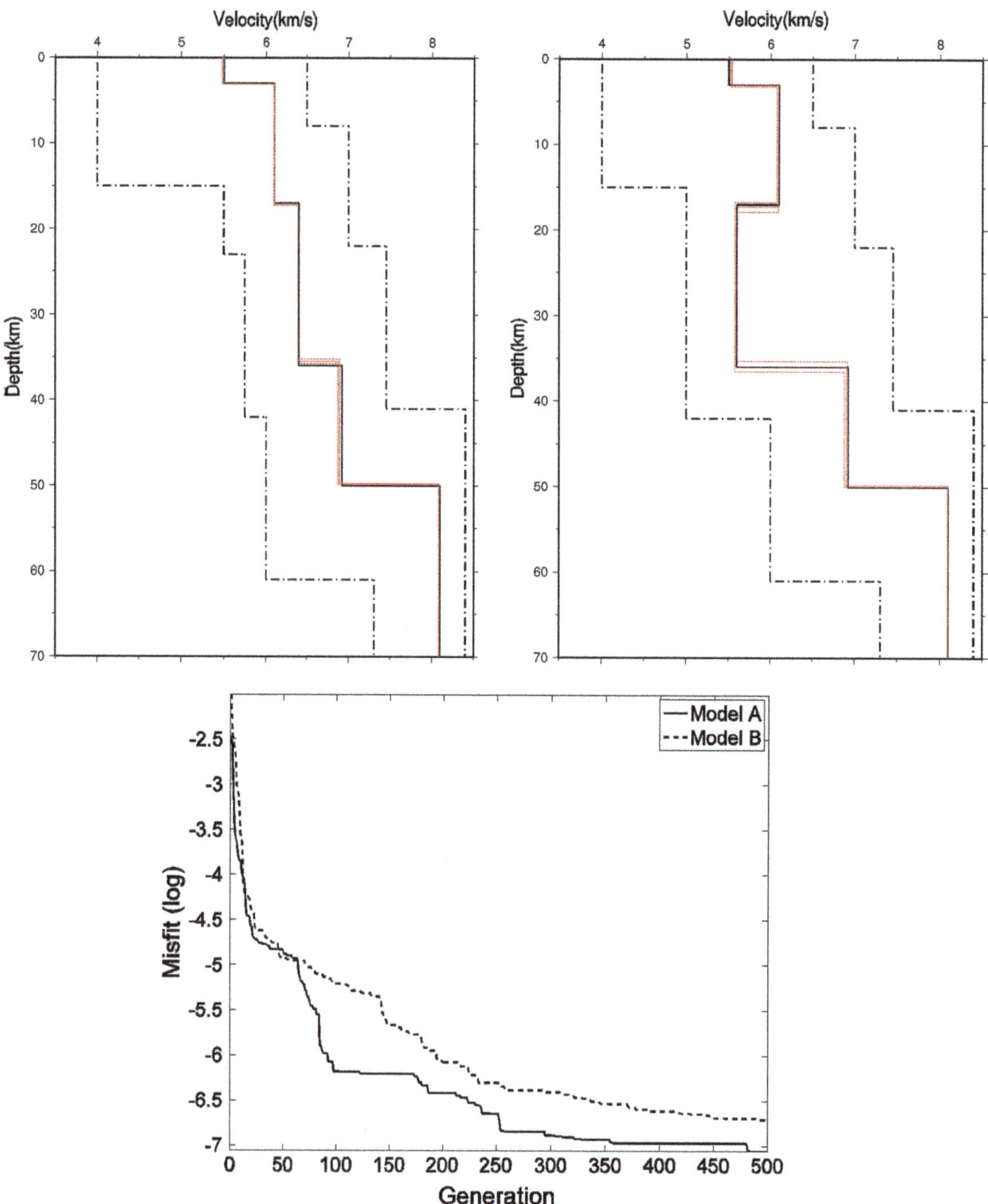

Fig. 2 Comparison between the FWA inverted model and a given true model in numerical tests. The *solid black line* denotes the given true model, while the *black dotted* and *dashed lines* mark the *upper* and *lower boundaries* of the velocity and depth search range. The *red solid line* denotes the inverted average velocity model, and the *light red dashed line* denotes ±1 SD of each model parameter. *Upper left* given true model A without a low-velocity layer; *upper right* given true model B incorporating a low-velocity layer. *Lower* the average misfit convergence value of model A (*solid line*) and model B (*dashed line*) along with generation increase

$$F = (1-\lambda) \times \frac{\sqrt{\sum_i \sum_j \left(\mathrm{Obs}_{ij} - \mathrm{Syn}_{ij}\right)^2}}{N_\mathrm{w} \times \sqrt{\sum_i \sum_j \mathrm{Obs}_{ij}^2}} + \lambda \times \left[1 - \frac{1}{N_\mathrm{w}} \times \sum_i \frac{\max(\mathrm{Obs}_i \times \mathrm{Syn}_i)}{\sqrt{\mathrm{Obs}_i \times \mathrm{Obs}_i}\sqrt{\mathrm{Syn}_i \times \mathrm{Syn}_i}}\right]. \quad (3)$$

Thus, (Obs × Syn) is:

$$(\mathrm{Obs} \times \mathrm{Syn})(\tau) = \int \mathrm{Obs}(t)\mathrm{Syn}(t-\tau)\mathrm{d}t$$

In this expression, N_w refers to the waveform data number, λ is the weight parameter used to balance the RMR and

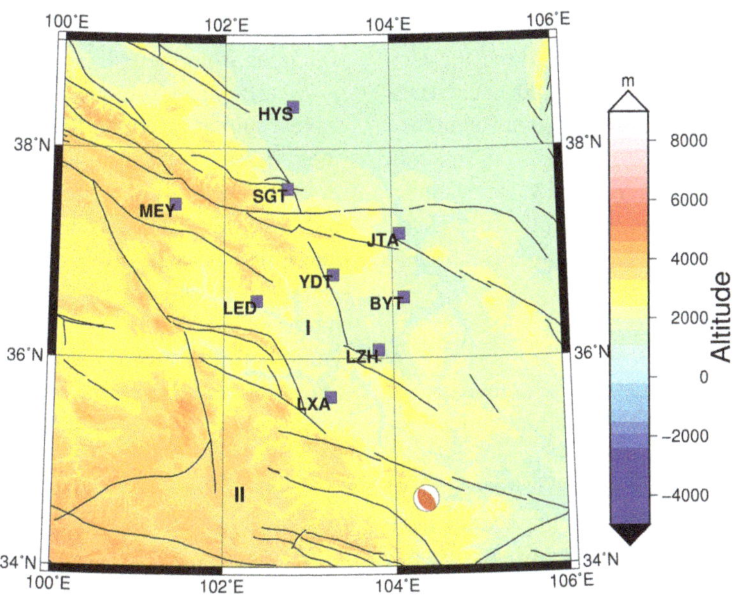

Fig. 3 Study area. The focal mechanism beach ball denotes the epicenter of the M_W 5.4 Minxian earthquake, the *blue solid squares* mark the positions of nine measurement stations, and *black solid lines* indicate faults. *Abbreviations* I, Qilian block; II, Qaidam block

Table 2 Earthquake parameters used in this study (http://www.globalcmt.org)

Epicenter	Centroid time (GMT)	Fault plane I			Fault plane II		
		Strike (°)	Dip (°)	Slip (°)	Strike (°)	Dip (°)	Slip (°)
34.65°N 104.35°E	2013-07-22 1:12:36.9	153	37	103	316	54	80

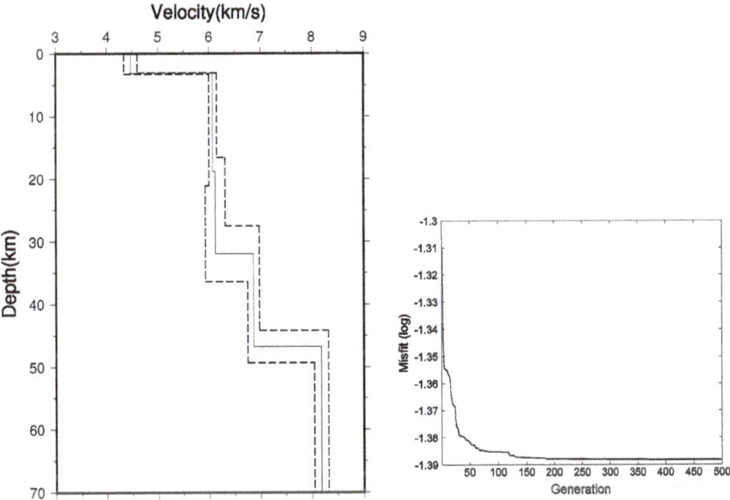

Fig. 4 *Left* average inversion models. The *solid line* in this figure denotes the final crustal velocity model, while the *two dashed lines* represent ±1 SD for model parameters. *Right* average misfit model value convergence plotted along with generation increase

waveform correlation error, while Obs_{ij} and Syn_{ij} refer to the amplitudes of the ith component of the observed and synthetic waveforms at the jth time point, respectively. In the objective function, the first term is the RMR used to determine the amplitude difference between synthetic and observation data, while the second term measures the similarity between observed and synthetic waveforms encompassing phase information. The weight parameter, λ, was set to 0.5 in this inversion so as to provide the same weight to the RMR and the waveform correlation coefficient (Li et al. 2016).

Like the numerical tests described above, we set the number of fireworks, *m*, and the sparks, *n*, to ten for each generation, and run FWA inversions for 500 generations using five different random seed values. This approach meant that the optimizer conducted up to 5×10^4 function evaluations in each run, and a final inverted velocity model was obtained by averaging the five most optimal models.

5 Results and discussion

Utilizing regional waveform inversions of seismic data, we determined an average 1D velocity structural model for central Gansu Province. Our average inversion model including standard errors is shown in Table 3. According to this inversion model, the thicknesses of the sedimentary layer, as well as the upper, middle, and lower crust are 3.15, 15.69, 13.08, and 14.83 km, respectively, while v_P values are 4.47, 6.07, 6.12, and 6.87 km/s, respectively, and the inverted v_P of the uppermost mantle is 8.18 km/s. In a previous study, Liu et al. (2006) used v_P intervals to determine the crustal composition of the northeastern margin of the Qinghai-Tibet plateau; the results of this study show that both the upper and middle crust contain felsic material, encompassing a total thickness of 28.77 km, while intermediate material is present in the lower crust, which has a thickness of 14.83 km.

The results of our study show that the thickness of the crust in central Gansu Province is 46.75 km, consistent with previously published receiver function results (Shen et al. 2013). These works calculated crustal thicknesses of 50.4 km at station BYT, 51.6 km at station LED, 46.2 km at station LXA, 49.2 km at station LZH, and 43.2 km at station YDT, translating to an average crustal thickness for the region covered by these five stations of 46.76 km. In this inversion, the velocity of upper mantle is 8.18 km/s, consistent with the upper mantle value used in the crust 1.0 model (Laske et al. 2013).

Comparisons of the synthetic waveforms generated by our final inverted 1D velocity model with those observed at the nine measurement stations are presented in Fig. 5. These data show a good overall fit between synthetic and observed waveforms and that the nine seismic stations in this study can be divided into three groups based on their azimuths from the epicenter.

Measurement stations LZH, YDT, SGT, and HYS all have a similar azimuth of about 340° from the epicenter. In these cases, both synthetic P- and S-waves for three components match the observed waveforms very closely, indicating that our inverted crustal velocity model accurately reproduces them in this orientation.

However, stations BYT and JTA are orientated at azimuths of about 355° from the epicenter; in these cases, while both the vertical and radial components match well with observations, there are obvious differences in the transverse component compared to observed waveforms. These differences may be the result of anisotropy and lateral heterogeneity in the crust along the BYT and JTA profiles with respect to the epicenter.

Stations LXA, LED, and MEY all have similar azimuths of about 319° with respect to the epicenter. Phase shifts between synthetic and observed S-waves are also clearly seen in vertical and radial components at these stations; although P-waves fit well. He et al. (2014) obtained a Moho depth map for China using the receiver function technique; the results of their study suggest that the depth of the Moho increases from about 45 km in the northeastern part of our study area to about 55 km in the southwest. Thus, this observed phase shift in S-waves maybe due to the Poisson solid, isotropy, and homogeneous medium assumptions that are inherent to the 1D velocity model. Research in South Korea (Chang and Baag 2006) and the Northeastern USA (Viegas et al. 2010) has also noted crustal anisotropy, which is beyond the scope of the 1D velocity model. This phenomenon should be further considered in future 2D and 3D crustal structural inversions.

The focal mechanism and location of an earthquake can also affect a final model. Thus, a station located on the nodal plane of a focal mechanism will record a seismogram with a smaller amplitude, so it is preferable to avoid the use of such station data for inversions. The distance from the epicenter can also affect models, so seismograms recorded at short epicentral distances also provide poor constraints for deeper parts of models as the turning point of the ray path is shallower. Because we utilized seismograms recorded simultaneously at nine stations, our model can be considered to be well constrained.

Table 3 A 1D average inverted crustal velocity model of central Gansu Province (including standard errors)

$v_P \pm \delta$ (km/s)	$h \pm \delta$ (km)
4.47 ± 0.13	3.15 ± 0.12
6.07 ± 0.08	15.69 ± 2.24
6.12 ± 0.19	13.08 ± 4.49
6.87 ± 0.11	14.83 ± 2.58
8.18 ± 0.14	

6 Conclusions

We present a 1D crustal velocity model for central Gansu Province that is based on regional waveform inversion using FWA. The three components of seismic waveform data used in this study were recorded by the regional

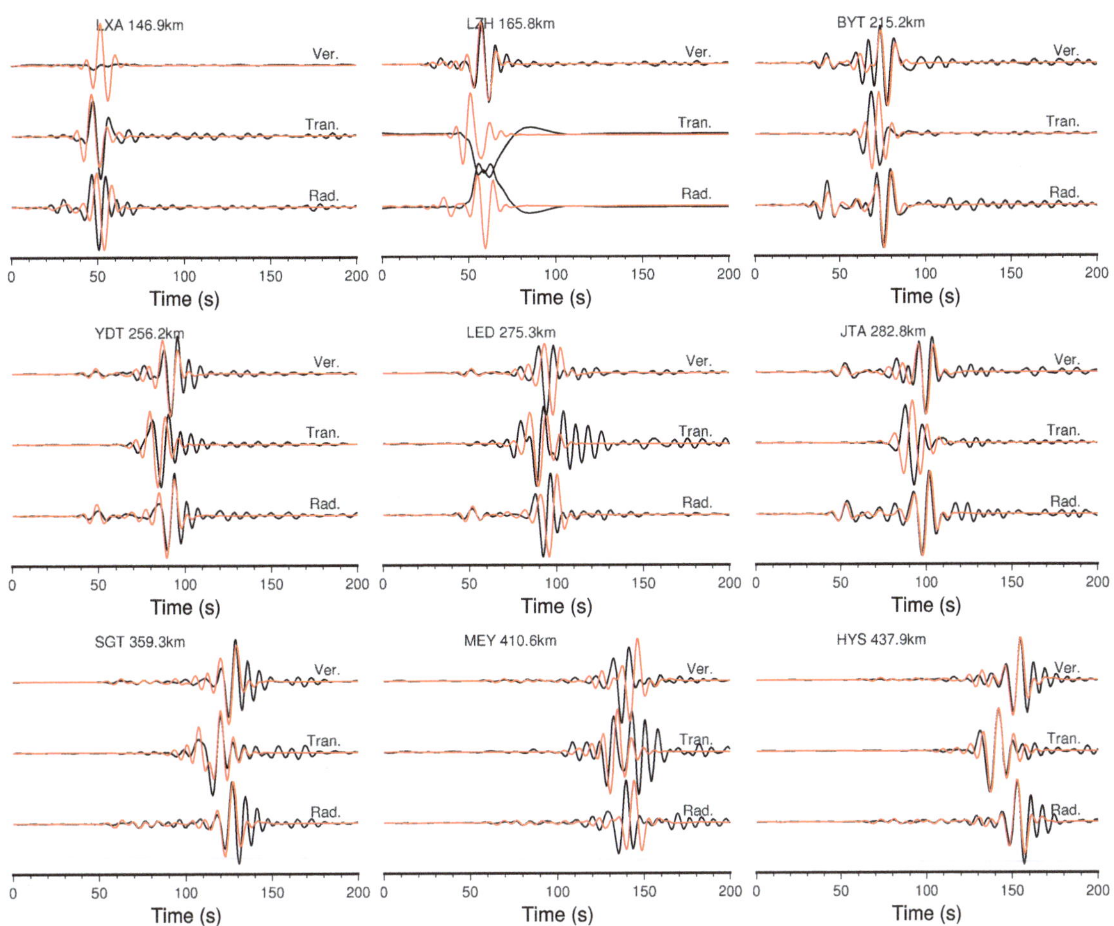

Fig. 5 Comparison of synthetic waveforms (*red*) from our final average inversion model with those observed (*black*) at nine stations in central Gansu Province. Station names and epicentral distances are listed at top of each figure. The *vertical* component for the LXA station, as well as the transverse and radial components for the LZH station were not used in the inversion as the observational records for these components contain high level of noises

seismic station network in Gansu Province. Numerical tests, the accuracy of our average inversion results, and standard deviations show that the FWA is suitable for regional waveform inversion. This inverted crustal velocity model indicates that the thicknesses of the sedimentary layer, as well as the upper, middle, and lower crust are 3.15, 15.69, 13.08, and 14.83 km, respectively, while calculated v_P values for these intervals are 4.47, 6.07, 6.12, and 6.87 km/s, respectively. Data show that the inverted v_P of the uppermost mantle is 8.18 km/s, while the crustal thickness is 46.75 km. The synthetic waveforms produced from this 1D model are in very close agreement with observations, although minor differences may be due to crustal anisotropy and lateral heterogeneous variations that were not taken into account during inversion. We suggest that these effects should be taken into account in future 2D and 3D velocity structural inversions.

Acknowledgements We thank Ke Ding for providing FWA code. This work was supported by the National Natural Science Foundation of China (No. 41174034). Some of the figures in this article were plotted using the Genetic Mapping Tools (GMT) software package distributed by Wessel et al. (2013). We are also grateful for the very helpful comments of two anonymous reviewers that enabled us to significantly improve this manuscript.

References

Abdelwahed MF, Zhao D (2014) Genetic waveform modeling for the crustal structure in Northeast Japan. J Asian Earth Sci 89:66–75

Bassin C, Laske G, Masters G (2000) The current limits of resolution for surface wave tomography in North America. EOS Trans AGU 81:F897

Chang SJ, Baag CE (2006) Crustal structure in southern Korea from joint analysis of regional broadband waveforms and travel times. Bull Seismol Soc Am 96(3):856–870

Ding K, Chen Y, Wang Y, Tan, Y (2015) Regional seismic waveform inversion using swarm intelligence algorithms, 2015. In: IEEE congress on evolutionary computation (CEC). pp 1235–1241

Ding K, Zheng S, Tan Y (2013) A GPU-based parallel fireworks algorithm for optimization. In: Proceedings of the 15th annual conference on Genetic and evolutionary computation. ACM, pp 9–16

Dziewonski AM, Anderson DL (1981) Preliminary reference Earth model. Phys Earth Plan Int 25:297–356

Fuchs K, Müller G (1971) Computation of synthetic seismograms with the reflectivity method and comparison with observations. Geophys J Int 23(4):417–433

He R, Shang X, Yu C, He R, Shang X, Yu C, Zhang H, Van der Hilst RD (2014) A unified map of Moho depth and V_p/V_s ratio of continental China by receiver function analysis. Geophys J Int 199(3):1910–1918

Jia S, Zhang X (2008) Study on the crust phases of deep seismic sounding experiments and fine crust structures in the northeast margin of Tibetan plateau. Chin J Geophys 51(5):1431–1443 **(in Chinese with English abstract)**

Laske G, Masters G, Ma Z, Pasyanos M (2013) Update on CRUST1.0—a 1-degree global model of Earth's crust. Geophys Res Abstr. 15 Abstract EGU2013-2658

Li C, Lei J (2014a) Numerical tests for effects of various parameters in niching genetic algorithm applied to regional waveform inversion. Earthq Sci 27(5):541–551

Li C, Lei J (2014b) Crustal velocity structure under southwestern Yunnan from regional waveform inversion. Chin Sci Bull 59(34):3398–3415. doi:10.1360/N972014-00407 **(in Chinese with English abstract)**

Li X, Wang Y, Chen YJ (2016) Inversion of ocean-bottom seismometer (OBS) waveforms for oceanic crust structure: a synthetic study. Earthq Sci 29(4):203–213

Li H, Michelini A, Zhu L, Li H, Michelini A, Zhu L, Bernardi F, Spada M (2007) Crustal velocity structure in Italy from analysis of regional seismic waveforms. Bull Seismol Soc Am 97(6):2024–2039

Li S, Wang Y, Liang Z, He S, Zeng W (2012) Crustal structure in southeastern Gansu from regional seismic waveform inversion. Chin J Geophys 55(2):206–218 **(in Chinese with English abstract)**

Liu M, Mooney WD, Li S, Liu M, Mooney WD, Li S, Okaya N, Detweiler S (2006) Crustal structure of the northeastern margin of the Tibetan plateau from the Songpan-Ganzi terrane to the Ordos basin. Tectonophysics 420(1):253–266

Maurice R, Stacey D, Wiens DA, Koper KD, Vera E (2003) Crustal and upper mantle structure of southernmost South America inferred from regional waveform inversion. J Geophys Res Solid Earth 108(B1):ESE15.1–ESE15.10

Shen X, Zhou Y, Zhang Y, Shen X, Zhou Y, Zhang Y, Liu X, Qin M, Li C (2013) Geodynamic significance of the crust structure beneath the eastern margin of Tibet. Progr Geophys 28(5):2273–2282 **(in Chinese with English abstract)**

Tan Y, Yu C, Zheng S, Ding K (2013) Introduction to fireworks algorithm. Int J Swarm Intell Res 4(4):39–70

Tan Y, Zhu Y (2010) Fireworks algorithm for optimization. In: Tan Y, Shi Y, Tan KC (eds) Advances in swarm intelligence, vol 175. Springer, Berlin, pp 355–364

Viegas GM, Baise LG, Abercrombie RE (2010) Regional wave propagation in New England and New York. Bull Seismol Soc Am 100(5A):2196–2218

Wessel P, Smith WH, Scharroo R, Wessel P, Smith WH, Scharroo R, Luis J, Wobbe F (2013) Generic mapping tools: improved version released. Eos Trans AGU 94(45):409–410

Zhou M, Zhang Y, Shi Y, Zhou M, Zhang Y, Shi Y, Zhang S, Fan B (2006) Three-dimensional crustal velocity structure in the northeastern margin of the Qinghai-Tibetan plateau. Progr Geophys 21(1):127–134 **(in Chinese with English abstract)**

8

S-wave velocity structure in the SE Tibetan plateau

Yan Cai · Jianping Wu · Weilai Wang · Lihua Fang · Liping Fan

Abstract We use observations recorded by 23 permanent and 99 temporary stations in the SE Tibetan plateau to obtain the S-wave velocity structure along two profiles by applying joint inversion with receiver functions and surface waves. The two profiles cross West Yunnan block (WYB), the Central Yunnan sub-block (CYB), South China block (SCB), and Nanpanjiang basin (NPB). The profile at ~25°N shows that the Moho interface in the CYB is deeper than those in the WYB and the NPB, and the topography and Moho depth have clear correspondence. Beneath the Xiaojiang fault zone (XJF), there exists a crustal low-velocity zone (LVZ), crossing the XJF and expanding eastward into the SCB. The NPB is shown to be of relatively high velocity. We speculate that the eastward extrusion of the Tibetan plateau may pass through the XJF and affect its eastern region, and is resisted by the rigid NPB, which has high velocity. This may be the main cause of the crustal thickening and uplift of the topography. In the Tengchong volcanic area, the crust is shown to have alternate high- and low-velocity layers, and the upper mantle is shown to be of low velocity. We consider that the magma which exists in the crust is from the upper mantle and that the complex crustal velocity structure is related to magmatic differentiation. Between the Tengchong volcanic area and the XJF, the crustal velocity is relatively high. Combining these observations with other geophysical evidence, it is indicated that rock strength is high and deformation is weak in this area, which is why the level of seismicity is quite low. The profile at ~23°N shows that the variation of the Moho depth is small from the eastern rigid block to the western active block with a wide range of LVZs. We consider that deformation to the south of the SE Tibetan Plateau is weak.

Y. Cai · J. Wu (✉) · W. Wang · L. Fang · L. Fan
Institute of Geophysics, China Earthquake Administration, Beijing 100081, China
e-mail: wjpwu@cea-igp.ac.cn

Y. Cai
e-mail: caiyan@cea-igp.ac.cn

J. Wu · L. Fang
Key Laboratory of Seismic Observation and Geophysical Imaging, Institute of Geophysics, China Earthquake Administration, Beijing 100081, China

Keywords SE Tibetan plateau · Velocity structure · Receiver function · Joint inversion · Tengchong volcano

1 Introduction

Collision between the Indian and Eurasian plates has induced intense orogenesis, produced many active faults, and caused material to escape from the Tibetan plateau (Molnar and Tapponnier 1975; Rowley 1996). To study the uplift and deformation mechanisms in this region, researchers have proposed many models, such as eastward extrusion of crustal material, crustal thickening and uplift of the plateau, or channel flow in a weak crust (Molnar and Tapponnier 1975; Tapponnier et al. 1982; Allegre et al. 1984; Royden et al. 1997). The SE Tibetan plateau is a region which is likely to be experiencing crustal flow and eastward extrusion, so it is quite important to study and test these models in this region. Thus, research into the deep crustal structure in the Tibetan plateau is needed to deepen our understanding of its evolution mechanisms.

Under forces from the NE subduction of the Indian plate and the eastward subduction of the Myanmar microplate, the Tibetan plateau has extruded eastward. Resisted by the rigid Yangtze platform, extrusion in the study area has become toward the southeast. Subjected to the intense tectonic activity, there has produced many major strike-slip faults in the SE Tibetan plateau. With the intense crustal movement and strike-slip motion of the active faults, there is a clockwise rotation around the eastern Himalaya syntaxis in the SE Tibetan plateau (Replumaz and Tapponnier 2003; Shen et al. 2001; Liang et al. 2013), and the strong earthquakes occurred frequently (Fig. 1). Combined with the GPS observation data, historical earthquakes, and focal mechanism, previous studies have divided the SE Tibetan plateau into many blocks (Fig. 1). The boundary faults among these blocks are shown to be compression-shear or tension-shear, and there is intense tectonic activity near these faults (Xu et al. 2005; Cheng et al. 2012). Thus, the velocity structure and deformation of the SE Tibetan plateau have attracted much attention. Researchers have used many techniques to study the crust-mantle structures beneath this region, such as receiver function inversion, surface wave dispersion curve inversion, joint inversion, travel time inversion, and deep seismic sounding (Wu et al. 2001, 2013; Wang et al. 2003, 2007, 2014; Bai and Wang 2004; Hu et al. 2008; Li et al. 2009; Sun et al. 2014; Fan et al. 2015). However, resolutions have been limited for the observations from permanent stations. Sun et al. (2014) used dense seismic data from Chinarry to obtain the velocity structure of a

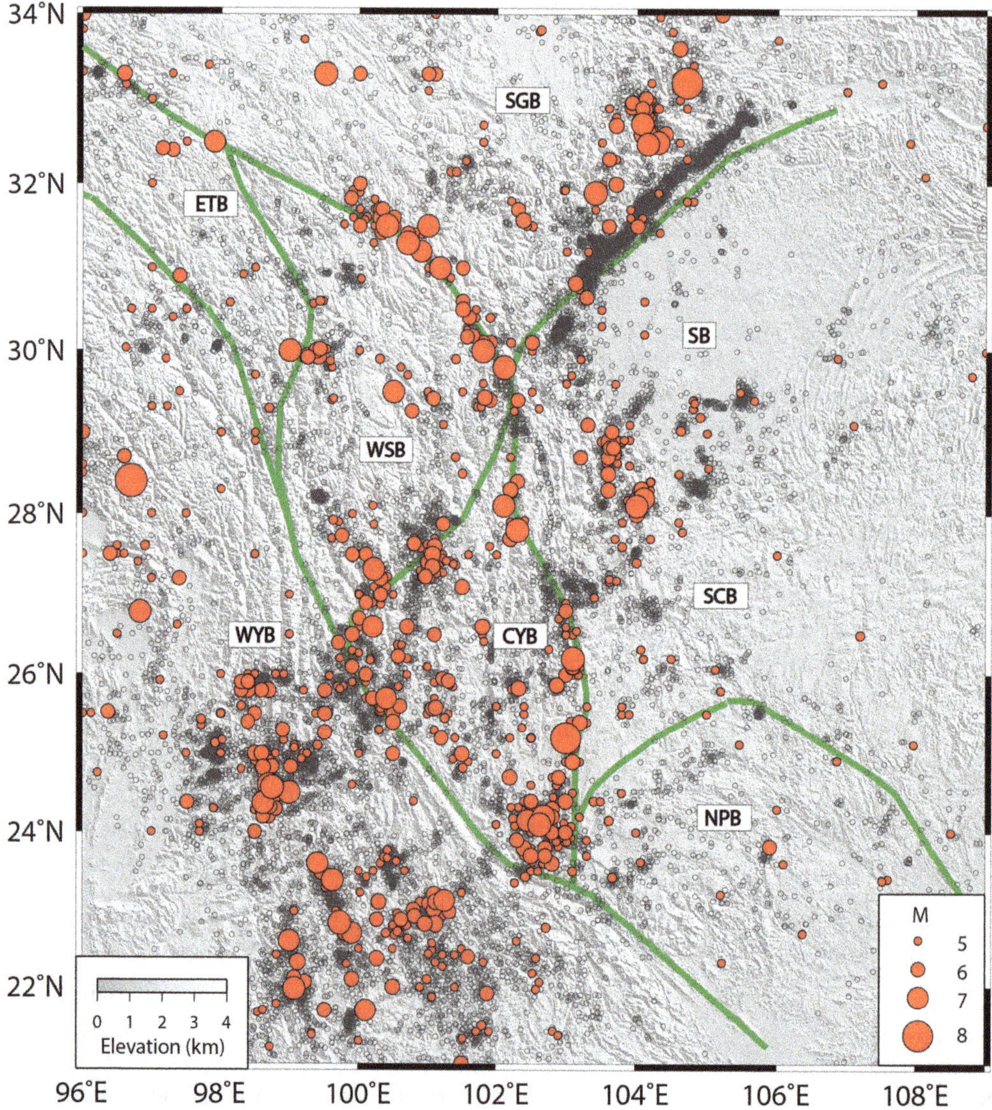

Fig. 1 Map of the blocks and tectonic boundaries in the SE Tibetan plateau. *Green lines* represent the boundaries between the tectonic units. The *red solid circles* represent the historical earthquakes with the magnitude larger than 5.0. The *black open circles* are the earthquakes with magnitude larger than 3.0 from 1970 to 2015 (the earthquake catalogue is from the China Earthquake Networks Center). *ETB* Eastern Tibet sub-block, *SGB* Songpan-Garze block, *WSB* Western Sichuan sub-block, *WYB* Western Yunnan sub-block, *CYB* Central Yunnan sub-block, *SCB* South China block, *NPB* Nanpanjiang basin, *SB* Sichuan basin

profile. They concluded that the two LVZs in the crust are crustal flow channels. Combining this with other research, we consider that the interpretation for the velocity structure should be a little different.

The youngest continental volcano in China—Tengchong volcano is located to the west of the West Yunnan block (WYB). Owing to the collision of the Indian and Eurasian plates, volcanic eruptions, geothermal activity, and crustal earthquakes are quite frequent and intense. Researchers have always been concerned with the deep structure of the volcano and its development (Qin et al. 1998; Wang et al. 2002; Wang and Gang 2004) and have obtained many important results and conclusions. These studies play an important role in predicting volcanic eruptions, understanding the relationships between the volcano and earthquakes, detecting geothermal storage and rationally utilizing geothermal resources.

In this paper, we use observation data recorded by not only the permanent stations but also the newly deployed dense seismic array in the SE Tibetan plateau to calculate receiver functions. We apply joint inversion with receiver functions and surface wave dispersion curves to obtain a high-resolution S-wave velocity structure. These new results provide detailed information for studying the crust-mantle structural features and tectonic boundary of the SE Tibetan Plateau.

2 Data and methods

The seismic data we used for inversion were recorded by 122 stations (Fig. 2): 23 permanent stations (Zheng et al. 2010) and 99 temporary stations. The permanent stations were in operation from 2007 to 2013. Of the temporary stations, 13 were in operation from December 2008 to December 2010 (ChinArray 2006) and 86 stations were in operation from June 2011 to November 2013 (ChinArray-Himalaya 2011).

We selected two seismic profiles with lengths of about 900 km and widths of 100 km. The teleseismic records of all stations with epicentral distances of between 30° and 90°, magnitudes >5.5, and SNR > 4.0 were selected to calculate receiver functions, and we chose 1436 teleseismic events (Fig. 3).

2.1 Extracting receiver functions

We extracted receiver functions from teleseismic P-waves using the maximum entropy deconvolution method (Burg 1972; Claerbout 1976; Wu et al. 2003). The original seismic records were processed by a band-pass filter of 0.02–2 Hz. We applied a Gaussian filter with an α-coefficient of 2.5 and a volume factor of 0.001 for the deconvolution method. To obtain a reliable velocity structure, we selected the receiver functions with a clear P-phase and stacked them according to their differences in slowness. Finally, we chose three stacked receiver functions to carry out the inversion.

2.2 Joint inversion of receiver functions and surface waves

The surface wave phase and group velocity dispersion curves are from Wang et al. (2014) and Fan et al. (2015). The origin of the surface wave dispersion technique is described in detail in their work. The dispersion curve is interpolated into each station. Generally, receiver function inversion has the advantage of obtaining the depth of seismic velocity

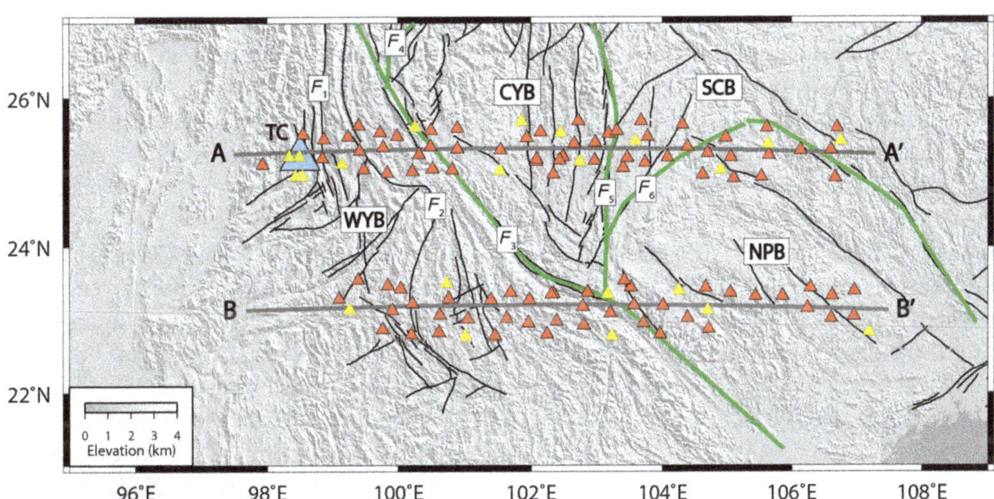

Fig. 2 Locations of the profiles and seismic stations. The *red* and *yellow triangles* represent the temporary and permanent stations, respectively. The *blue triangle* indicates the Tengchong volcanic area. *Gray lines* are the profiles *AA′* and *BB′* that were studied in this work. *Black lines* show the faults in the study area. F_1 Nujiang fault, F_2 Lancangjiang fault, F_3 Red River fault, F_4 Xiaojinhe fault, F_5 Xiaojiang fault, F_6 Mile fault, *TC* Tengchong volcano

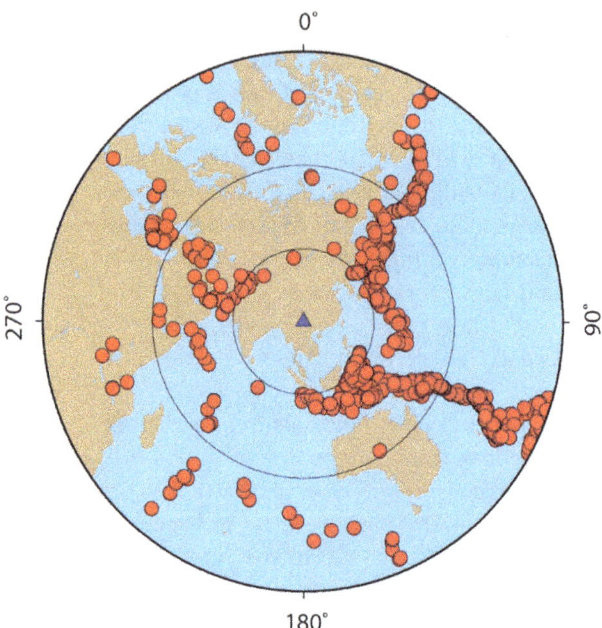

Fig. 3 Distribution of teleseismic events used for receiver functions. The *red circles* indicate teleseismic events. The *blue triangle* is the central point of the seismic stations used in this study

discontinuity. However, due to noise interference and multiple waves in the low-velocity layer near the surface, the calculated velocity structure is inaccurate. To solve this problem, we applied the surface wave dispersion curves to constrain the receiver function inversion. Surface wave dispersion curves should effectively reveal variation in S-wave velocity, but it is difficult to determine the exact distribution of velocity discontinuity. Joint inversion of receiver functions and surface wave dispersion curves combines the advantages of the two methods and could allow us to obtain a reliable S-wave velocity structure.

During joint inversion, the weight values of the fitting error for the receiver functions and the surface wave were 0.2 and 0.8, respectively. The average v_P/v_S ratio was also inverted in the inversion procedure. We used the same initial velocity model because the inversion results were not very sensitive to the initial model. With different initial models, the inversion results were similar. The initial v_S was set to 3.0 km/s at 0–10 km, 3.3 km/s at 10–26 km, 3.5 km/s at 26–34 km, 4.1 km/s at 34–46 km, and 4.5 km/s at 46–200 km. There were ten iterations for each station. From Fig. 4, it can be seen that the calculated receiver functions and surface wave dispersion curves at station 53,220 fit well with the observed data.

3 Results

With the dense seismic array deployed in this area, we obtained the S-wave velocity structures of the two profiles to a depth of 100 km (Figs. 5, 6). From profile *AA′* (Fig. 5), we could see that the Moho interface at the middle of the profile is deeper than that at both ends. The Moho depth decreases from about 46 km in the Central Yunnan block (CYB) to 34 km in the Nanpanjiang basin (NPB) and to 38 km in the WYB; and the topography and Moho depth have clear and good correspondence. Beneath the Xiaojiang fault (XJF), there is a crustal low-velocity zone (LVZ) located at 25–38 km depth. The LVZ has a length of 300 km, crosses the XJF zone, and expands eastward about 160 km into the south China block (SCB). To its east, the NPB is shown to be of relatively high velocity within the crust and upper mantle. The LVZ passes through the XJF zone and seems to be resisted by the NPB, which is of high velocity; this result differs little from previous understanding that the XJF is the boundary of the high-velocity eastern region and low-velocity western region. In the Tengchong volcanic area, the velocity variation in the crust is more complex than that of its surroundings. On the whole, the crust shows alternate thin high-velocity zones (HVZ) and LVZs, and the upper mantle velocity is low. Between the Tengchong volcanic area and the XJF, the crustal velocity is relatively high, and the HVZ crosses the Red River fault (RRF). Profile *BB′* (Fig. 5) shows a conspicuous variation in velocity within the WYB and the NPB: the WYB has a wide range of crustal LVZs but NPB is shown to be of relatively high velocity. Compared with the northern profile, the Moho depths at the eastern stable block and western active block have quite gradual variations.

4 Discussion

Near the middle branch of the XJF, there exists a LVZ in the middle and lower crust and upper mantle. The crustal LVZ is located not only in the XJF; but also crosses the XJF and expands eastward into the SCB, which shows that the XJF is not the boundary of the eastern HVZ and western LVZ. Receiver function studies found that Poisson's ratio is high around the XJF (Wang et al. 2014). The surface heat flow is about 70 mW/m^2 near the middle branch of the XJF and its eastern area (Xu et al. 1992). Combined with the low velocity in the upper mantle, this implies that the crustal LVZ may be related to the high temperature in the upper mantle. When the temperature rises, the strength of the crustal material weakens, and then the crust easily deforms. Thus, we could speculate that with the high temperature near the XJF and its eastern area, the eastward extrusion of the Tibetan plateau may have passed through the XJF and affected its eastern region. The NPB is of relatively high velocity and has a low temperature (Xu et al. 1992) and low Poisson's ratio (Wang et al. 2014). We

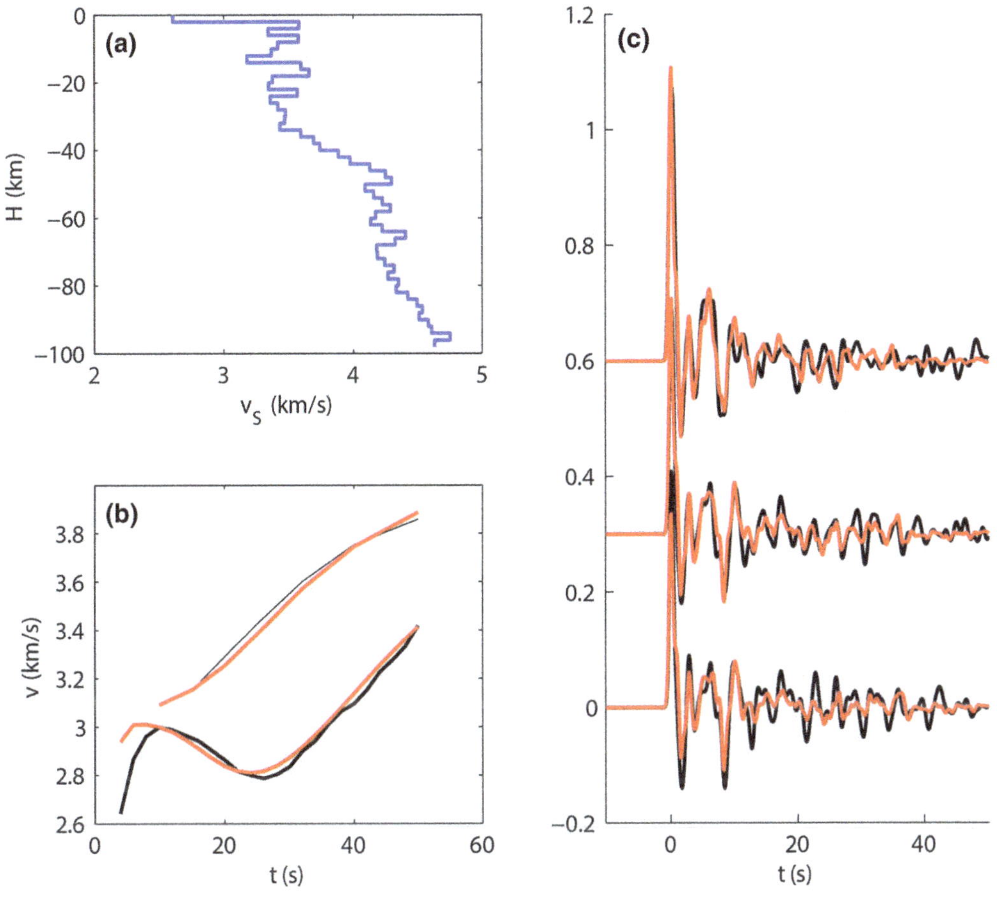

Fig. 4 The joint inversion results at station 53,220. **a** The S-wave velocity from joint inversion at station 53,220. **b** The surface wave dispersion curve (*black lines*) and their fit curves (*red lines*). **c** The original receiver functions (*black lines*) and their fit curves (*red lines*)

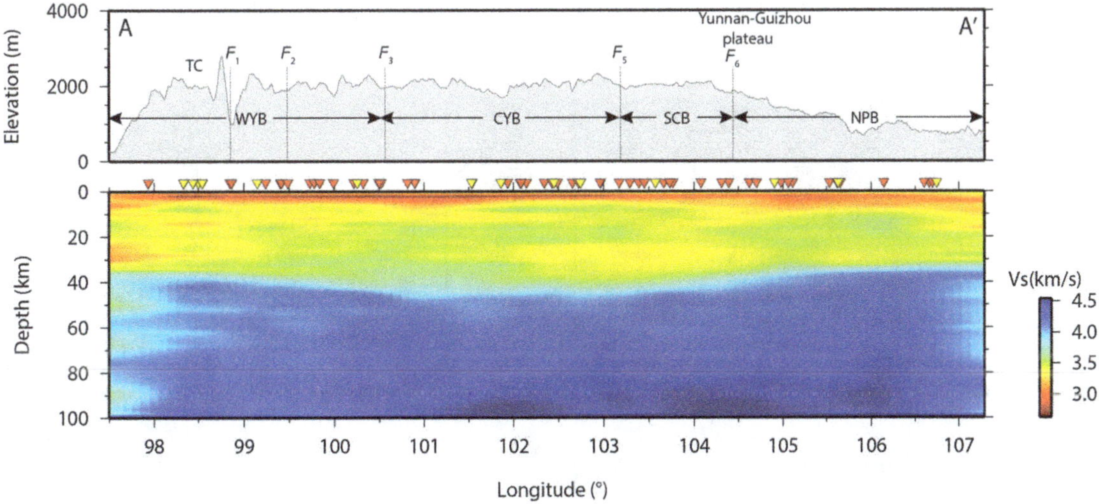

Fig. 5 The S-wave velocity structure beneath profile *AA′*

consider that the rocks in the NPB have high strength, and so they block the eastward extrusion of the Tibetan plateau, causing topography uplift in the Yunnan-Guizhou plateau and crustal thickening. Wang et al. (2009) obtained the P wave velocity structure crossing XJF zone from the deep seismic sounding profile. The results implied that there

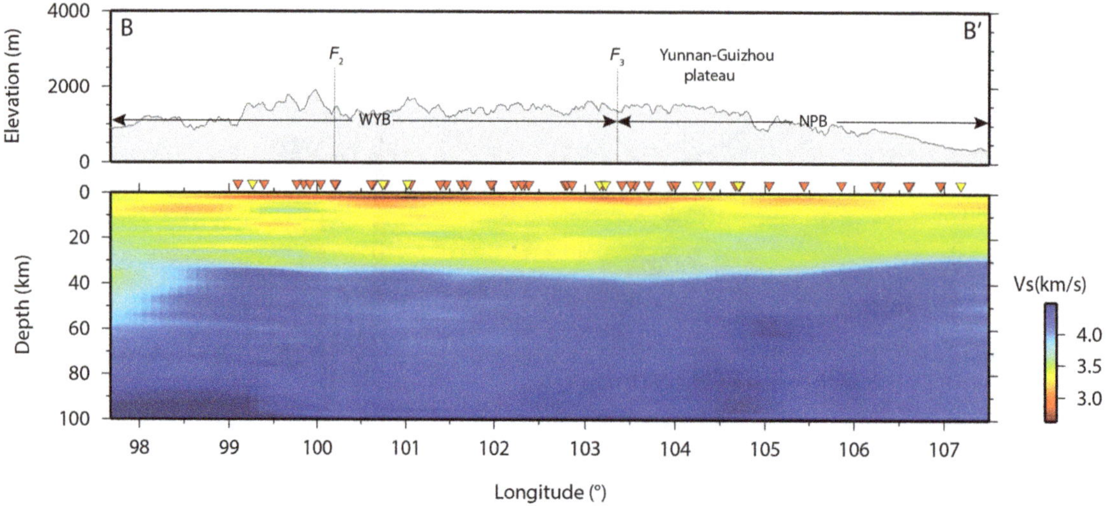

Fig. 6 The S-wave velocity structure beneath profile *BB′*

exists LVZ in the lower crust beneath the east of XJF. Our results show that the crustal LVZ crosses XJF and presents in both sides of XJF zone, which is also identified from the P wave tomography (Wang et al. 2003; Wu et al. 2013).

In the Tengchong volcanic area, the crustal velocity structure is shown to be more complex than that in other areas (Fig. 5). As shown in Fig. 7, the TNC and MIZ stations are both located in the Tengchong volcanic area. We

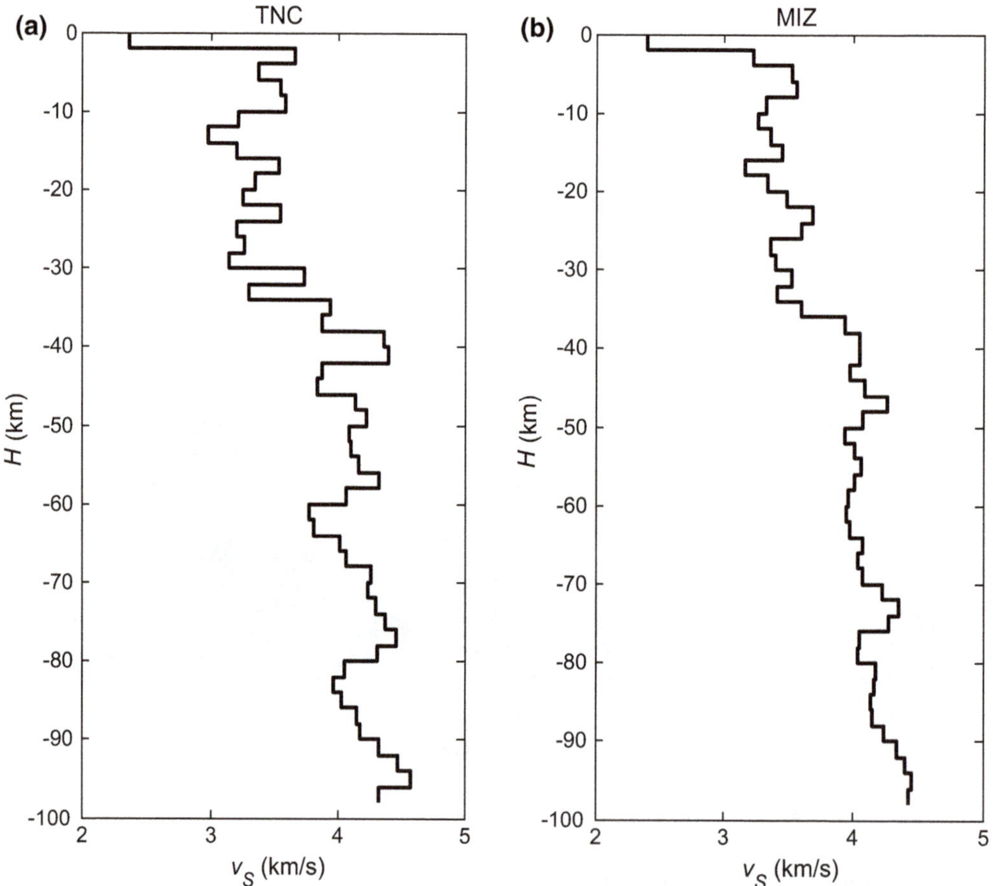

Fig. 7 The joint inversion results at TNC **a** and MIZ **b** stations

can see that the shallow crust within 5–10 km is of high velocity, with $v_P = 3.6$–3.7 km/s. Under the HVZ, the velocity decreases rapidly and reaches to 3.0 km/s. From 16 to 23 km, the crustal velocity rises again to about 3.6 km/s, and then the crust below 25 km is shown to have a relatively low velocity. Overall, the crustal velocity structure shows that HVZs alternate with LVZs, and the upper mantle has a low velocity. Moreover, the Tengchong volcanic area has a feature of high heat flow (Xu et al. 1992). Thus, it is speculated that the crustal LVZ is related to the high temperature and that the low-velocity hot material is mainly from the upper mantle. Several magma chambers may have developed in the crust after the intrusion of magma. When the magma cooled and solidified, the velocity of that rock became higher. We infer that the crustal high velocity anomaly is the main result of the intrusion of the basic or ultrabasic rocks from the deep lithosphere, and the alternation of HVZs and LVZs in the crust is the result of magmatic differentiation. Sun et al. (2014) obtained the S-wave velocity structure near our profile AA' with receiver functions and group velocity of Rayleigh wave. Both the results of our study and their work show that two crustal LVZs present beneath the Tengchong area and XJF zone. Sun et al. (2014) argued that the two LVZs are the channels of the crustal flow in the SE Tibetan plateau. In contrast to their results, our results show that there also exist HVZs clearly in the crust beneath Tengchong volcanic area. The results from deep seismic sounding profile and P wave tomography (Wang et al. 2002, 2003; Ma et al. 2008) also show the existence of both high velocity anomaly and low velocity anomaly in the crust beneath Tengchong volcanic area. Combining with the high temperature and low velocity within the upper mantle, we are inclined to conclude that the crustal low-velocity material beneath Tengchong volcanic area is mainly from the deep lithosphere.

As shown in Fig. 5, between the Tengchong volcanic area and the XJF zone, the crustal velocity is shown to be relatively high. The level of seismicity in this area is weak (Xu et al. 2005). The heat flow (Xu et al. 1992), Poisson's ratio (Wang et al. 2014), and electrical conductivity (Bai et al. 2010) are lower than those in the surrounding areas. This implies that the rocks in this area have high strength and weak deformation. However, it is difficult to predict whether strong earthquakes could occur in such an area in the future. From Fig. 6, we could see that the average crustal velocity in the WYB is lower than that in the NPB, and a wide range of LVZs are distributed in the WYB, which is quite different from the situation in the NPB. The Moho depth varies more gradually than that in profile AA' (Fig. 5). Similar crustal thickness in the eastern stable block and western active block indicates that the deformation in the south of the study region is relatively weak.

5 Conclusions

We investigated the S-wave velocity structure beneath the SE Tibetan plateau by applying joint inversion of receiver functions and surface wave phase velocity and group velocity dispersion curves. The observation data are recorded at most of the permanent stations and from a dense seismic array. Combining our results with previous studies, we have shown some detailed and novel velocity anomalies. The major conclusions are as follows:

Beneath the XJF, there exists a crustal LVZ, which crosses the XJF and expands eastward into the SCB. The LVZ is related to hot material in the upper mantle. The high temperature has weakened the strength of the crust, contributing to eastward extrusion of the Tibetan plateau to passing through the XJF and affecting its eastern region. The rigid high-velocity NPB has resisted the eastward extrusion of the Tibetan plateau, causing crustal thickening and topography uplift in the Yunnan-Guizhou plateau.

In the Tengchong volcanic area, the crustal velocity structure is more complex than that of other regions. Crustal velocity is shown to be high at depths of 5–10 and 18–22 km, and low at 10–18 and 22–38 km, and velocity is low in the upper mantle. We speculate that the crustal LVZ is related to the magma chambers in the crust. When the magma cooled and solidified, the velocity of the rocks became higher.

Between the Tengchong volcanic area and the XJF, the crustal velocity is relatively high. The seismicity level is clearly low compared with other regions. We consider that the strength of the rocks in this area is high and the deformation is weak. However, due to the complex tectonic activity in the SE Tibetan plateau, it is difficult to predict whether strong earthquakes will occur in this region in the future.

Acknowledgments The waveform data were provided by Data Management Center of China National Seismic Networks and China Seismic Array Data Management Center at the Institute of Geophysics, China Earthquake Administration. This work was supported by a National Natural Science Foundation of China (Grant No. 41374097) and China National Special Fund for Earthquake Scientific Research in Public Interest (Grant No. 201008001).

References

Allegre CJ, Courtill V, Tapponnier P, Hirn A, Mattauer M, Coulon C, Jaeger JJ, Achache J, Scharer U, Marcoux J (1984) Structure and evolution of the Himalaya-Tibet orogenic belt. Nature 307(5946):17–22

Bai ZM, Wang CY (2004) Tomography research of the Zhefang-Bingchuan and Menglian-Malong wide-angle seismic profile in Yunnan province. Chin J Geophys 47(2):257–267 **(in Chinese with English abstract)**

Bai DH, Unsworth MJ, Meju MA, Ma XB, Teng JW, Kong XR, Sun Y, Sun J, Wang LF, Jiang CS (2010) Crustal deformation of the eastern Tibetan Plateau revealed by magnetotelluric imaging. Nat Geosci 3(5):358–362

Burg JP (1972) The relationship between maximum entropy spectra and maximum likelihood. Spectra Geophys 37(2):375–376

Cheng J, Xu XW, Gan WJ, Ma WT, Chen WT, Zhang Y (2012) Block model and dynamic implication from the earthquake activities and crustal motion in the southeastern margin of Tibetan Plateau. Chin J Geophys 55(4):1198–1212 **(in Chinese with English abstract)**

ChinArray (2006) China seismic array waveform data. China Earthq Adm. doi:10.12001/ChinArray.Data

ChinArray-Himalaya (2011) China seismic array waveform data of Himalaya project. Inst Geophys China Earthq Adm. doi:10.12001/ChinArray.Data.Himalaya

Claerbout JF (1976) Fundamentals of geophysical data processing. McGraw-Hill, New York, pp 1–357

Fan LP, Wu JP, Fang LH (2015) The characteristic of Rayleigh wave group velocities in the southeastern margin of the Tibetan Plateau and its tectonic implications. Chin J Geophys 58:1555–1567 **(in Chinese with English abstract)**

Hu JF, Yang HY, Zhao H (2008) Structure and significance of S-wave velocity and Poisson's Ratio in the crust beneath the eastern side of the Qinghai-Tibet Plateau. Pure appl Geophys 165(5):829–845

Li YH, Wu QJ, Tian XB, Zhang RQ, Pan JT (2009) Crustal structure in the Yunnan region determined by modeling receiver functions. Chin J Geophys 52(1):67–80 **(in Chinese with English abstract)**

Liang S, Gan W, Shen C, Xiao G, Liu J, Chen W, Xiao D, Zhou D (2013) 3D velocity field of present-day crustal motion of the Tibetan Plateau derived from GPS measurements. J Geophys Res 118(10):5722–5732

Ma HS, Zhang GM, Wen XZ, Zhou LQ, Shao ZG (2008) 3-D *P* wave velocity structure tomographic inversion and its tectonic interpretation in southwest China. Earth Sci 33(5):591–602 **(in Chinese with English abstract)**

Molnar P, Tapponnier P (1975) Cenozoic tectonics of Asia: effects of a continental collision. Science 189(4201):419–426

Qin JZ, Huang PG, Zhang JW (1998) Characteristics of *Q* values around Tengchong volcano and adjacent areas. J Seismol Res 21:358–361 **(in Chinese with English abstract)**

Replumaz A, Tapponnier P (2003) Reconstruction of the deformed collision zone between India and Asia by backward motion of lithospheric blocks. J Geophys Res 108(B6):2285. doi:10.1029/2001JB000661

Rowley DB (1996) Age of initiation of collision between India and Asia: a review of stratigraphic data. Earth Planet Sci Lett 145(1–4):1–13

Royden LH, Burchfiel BC, King RW, Wang E, Chen Z, Shen F, Liu Y (1997) Surface deformation and lower crustal flow in Eastern Tibet. Science 276(5313):788–790

Shen F, Royden LH, Burchfiel BC (2001) Large-scale crustal deformation of the Tibetan Plateau. J Geophys Res 106(B4):6793–6816

Sun XX, Bao XW, Xu MJ, Eaton DW, Song XD, Wang LS, Ding ZF, Ning M, Yu DY, Hua L (2014) Crustal structure beneath SE Tibet from joint analysis of receiver functions and Rayleigh wave dispersion. Geophys Res Lett 41(5):1479–1484

Tapponnier P, Peltzer G, Dain AYL, Armijo R, Cobbold P (1982) Propagating extrusion tectonics in Asia: new insights from simple experiments with plasticine. Geology 10(12):611

Wang CY, Gang HF (2004) Crustal structure in Tengchong volcano-geothermal area, western Yunnan, China. Tectonophysics 380(1):69–87

Wang CY, Chan WW, Mooney WD (2003) Three-dimensional velocity structure of crust and upper mantle in southwestern China and its tectonic implications. J Geophys Res Solid Earth 108(B9):ESE 13–ESE 11

Wang CY, Han WB, Wu JP, Lou H, Chan WW (2007) Crustal structure beneath the eastern margin of the Tibetan Plateau and its tectonic implications. J Geophys Res Solid Earth 112(B7):3672

Wang CY, Lou H, Wang XL, Qin JZ, Yang RH, Zhao JM (2009) Crustal structure in Xiaojiang fault zone and its vicinity. Earthq Sci 22(4):347–356

Wang CY, Lou H, Wu JP, Bai ZM, Gang HF, Zheng QJ (2002) Seimological study on the crustal structure of Tengchong volcano-geothermal area. Acta Seismol Sin 24(3):231–242

Wang WL, Wu JP, Fang LH, Lai GJ, Yang T, Cai Y (2014) S wave velocity structure in southwest China from surface wave tomography and receiver functions. J Geophys Res Solid Earth 119(2):1061–1078. doi:10.1002/2013jb010317

Wu JP, Ming YH, Wang CY (2001) The S wave velocity structure beneath digital seismic stations of Yunnan province inferred from teleseismic receiver function modelling. Chin J Geophys 44(2):228–237 **(in Chinese with English abstract)**

Wu QJ, Tian XB, Zhang NL, Li WP, Zeng RS (2003) Receiver function estimated by maximum entropy deconvolution. Acta Seismol Sin 16(4):404–412 **(in Chinese)**

Wu JP, Yang T, Wang WL, Ming YH (2013) Three dimensional P-wave velocity structure around Xiaojiang fault system and its tectonic implications. Chin J Geophys 56(7):2257–2267 **(in Chinese with English abstract)**

Xu Q, Wang JA, Wang JC, Zhang WR (1992) Terrestrial heat flow and its tectonic significance in Yunnan, China. Geotectonica Et Metallogenia 16(3):285–299 **(in Chinese with English abstract)**

Xu Y, Yang JQ, Su YJ, Liu J (2005) Analysis on accurate location of earthquakes in Yunnan area and its tectonic meaning. J Seismol Res 28(4):340–344 **(in Chinese with English abstract)**

Zheng XF, Yao ZX, Liang JH, Zheng J (2010) The role played and opportunities provided by IGP DMC of China national seismic network in Wenchuan earthquake disaster relief and researches. Bull Seismol Soc Am 100(5):2866–2872

9

Internal co-seismic deformation and curvature effect based on an analytical approach

Jie Dong · Wenke Sun

Abstract In this study, we present a new method to compute internal co-seismic deformations of a homogeneous sphere, based on our previous approach (Dong et al. 2016). In practical numerical computations, we consider a strike-slip point source as an example, and compute the vertical co-seismic displacement on different internal spherical surfaces (including the Earth surface). Numerical results show that the internal co-seismic deformations are generally larger than that on the Earth surface; especially, the maximum co-seismic displacement appears around the seismic source. The co-seismic displacements are opposite in sign for the areas over and beneath the position of the seismic source. The results also indicate that the curvature effect of the internal deformation is pretty large, and larger than that on the Earth surface. The results indicate that the dislocation theory for a sphere is necessary in computing internal co-seismic deformations.

Keywords Internal displacement · Curvature effect · Spherical model

1 Introduction

Various dislocation theories have been developed for different geometrical Earth models, such as half-space media (Okada 1985; Okubo 1992), homogeneous sphere (Sun and Okubo 1993), or inhomogeneous sphere (Sun and Okubo 1993). The theory for a spherical Earth model is considered better than that for a half-space model, because the former takes the Earth curvature into account. However, due to the mathematical simplicity, the dislocation theory for a half-space model is still widely applied.

As modern geodetic technique developed, such as GPS and gravity missions, global co-seismic deformation could be detected. In this case, a more precise dislocation theory is actually necessary. We must be sure it is safe to apply a theory for a simple half-space Earth model in computing co-seismic deformation, especially for a far-field even a global scale deformation. Okubo et al. (2002) found the far-field displacement should be analyzed in the framework of spherical Earth theory by comparing the results for an elastic homogeneous half-space model and a radially stratified elastic Earth. For this purpose, some researchers made efficient studies to try to understand how large the effects of the curvature and layered structure are. So far, almost all the investigations are made for the deformation on the surface of the Earth, since the geodetic observation, such as GPS and gravity measurements, is usually performed on the Earth surface. Pollitz (1996) presented a method to illustrate the effects of Earth's sphericity and layering on the calculated deformation field, whose results showed the curvature effect is generally <2% within 100 km of the point source depth. Sun and Okubo (2002) found that both the layering and curvature effects on the co-seismic surface deformation are very large. The layered structure effect reaches a discrepancy of more than 25%. Fu et al. (2010) studied the total effects of curvature and radial heterogeneity in the case of the 2008 Wenchuan earthquake and the 2004 Sumatra earthquake. Wang et al. (2010) found the total effects of the curvature and layer structures are large, without separating the two effects. Recently, Dong et al. (2014, 2016) systematically studied the effects of Earth's layered structure, gravity and

J. Dong · W. Sun (✉)
Key Laboratory of Computational Geodynamics, University of Chinese Academy of Sciences, Beijing 100049, China
e-mail: sunw@ucas.ac.cn

curvature on co-seismic deformation. Their results show that those effects are very larger and cannot be neglected.

Notice again that all above studies were performed on the Earth surface. There are less study referring to the internal deformation, including the internal co-seismic deformation and the curvature effect. Although Okada (1992) presented a set of expressions of the internal deformation based on a homogeneous elastic half-space mode, the study about the internal deformation is still basically in the stage of theoretical discussion. Computing the internal co-seismic deformation can enhance our understanding of the stress status, mass redistribution, seismic mechanism, and so on. Recently, Takagi and Okubo (2017) presented a new method of computing internal displacement, stress, strain, and gravitational changes caused by a point dislocation in a spherical Earth model. However, in their method the asymptotic solutions of the radial functions are introduced. Actually, for a homogeneous sphere, it can be proved that the asymptotic solutions are not necessary; a more accurate and straightforward approach can be applied.

Therefore, in this study, we present a set of formulas to compute the internal deformation for the homogeneous spherical model, based on our previous study (Dong et al. 2016). Then we compare the internal co-seismic deformation computed by the new formula, and the corresponding deformation calculated by applying Okada's (1992) formulas, to investigate the curvature effect of the internal co-seismic deformation.

2 Expressions of internal co-seismic deformations for a homogeneous sphere

Conventionally, researchers study the surface co-seismic deformation (the blue sphere in Fig. 1) for a homogeneous sphere or an inhomogeneous sphere. In this section, we try to derive expressions to calculate internal deformations for a specified inner surface (such as the red sphere in Fig. 1). Although Okada (1992) presented analytical formulas for computing the internal deformation based on a homogeneous elastic half-space model, it is impossible to apply these formulas in the case of a sphere since the Earth's spherical curvature is neglected. Here, we derive the expressions of internal co-seismic deformations for a homogeneous sphere in spherical coordinates (r, θ, φ), where r is the geocentric distance, and (θ, φ) express the co-latitude and longitude, respectively, based on the approach of Dong et al. (2016).

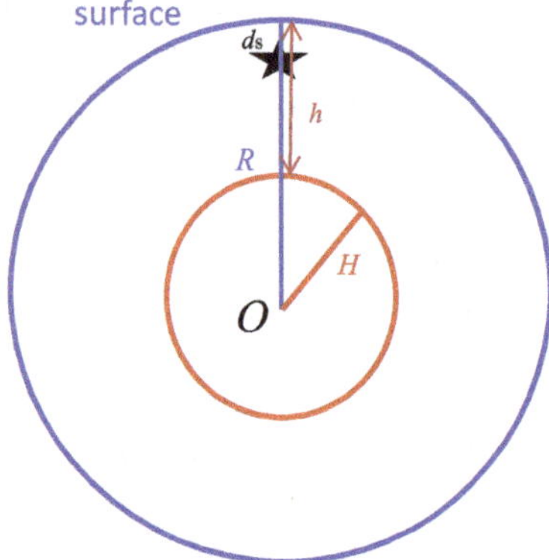

Fig. 1 Sketch showing the positions of Earth (*blue one*) and the internal layer sphere (*red one*). R is the radius of the Earth, d_s is the source depth, h is the internal layer depth under the surface, and H is the radius of the internal sphere

Since we consider a homogeneous sphere without gravity, the co-seismic displacement ($\boldsymbol{u}$), and stress ($\boldsymbol{\tau}$) excited by a unit point source ($\boldsymbol{f}$) at a location $(r_0, \theta_0, \varphi_0)$ satisfies the equations of equilibrium and stress-strain relation (Alterman et al. 1959; Takeuchi and Saito 1972):

$$\nabla \cdot \boldsymbol{\tau} + \rho \boldsymbol{f} = 0 \tag{1}$$

$$\boldsymbol{\tau} = \lambda \boldsymbol{I} \nabla \cdot \boldsymbol{u} + \mu[\nabla \boldsymbol{u} + (\nabla \boldsymbol{u})^{\mathrm{T}}] \tag{2}$$

where $\boldsymbol{I}$ is the unit tensor, superscript T stands for transpose, and μ and λ are the Lame constants of the Earth.

Generally, any function can be expressed as spherical harmonics on a unit sphere. To solve Eqs. (1) and (2), the co-seismic displacement $\boldsymbol{u}(r, \theta, \varphi)$ and stress $\boldsymbol{\tau}(r, \theta, \varphi)$ can be expressed as:

$$\begin{aligned} \boldsymbol{u}(r, \theta, \varphi) = \sum_{n,m} \big[& y_1(r) \boldsymbol{R}_n^m(\theta, \varphi) + y_3(r) \boldsymbol{S}_n^m(\theta, \varphi) \\ & + y_1^t(r) \boldsymbol{T}_n^m(\theta, \varphi) \big] \end{aligned} \tag{3}$$

$$\begin{aligned} \boldsymbol{\tau} \cdot \boldsymbol{e}_r(r, \theta, \varphi) = \sum_{n,m} \big[& y_2(r) \boldsymbol{R}_n^m(\theta, \varphi) + y_4(r) \boldsymbol{S}_n^m(\theta, \varphi) \\ & + y_2^t(r) \boldsymbol{T}_n^m(\theta, \varphi) \big] \end{aligned} \tag{4}$$

where $\boldsymbol{e}_r(r, \theta, \varphi)$ is the radial unit vector, and $\boldsymbol{\tau} \cdot \boldsymbol{e}_r(r, \theta, \varphi)$ represents the radial component of the stress. $\boldsymbol{R}_n^m(\theta, \varphi)$, $\boldsymbol{S}_n^m(\theta, \varphi)$, and $\boldsymbol{T}_n^m(\theta, \varphi)$ are conventional spherical harmonic functions,

$$\begin{aligned} \boldsymbol{R}_n^m(\theta, \varphi) &= \boldsymbol{e}_r Y_n^m(\theta, \varphi) \\ \boldsymbol{S}_n^m(\theta, \varphi) &= \left(\boldsymbol{e}_\theta \frac{\partial}{\partial \theta} + \boldsymbol{e}_\varphi \frac{1}{\sin\theta} \frac{\partial}{\partial \varphi} \right) Y_n^m(\theta, \varphi) \\ \boldsymbol{T}_n^m(\theta, \varphi) &= \left(\boldsymbol{e}_\theta \frac{1}{\sin\theta} \frac{\partial}{\partial \varphi} - \boldsymbol{e}_\varphi \frac{\partial}{\partial \theta} \right) Y_n^m(\theta, \varphi) \end{aligned} \tag{5}$$

$$\begin{aligned} & Y_n^m(\theta, \varphi) = P_n^m(\cos\theta) \mathrm{e}^{\mathrm{i}m\varphi} \\ & Y_n^{-|m|}(\theta, \varphi) = (-1)^m P_n^{|m|}(\cos\theta) \mathrm{e}^{-\mathrm{i}|m|\varphi} \\ & m = 0, \pm 1, \pm 2, \ldots \pm n \end{aligned}$$

$Y_n^m(\theta, \varphi)$, $Y_n^{-|m|}(\theta, \varphi)$ are functions of the associated Legendre's functions P_n (cos θ). $(\boldsymbol{e}_r, \boldsymbol{e}_\theta, \boldsymbol{e}_\varphi)$ is the base vectors in spherical coordinate for radial, co-latitude and longitude directions, respectively. The superscript t stands for the toroidal deformation, which is parallel to the spherical surface. For the spheroidal deformation, y_1 and y_3 are radial and horizontal components of displacement; y_2 and y_4 are radial and horizontal components of stress, while y_1^t and y_2^t are horizontal displacement and stress of toroidal deformation, respectively. Similarly, the point force f can be expressed as spherical harmonics, and details are omitted here but can refer to Sun et al. (2009) or Dong et al. (2016).

Substituting the formulae (3)–(4) into (1) and (2), and neglecting the gravity effect ($g = 0$), we obtain four ordinary spheroidal differential Eq. (6) and two toroidal Eq. (7) as:

$$\begin{cases} \dfrac{dy_1}{dr} = \dfrac{1}{\beta}\left\{y_2 - \dfrac{\lambda}{r}[2y_1 - n(n+1)y_3]\right\} \\ \dfrac{dy_2}{dr} = \dfrac{4}{r}\left(\dfrac{3\kappa\mu}{r\beta}\right)y_1 - \dfrac{4\mu}{r\beta}y_2 - \dfrac{n(n+1)}{r}\left(\dfrac{6\mu\kappa}{r\beta}\right)y_3 + \dfrac{n(n+1)}{r}y_4 - F_2\dfrac{\delta(r-r_0)}{r_0^2} \\ \dfrac{dy_3}{dr} = \dfrac{1}{\mu}y_4 - \dfrac{1}{r}(y_1 - y_3) \\ \dfrac{dy_4}{dr} = -\dfrac{6\mu\kappa}{r^2\beta}y_1 - \dfrac{\lambda}{r\beta}y_2 + \left\{\dfrac{2\mu}{r^2\beta}\left[(2n^2+2n-1)\lambda + 2(n^2+n-1)\mu\right]\right\}y_3 \\ \quad -\dfrac{3}{r}y_4 - F_4\dfrac{\delta(r-r_0)}{r_0^2} \end{cases} \tag{6}$$

Similarly, the toroidal solutions (one regular solution and one irregular solution) are:

$$\begin{pmatrix} y_{11}^t(r) & y_{12}^t(r) \\ y_{21}^t(r) & y_{22}^t(r) \end{pmatrix} = \begin{pmatrix} r^n & -r^{-(n+1)} \\ \mu(n-1)r^{n-1} & \mu(n+2)r^{-(n+2)} \end{pmatrix} \tag{9}$$

Then the general solution ($\boldsymbol{X}$) can be expressed by a combination of the fundamental spheroidal solutions as

$$x_j(r) = \sum_{i=1}^{4} \beta_i y_{ji}(r), \quad j = 1,2,3,4 \tag{10}$$

$$\begin{cases} \dfrac{dy_1^t}{dr} = \dfrac{1}{r}y_1^t + \dfrac{1}{\mu}y_2^t \\ \dfrac{dy_2^t}{dr} = \dfrac{\mu(n-1)(n+2)}{r^2}y_1^t - \dfrac{3}{r}y_2^t - F_2^t\dfrac{\delta(r-r_0)}{r_0^2} \end{cases} \tag{7}$$

The general solution ($\boldsymbol{X}$) of Eqs. (6) and (7) can be analytically obtained according to Love (1911). Although Love (1911) studied this problem, it is difficult to find a suitable solution of $\boldsymbol{X}$ from his publication; therefore, we derive the expressions of $\boldsymbol{X}$ in this study. Omitting the tedious mathematical work, we present four sets of fundamental spheroidal solutions $y_{ji}(i,j = 1, 2, 3, 4)$ and two sets of toroidal solutions $y_{ji}^t(i,j = 1, 2)$.

The spheroidal solutions of the homogeneous Eq. (6), including two sets of regular solutions and two sets of irregular solutions, can be obtained analytically:

$$\begin{pmatrix} y_{11}(r) & y_{12}(r) & y_{13}(r) & y_{14}(r) \\ y_{21}(r) & y_{22}(r) & y_{23}(r) & y_{24}(r) \\ y_{31}(r) & y_{32}(r) & y_{33}(r) & y_{34}(r) \\ y_{41}(r) & y_{42}(r) & y_{43}(r) & y_{44}(r) \end{pmatrix} =$$

$$\begin{pmatrix} -(n+1)r^{-n-2} & -\left[\dfrac{(n+1)\lambda + (n+3)\mu}{(n-2)\lambda + (n-4)\mu}\right]nr^{-n} & \dfrac{n\lambda + (n-2)\mu}{(n+3)\lambda + (n+5)\mu}(n+1)r^{n+1} & nr^{n-1} \\ 2\mu(n+1)(n+2)r^{-n-3} & \dfrac{(n^2+3n-1)\lambda + n(n+3)\mu}{(n-2)\lambda + (n-4)\mu}2\mu nr^{-n-1} & \dfrac{(n^2-n-3)\lambda + (n^2-n-2)\mu}{(n+3)\lambda + (n+5)\mu}2\mu(n+1)r^n & 2\mu n(n-1)r^{n-2} \\ r^{-n-2} & r^{-n} & r^{n+1} & r^{n-1} \\ -2\mu(n+2)r^{-n-3} & -\dfrac{(n^2-1)\lambda + (n^2-2)\mu}{(n-2)\lambda + (n-4)\mu}2\mu r^{-n-1} & \dfrac{(n^2+2n)\lambda + (n^2+2n-1)\mu}{(n+3)\lambda + (n+5)\mu}2\mu r^n & 2\mu(n-1)r^{n-2} \end{pmatrix} \tag{8}$$

where β_i are unknown constants. To determine the solution on the Earth surface, we introduce the boundary conditions,

$$y_2(r)|_{r=R} = y_4(r)|_{r=R} = 0 \tag{11}$$

$$\boldsymbol{y}_j(r)|_{r=r_s^+} - \boldsymbol{y}_j(r)|_{r=r_s^-} = \boldsymbol{s}_j, \quad j = 1,2,3,4 \tag{12}$$

$$\boldsymbol{y}(r)|_{r=0} < +\infty \tag{13}$$

where $\boldsymbol{s}$ is seismic source function, which is given by Takeuchi and Saito (1972).

Thus, we may obtain the following equations for the spheroidal solution. Here we take the vertical strike-slip source as an example. The strike-slip source is formed in shear force, with a relative movement across the strike of the fault:

$$\begin{pmatrix} y_{21}(R) & y_{22}(R) & y_{23}(R) & y_{24}(R) & 0 & 0 \\ y_{41}(R) & y_{42}(R) & y_{43}(R) & y_{44}(R) & 0 & 0 \\ y_{11}(r_s^+) & y_{12}(r_s^+) & y_{13}(r_s^+) & y_{14}(r_s^+) & -y_{13}(r_s^-) & -y_{14}(r_s^-) \\ y_{21}(r_s^+) & y_{22}(r_s^+) & y_{23}(r_s^+) & y_{24}(r_s^+) & -y_{23}(r_s^-) & -y_{24}(r_s^-) \\ y_{31}(r_s^+) & y_{32}(r_s^+) & y_{33}(r_s^+) & y_{34}(r_s^+) & -y_{33}(r_s^-) & -y_{34}(r_s^-) \\ y_{41}(r_s^+) & y_{42}(r_s^+) & y_{43}(r_s^+) & y_{44}(r_s^+) & -y_{43}(r_s^-) & -y_{44}(r_s^-) \end{pmatrix} \times \begin{pmatrix} \beta_1 \\ \beta_2 \\ \beta_3 \\ \beta_4 \\ \beta_5 \\ \beta_6 \end{pmatrix} = \begin{pmatrix} 0 \\ 0 \\ s_1^{12}(r_s) \\ s_2^{12}(r_s) \\ s_3^{12}(r_s) \\ s_4^{12}(r_s) \end{pmatrix} \tag{14}$$

where R is the radius of the Earth, and $r_s = (R - r_0)/R$ denotes the normalized radius distance of the source. After solving Eq. (14), β_i can be determined analytically, but the tedious calculations are not presented. Then, we can obtain the radial and horizontal components of displacement on the surface, and even inside the Earth.

Similarly, for the toroidal solution, we have:

$$\begin{pmatrix} y_{21}^t(R) & y_{22}^t(R) & 0 \\ y_{11}^t(r_s^+) & y_{12}^t(r_s^+) & -y_{11}^t(r_s^-) \\ y_{21}^t(r_s^+) & y_{22}^t(r_s^+) & -y_{21}^t(r_s^-) \end{pmatrix} \begin{pmatrix} \beta_1 \\ \beta_2 \\ \beta_3 \end{pmatrix} = \begin{pmatrix} 0 \\ s_1^{t,12}(r_s) \\ s_2^{t,12}(r_s) \end{pmatrix} \tag{15}$$

The constants β_i can be obtained in an analytical form as

$$\begin{pmatrix} \beta_1 \\ \beta_2 \\ \beta_3 \end{pmatrix} = \frac{r_s^{n-1}}{8\pi n(n+1)} \begin{pmatrix} \frac{n+2}{n-1} \\ -1 \\ \frac{n+2}{n-1} + r_s^{-2n-1} \end{pmatrix} \tag{16}$$

Then we obtain the toroidal solution as

$$y_1^{t,n,12}(r) = \frac{r_s^{n-1}}{8\pi n(n+1)} \left(\frac{n+2}{n-1} r^n + r^{-n-1} \right) \tag{17}$$

Similarly, we can obtain the solutions for other sources. Finally, we may compute co-seismic displacement $\boldsymbol{u}(r, \theta, \varphi)$ by harmonics summation.

According to the above approach, we may easily derive the corresponding explicit expressions of internal deformations, by applying the four sets of fundamental spherical solutions ($\boldsymbol{y}$) and two sets of toroidal solutions for a homogeneous Earth model. In order to derive expressions of the internal co-seismic deformations for any layer (h) inside the Earth, we should consider two conditions:

Case I When $h < d_s$, i.e., the internal surface to be computed is over the seismic source, located between the Earth surface and the seismic source, and we get the y-variables by solving the following equations

$$\begin{pmatrix} y_1 \\ y_3 \end{pmatrix} = \begin{pmatrix} y_{11}(r) & y_{12}(r) & y_{13}(r) & y_{14}(r) \\ y_{31}(r) & y_{32}(r) & y_{33}(r) & y_{34}(r) \end{pmatrix} \begin{pmatrix} \beta_1 \\ \beta_2 \\ \beta_3 \\ \beta_4 \end{pmatrix}$$

$$y_1^t = \begin{pmatrix} y_{11}^t(r) & y_{12}^t(r) \end{pmatrix} \begin{pmatrix} \beta_1^t \\ \beta_2^t \end{pmatrix} \tag{18}$$

Case II when $h > d_s$, i.e., the internal surface to be computed is beneath the seismic source, located between the seismic source and the mantle-core boundary, and we get the y-variables by solving the following equations

$$\begin{pmatrix} y_1 \\ y_3 \end{pmatrix} = \begin{pmatrix} y_{13}(r) & y_{14}(r) \\ y_{33}(r) & y_{34}(r) \end{pmatrix} \begin{pmatrix} \beta_5 \\ \beta_6 \end{pmatrix}$$

$$y_1^t = y_{11}^t(r)\beta_3^t \tag{19}$$

After deriving all these y-symbols based on the above mathematical processing, we may finally obtain the expressions of internal co-seismic displacement Green functions as

$$\begin{aligned} u_r(r,\theta,\varphi) &= \sum_{n,m} y_{1,m}^n(r) Y_n^m(\theta,\varphi) \cdot R^2 \\ u_\theta(r,\theta,\varphi) &= \sum_{n,m} y_{3,m}^n(r) \frac{\partial Y_n^m(\theta,\varphi)}{\partial\theta} \cdot R^2 \\ &\quad + \sum_{n,m} y_{1,m}^{t,n}(r) \frac{1}{\sin\theta} \frac{\partial Y_n^m(\theta,\varphi)}{\partial\varphi} \cdot R^2 \\ u_\varphi(r,\theta,\varphi) &= \sum_{n,m} y_{3,m}^n(r) \frac{1}{\sin\theta} \frac{\partial Y_n^m(\theta,\varphi)}{\partial\varphi} \cdot R^2 \\ &\quad - \sum_{n,m} y_{1,m}^{t,n}(r) \frac{\partial Y_n^m(\theta,\varphi)}{\partial\theta} \cdot R^2 \end{aligned} \tag{20}$$

Note that although these formulas in Eq. (20) are given by spherical harmonic functions, they are still analytical solutions. These summations in Eq. (20) can be evaluated analytically using the mathematical skill as used in Sun et al. (2009). It means that we can compute the co-seismic deformations for any layer inside the Earth. In a practical computation, in the above scheme, the inputs we need are the source depth d_s, internal surface depth h, and radius R.

3 Internal co-seismic deformations of a homogeneous sphere

In order to display the internal deformations, we assume a strike-slip point source to locate at north pole (Sun and Okubo 1993). Then we consider two source depths at 30 and 637 km, respectively. The 30-km source represents a shallow event, and the 637-km (10% of the Earth radius) source stands for a deep event, so that we can observe the

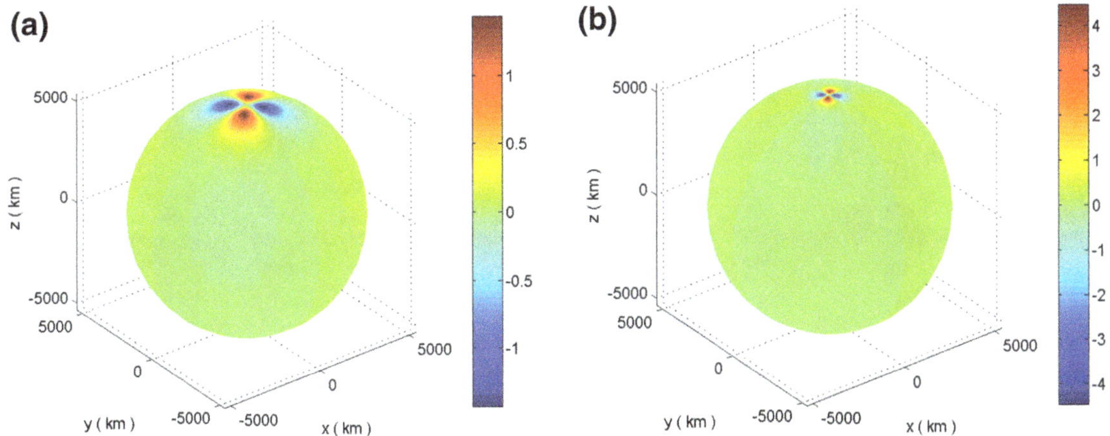

Fig. 2 Vertical displacement on the internal sphere $h = 1000$ km caused by the strike-slip point source ($UdS/R^2 = 1$) at depths of 30 km (**a**) and 637 km (**b**), the vertical displacement is dimensionless

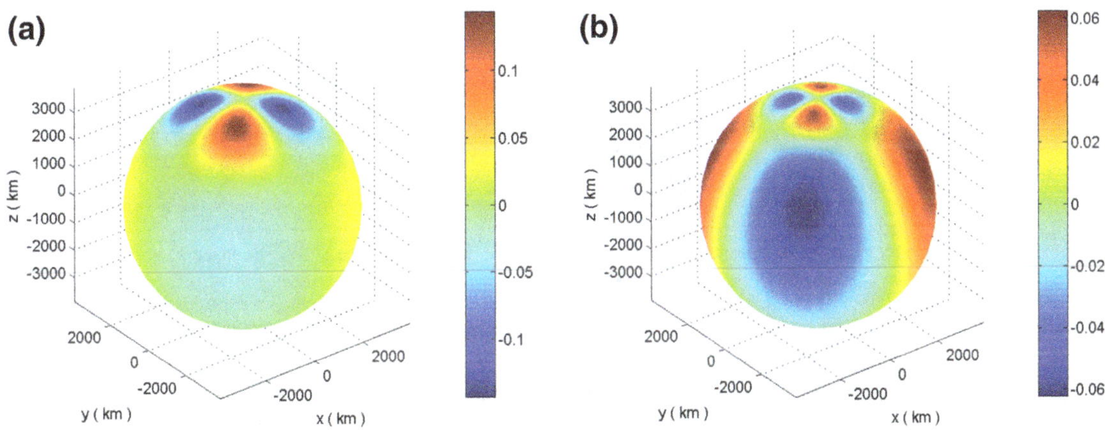

Fig. 3 Same as Fig. 2 but for the internal sphere $h = 2500$ km, the vertical displacement is dimensionless

property of the inner deformation and the curvature effect for different source depth.

Applying the above computing scheme, we calculate co-seismic displacements on two internal spheres caused by the two sources, with depth of $h = 1000$ km and $h = 2500$ km. Figure 2 shows the vertical co-seismic displacements on an internal sphere with depth of $h = 1000$ km, caused by the two sources, while Fig. 3 gives the same results as Fig. 2 but for an internal sphere of $h = 2500$ km. Both Figs. 2 and 3 show that the co-seismic deformation behaves in a quadrant pattern, similar to the deformation on the Earth surface. Since the Earth's surface in the quadrants 1 and 3 sink down, and the other two quadrants rise (further discussions on the distribution of the surface deformation refer to the Figs. 4, 5 below); while the deformation at the two internal spheres ($h = 1000$ km and $h = 2500$ km) shown in Figs. 2 and 3 indicates that the deformation in the quadrants 1 and 3 rise, and the quadrants 2 and 4 sink down. The reason for this opposite deformation between the Earth surface and the inner surface is due to the position of the inner surface. Generally, we found that if the internal surface is located beneath the source, the deformation appears opposite in sign with that over the source (including the earth surface).

Comparing Fig. 2a, b shows that the amplitude of the vertical displacement caused by the source at radius of 637 km is larger than that at radius of 30 km, because the source at $d_s = 637$ km is nearer to the internal surface of $h = 1000$ km. In addition, the magnitude of the vertical displacements for both depths decays quickly as the epicentral distance (θ) increases, meaning that the local co-seismic deformation dominates.

From Fig. 3, we see that the amplitude of the co-seismic displacements on the internal sphere of $h = 2500$ km is smaller than that on the sphere of $h = 1000$ km. This phenomenon is normal, because the former is farther from seismic source and the deformation decays with distance. In addition, the deformation on sphere of $h = 2500$ km appears wide distribution covering a more large area, because the farther the distance apart from the seismic

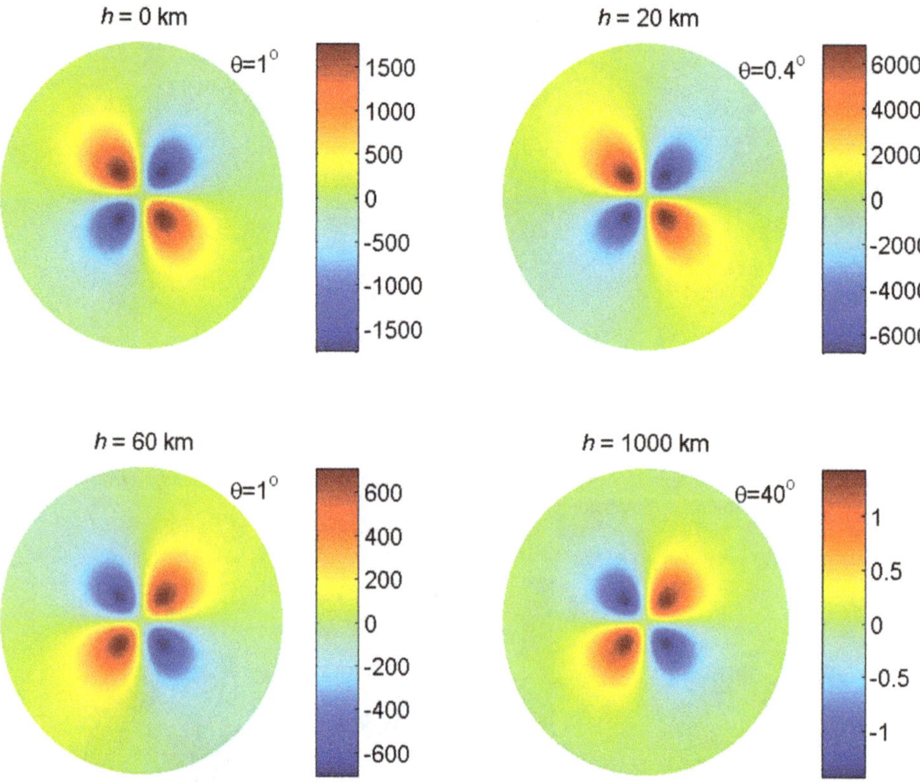

Fig. 4 Vertical co-seismic displacements on different internal spheres (h is the depth under the surface) caused by the strike-slip point source ($UdS/R^2 = 1$) at a source depth of 30 km, the vertical displacement is dimensionless

source, the weaker the co-seismic deformations. It means that the high frequency components of the deformation become weaker and weaker, while the low frequency components become dominating.

To observe the property of the co-seismic displacements changing for different depth, we compute and plot co-seismic vertical displacements on several internal spheres with different depths (0, 20, 60 and 1000 km) and different epicentral distance θ. Results are plotted in Fig. 4 for a seismic source depth of 30 km, and in Fig. 5 for a source depth of 637 km. In Fig. 4, when $h = 0$ km (on the surface of the Earth), we find that the larger displacements appear within $\theta = 1°$, almost the same as that of the homogeneous sphere's numerical solutions of Sun et al. (2009). When $h = 20$ km, the larger displacements appear within $\theta = 0.4°$ and show largest deformation in magnitude among the four depths, more concentrate to the epicenter, because the deformation on the internal sphere of $h = 20$ km is nearer to the source of 30 km than the Earth surface ($h = 0$ km). When $h = 60$ km, we find that the co-seismic displacement is much smaller than that on the Earth surface, even though the distances between the seismic source ($h = 30$ km) and the two spheres ($h = 0$ km and $h = 60$ km) are the same. This phenomenon is understandable, because the co-seismic deformation depends on the geocentric distance, and the distances from the Earth's center to the Earth surface and depth of 60 km are different. Finally when $h = 1000$ km, the larger co-seismic displacement appears in wider area, covering large epicentral distance. On the other hand, the magnitude of the displacement is much smaller than that on other spheres, because this sphere ($h = 1000$ km) is very far from the source and the deformation decays quickly.

In addition, we find that the coverage (epicentral distance) of the largest deformation is proportional to the distance of ($|h - d_s|$). That is, the epicentral distance for the large deformation becomes larger as the internal sphere farther from the source. Furthermore, we find that, as pointed out above, the co-seismic displacements are opposite in sign for those areas over and beneath the seismic source. Generally, the displacement becomes large when the computing point is near the source, as shown in Figs. 4 and 5.

Comparing Figs. 4 and 5 shows that the internal deformation obviously changes for different source depth. The vertical displacements on the Earth surface become much smaller when the source is located in depth of 637 km, while the deformation on sphere of $h = 1000$ km becomes larger due to the relatively near source. Note that all properties shown in Fig. 4 also apply to Fig. 5.

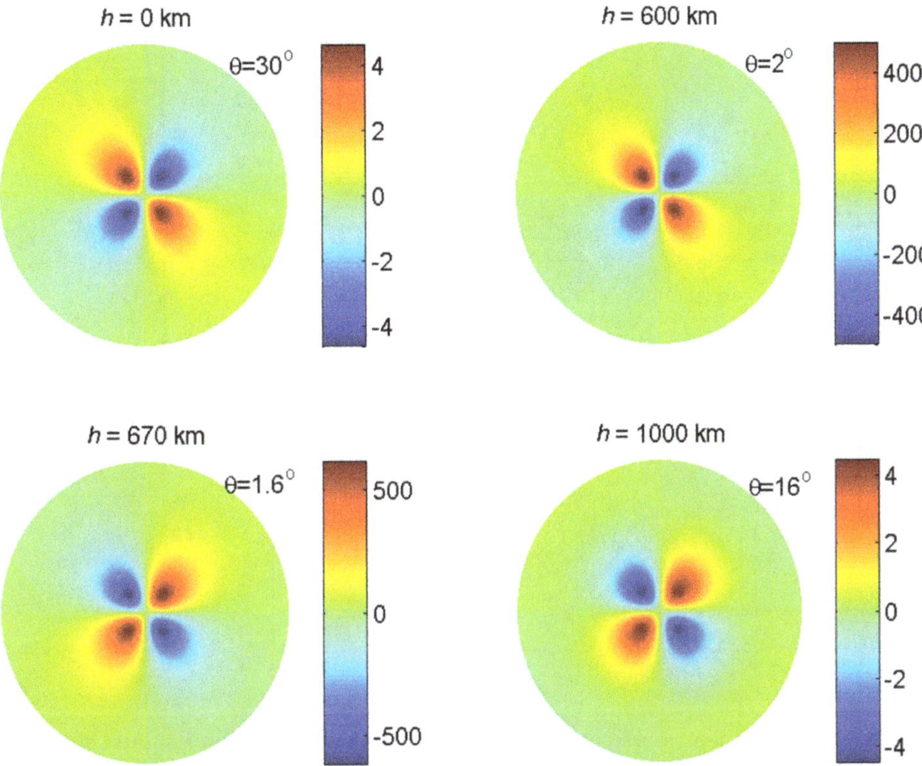

Fig. 5 Same as Fig. 4 but a source depth of 637 km

4 Curvature effect in computing internal co-seismic deformations

In this section, we investigate the curvature effect for internal co-seismic deformation based on the above approach. For this purpose, we calculate internal vertical co-seismic displacements for different depths by using the above approach and the theory of Okada (1992). In computation, we consider two strike-slip seismic sources at depths of 30 and 637 km, respectively. Similarly, the point source is normalized by factor of $UdS/R^2 = 1$. Then we compute the vertical displacements for different depths (or internal spheres) and plot them in Figs. 6 and 7, for the two sources, respectively. A comparison of vertical co-seismic displacements on different internal spheres (h) computed by half-space model (red line) and spherical model (blue dashed line) is clearly shown in the figures. The difference between the results represents the curvature effect of the internal deformation.

Figure 6 shows that the difference (curvature effect) between the two theories on Earth surface and shallow areas over the source is not obvious and difficult to be identified since both curves overlap, although the numerical result shows the curvature effect is about 1.2% on the surface. However, the curvature effect becomes larger and larger when the depth of the internal sphere goes deeper and deeper. However, when the source locates deep as shown in Fig. 7 for depth of 637 km, the curvature effect becomes larger, even we can identify them directly. It means that the curvature effect becomes larger for deep seismic events.

To evaluate the curvature effect in quantity, we represent the curvature effect in form of a relative error defined as

$$\varepsilon = \frac{\left|\boldsymbol{u}^{(\mathrm{s})}\right| - \left|\boldsymbol{u}^{(\mathrm{h})}\right|}{\left|\boldsymbol{u}^{(\mathrm{h})}\right|_{\max}} \tag{21}$$

where $\boldsymbol{u}^{(\mathrm{s})}$ is the displacement computed for a spherical model, $\boldsymbol{u}^{(\mathrm{h})}$ is that computed for a half-space model (Okada 1992), while the term $\left|\boldsymbol{u}^{(\mathrm{h})}\right|_{\max}$ stands for the maximum value of the co-seismic deformation. Then the curvature effects in Figs. 6 and 7 can be represented in Fig. 8a, b. Figure 8 shows that the curvature effect on the Earth surface is small, which is about 1.2%, in agreement with the conclusion of Dong et al. (2016). It also shows that the curvature effect (ε) becomes larger as the distance between the internal sphere and the source goes larger. Figure 8 also shows a proportional relation between the curvature effect and the source depth, i.e., $\varepsilon \propto |h - d_s|$. This relation should be further proved in theory or more numerical work.

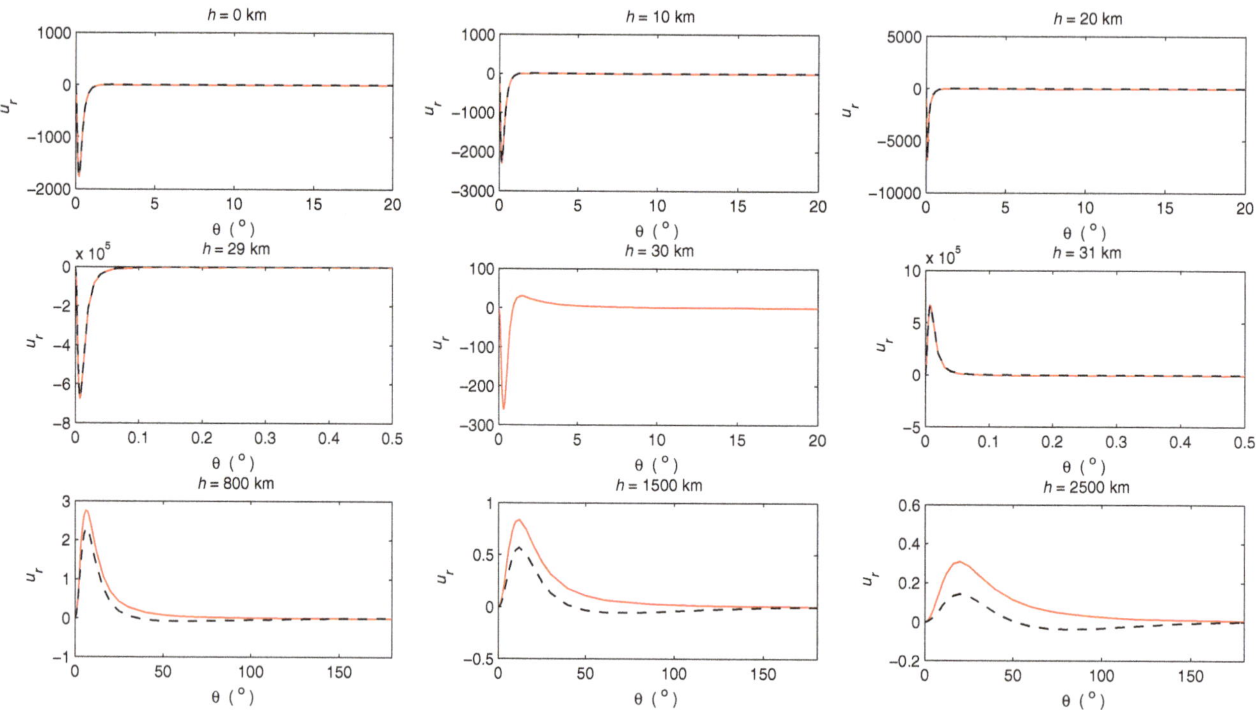

Fig. 6 Comparison of vertical co-seismic displacements on different internal spheres (h) computed by half-space model (*red line*) and spherical model (*blue dashed line*) caused by the strike-slip point source ($U\mathrm{d}S/R^2 = 1$) at a source depth of 30 km, the vertical displacement is dimensionless

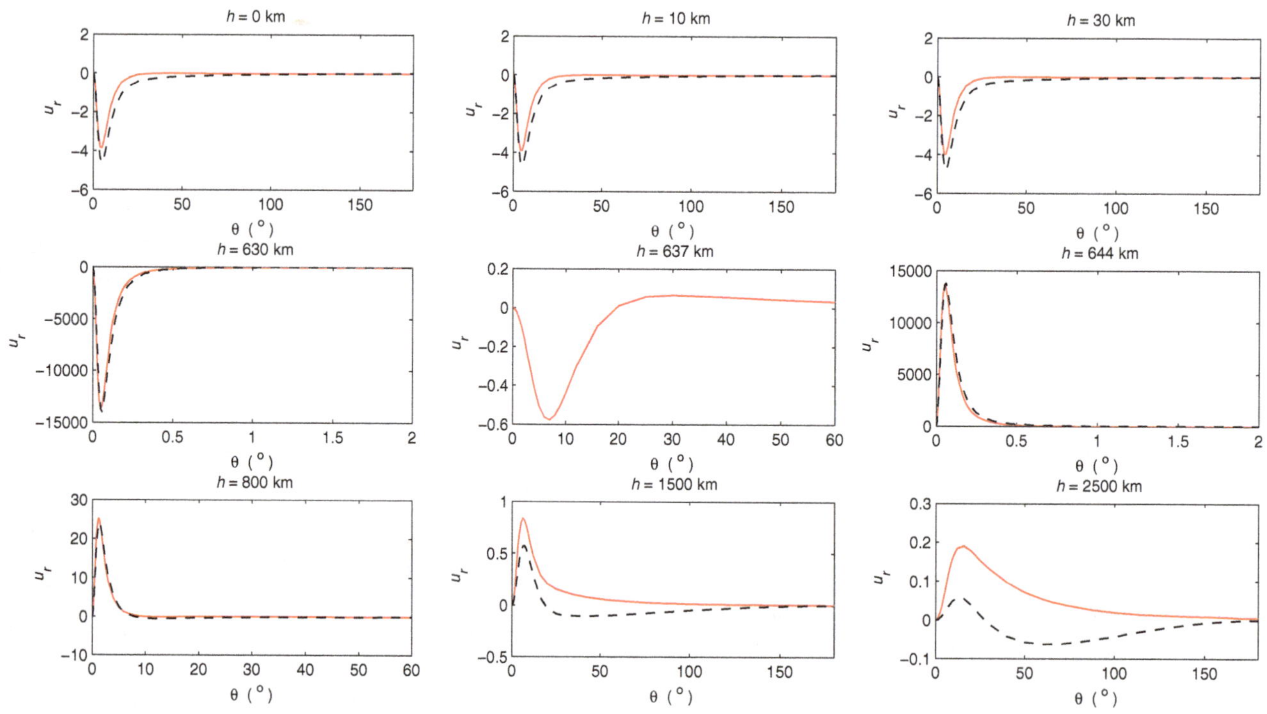

Fig. 7 Same as Fig. 6 but for a source depth of 637 km

5 Discussion and conclusions

In this study, we present expressions for computing the internal deformation of a homogeneous sphere, based on our previous approach (Dong et al. 2016). These expressions are given in form of analytical solutions, similar to that of Okada (1992), which is given for a homogeneous half-space model. In practical numerical computations, we

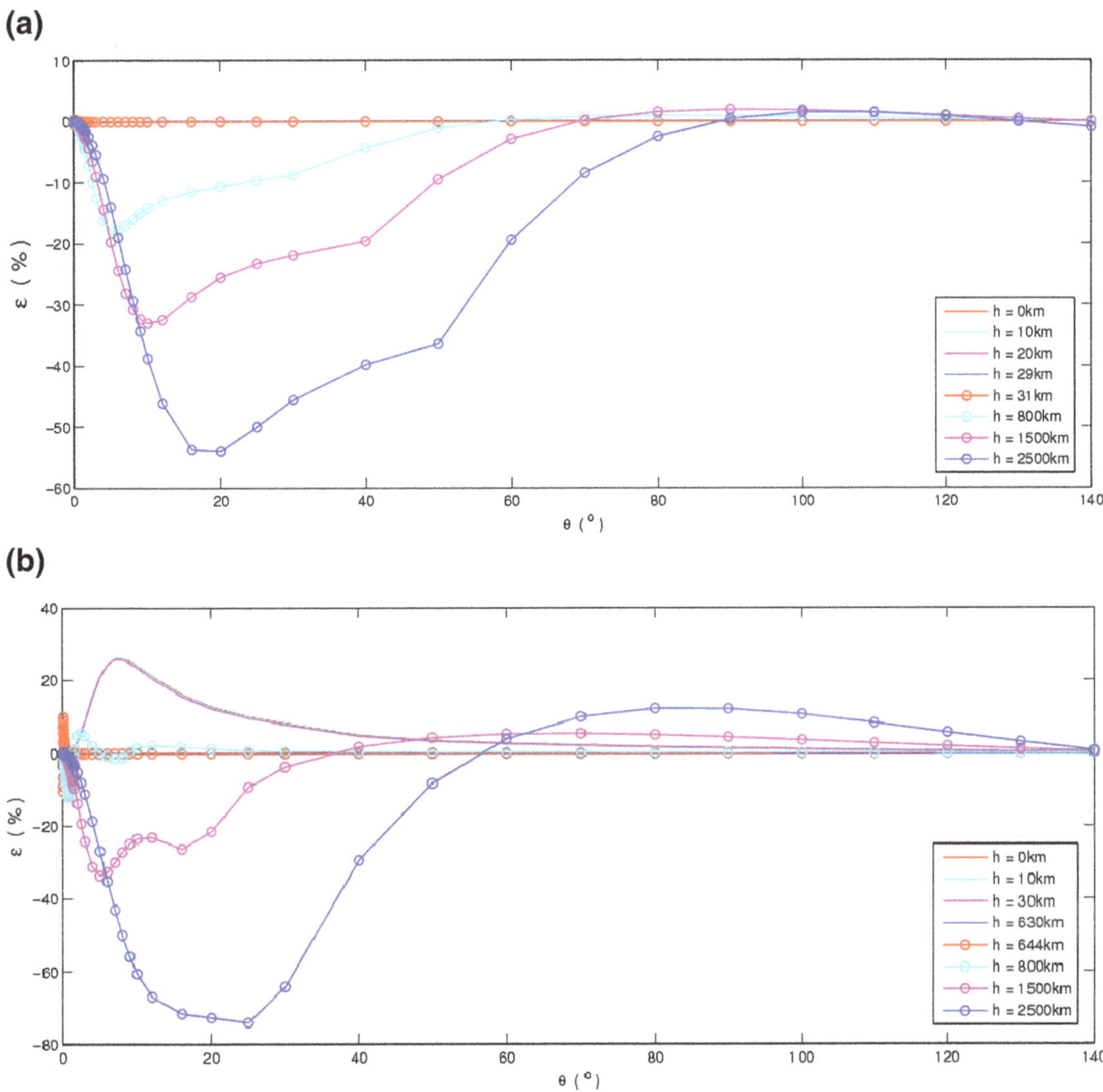

Fig. 8 Curvature effects on different inner spheres caused by the strike-slip point source ($UdS/R^2 = 1$) at depths of 30 km (**a**) and 637 km (**b**). The Subplot **a** shows that the curvature effects for those internal spheres are 1.2%, 1.1%, 0.7%, 0.4%, 0.4%, 18%, 33%, and 54%, respectively; **b** shows that the corresponding curvature effects are 26.1%, 26%, 25.8%, 11.6%, 10.4%, 12.2%, 34%, and 75%, respectively

consider a strike-slip point source as an example, and compute the vertical co-seismic displacement on different internal spherical surfaces (including the Earth surface).

Numerical results of the internal deformations show that the internal co-seismic deformations are generally larger than that on the Earth surface; especially, the maximum co-seismic displacement appears around the seismic source. For the point source at depth of 30 km, the displacement magnitude reaches about 10^5(normalized by factor of $UdS/R^2 = 1$) near the source, while the displacement amplitude in other places decreases quickly as the distance apart from the source increases. The results also show that the displacements are opposite in sign for the areas over and beneath the position of the seismic source. The results of the study also indicate that the curvature effect of the internal deformation is pretty large. Our previous study (Dong et al. 2016) on the Earth surface showed that the curvature effect is generally <5% for point source with depth less than 100 km. The current study shows that the curvature effect of internal deformation becomes larger comparing to the surface, e.g., the curvature effect on the CMB (core–mantle boundary) spherical surface can reach 100%. The above results indicate that the dislocation theory for a sphere is necessary in computing internal co-seismic deformations, comparing to the theory for a half-space media (Okada 1992).

Note that since the numerical computations are made for only the strike-slip source and the vertical co-seismic displacement in the study, the corresponding conclusions are actually limited. Because the co-seismic deformations include different geophysical phenomena, such as displacement, strain, potential (geoid) and gravity changes, the corresponding deformation property and pattern are different. The displacement is a vector, including two components, vertical displacement and horizontal displacement, and its spatial distribution pattern and magnitude are different. In addition, all these co-seismic deformations appear different spatial distribution property, depending on source types, source depth. Therefore, the numerical computation and discussion and conclusions are considered as a case study. For different source types and different co-seismic deformations, there may be somehow changeable or adjustable conclusions, maybe slightly. However, the method and conclusions in this study are still important in enhancing our understanding of the property of the internal deformation and curvature effect.

The method presented in this study can be used to compute Green's functions of the internal co-seismic deformation, including all types of physical variables, such as displacement, strain, tilt, and so on. Then we may apply the Green's functions to compute co-seismic deformations at any source depth by arbitrary source types, though simple numerical integrations over limited fault plane. These practical applications remain in our future work.

Acknowledgements We thank two anonymous reviewers for their careful reading in the previous version of the paper and for their helpful comments. This research was supported financially by the National Natural Science Foundation of China (Nos. 41331066, 41604067 and 41474059), China Postdoctoral Science Foundation Funded Project (No. 119103S268), CAS Key Study Program QYZDY-SSW-SYS003 and the CAS/CAFEA International Partnership Program for Creative Research Teams (No. KZZD-EW-TZ-19).

References

Alterman Z, Jarosch H, Pekris CL (1959) Oscillation of the Earth. Proc R Soc. Lond A252:80–95

Dong J, Sun W, Zhou X, Wang R (2014) Effects of Earth's layered structure, gravity and curvature on coseismic deformation. Geophys J Int 199:1442–1451

Dong J, Sun W, Zhou X, Wang R (2016) An analytical approach to estimate curvature effect of coseismic deformations. Geophys J Int 206:1327–1339

Fu G, Sun W, Fukuda Y, Gao S, Hasegava T(2010) Effects of Earth's curvature and radial heterogeneity in dislocation studies: case studies of the 2008 Wenchuan earthquake and the 2004 Sumatra earthquake. Earthq Sci 23:301–308

Love AEH (1911) Some problem of geodynamics. Cambridge University Press, Cambridge, pp 89–104

Okada Y (1985) Surface deformation due to shear and tensile faults in a half-space. Bull Seismol Soc Am 75:1135–1154

Okada Y (1992) Deformation due to shear and tensile faults in a half-space. Bull Seismol Soc Am 82:1018–1040

Okubo S (1992) Potential and gravity changes due to shear and tensile faults. J Geophys Res 97:7137–7144

Okubo S, Sun W, Yoshino T, Kondo T, Amagai J, Kiuchi H, Koyama Y, Ichikawa R, Sekido M (2002) Far-field deformation due to volcanic activity and earthquake Swarm, Vistas for geodesy in the New Millennium. In: Adam J, Schwarz KP (eds) International Association of Geodesy Symposia, vol 125, pp 518–522

Pollitz FF (1996) Coseismic deformation from earthquake faulting in a layered spherical Earth. Geophys J Int 125:1–14

Sun W, Okubo S (1993) Surface potential and gravity changes due to internal dislocations in a spherical Earth, I. Theory for a point dislocation. Geophys J Int 114(3):569–592

Sun W, Okubo S (2002) Effects of Earth's spherical curvature and radial heterogeneity in dislocation studies-for a point dislocation. Geophys Res Lett 29(12):1605. doi:10.1029/2001GL014497

Sun W, Okubo S, Fu G, Araya A (2009) General formulations of global co-seismic deformations caused by an arbitrary dislocation in a spherically symmetric Earth model—applicable to deformed Earth surface and space-fixed point. Geophys J Int 177:817–833. doi:10.1111/j.1365-246X.2009.04113.x

Takagi Y, Okubo S (2017) Internal deformation caused by a point dislocation in a uniform elastic sphere. Geophys J Int 208:973–991

Takeuchi H, Saito M (1972) Seismic surface waves. Methods Comput Phys 11:217–295

Wang W, Sun W, Jiang Zaisen (2010) Comparison of fault models of the 2008 Wenchuan earthquake (M_S8.0) and spatial distributions of co-seismic deformations. Tectonophysics 491:85–95

10

Waveform inversion and analysis of an unusual earthquake swarm in the Boshan mining area, Shandong Province, China

Jian-Chang Zheng · Jin-Hua Zhao · Chang-Peng Xu · Peng Wang

Abstract Moment tensor solutions were retrieved for the earthquake swarm that occurred during November and December 2010 in the Boshan mining area, Shandong Province, China. The results showed that the double-couple components in the source mechanisms were higher at the beginning of the swarm and consisted mainly of shear faulting controlled by tectonic stress. The subsequent events had significant non-double-couple components, indicating tensile faulting. The double-couple components predominately presented as normal faulting and the P axes were orientated almost vertically. The slip vectors of the swarm events were relatively stable. With reference to the tectonic features near the epicenter, we concluded that the swarm was a result of subordinate fault motion related to the Wangmu Mountain fault and that high-pressure pore fluids played a crucial role in the activity of the earthquake swarm.

Keywords Moment tensor · Tensile crack · Mechanism · Boshan earthquake swarm · Pore pressure

1 Introduction

In 2010, a series of earthquake occurred near Boshan, a mining area city in the past in Shandong Province, China. The earthquake sequence continued over 1 month, from November 24, 2010 to December 27, 2010 and were measured by the Shandong seismic network. The tremors from these earthquakes propagated widely through these mining areas and affected almost all the villages in Boshan, with consequent social impacts. There were six earthquakes of $M_L \geq 2.0$ and the maximum event of M_L 3.4 occurred on December 20, 2010. This event had the highest magnitude of all the earthquakes that have occurred in the mining areas of Shandong Province in recent years.

The term "mine earthquake" is used here to refer to all earthquakes occurring in mining areas, whether they were caused by rock bursts, mine collapses, or tectonics. Mine earthquakes can be divided into three types: (1) earthquakes in which a rupture occurs in the surrounding rock mass of the mining areas—these are caused by the adjustment of stress in the Earth's crust and result in slippage along a few minor faults; (2) shocks caused by mine collapse, roof falls, or rock bursts; and (3) shocks caused by the detonation of gas or powdered coal in mine tunnels (Zhang et al. 1993). The source of the earthquake is obviously different in each of these three categories. The source type can be indicated by the moment tensor. For earthquakes caused by tectonic activity resulting from stress adjustment in mining areas, the source mechanism follows the double-couple (DC) shear dislocation model and is close to that of natural earthquakes. The source of earthquakes caused by mine collapse, roof falls, or rock bursts is primarily uniaxial compression, and there is a large compensated linear vector dipole (CLVD) component in the moment tensor solutions and a characteristic implosion mechanism resulting from the sudden closure of goafs. For mine earthquakes resulting from the detonation of gas or powdered coal, the explosive source (i.e., the isotropic volume component) plays a large part in the source mechanism. Some studies have shown that many mine earthquakes have non-DC mechanisms and may contain a large volume or linear component under some

J.-C. Zheng (✉) · J.-H. Zhao · C.-P. Xu · P. Wang
Earthquake Administration of Shandong Province, Jinan 250014, China
e-mail: zjcmail@yeah.net

circumstances (Gibowicz and Kijko 1998; McGarr 1992a, b; Feignier and Young 1992; Phillips et al. 1999; Fletcher and McGarr 2005; Stec 2007; Julià et al. 2009). There have been a number of studies in China in recent years of the source mechanisms of mining-induced earthquakes (Li et al. 2005; He et al. 2007; Chen et al. 2009), although there has been little specific research into the moment tensor of mine earthquakes.

Non-tectonic earthquakes and natural earthquakes are technically distinguishable as they differ in many aspects, such as the first motion of the seismic waves, the ground motion frequency, and the rate of attenuation. Many researchers have conducted in-depth research into the differences between non-tectonic and natural earthquakes using seismic waveforms. For example, Lin et al. (1990) compared the seismic phase characteristics of collapse earthquakes and tectonic earthquakes and suggested that collapse earthquakes have a different wave propagation velocity, attenuation, and amplitude. Gibowicz (1995) studied the relationship between the seismic moment of mining-induced events and the P and S wave corner frequencies in various regions using statistical comparisons. Oye et al. (2005) studied seismic source parameters such as the coda attenuation and apparent stress of a mine earthquake swarm in Finland using the spectral integral and multiple empirical Green's functions. He et al. (2006) and Liu et al. (2005) solved the recognition factors for mine and tectonic earthquakes based on wavelet packet decomposition of the waveform. Shen et al. (2006) studied the rupture and nucleation processes of mine earthquakes based on micro-tremor records and micro-seismic waveforms in mining areas. Zhang et al. (2009) compared and analyzed the time-frequency characteristics of the spectra of mine and tectonic earthquakes in the Three Gorges area of China.

During emergency work after this swarm of earthquakes in 2010, we noticed that these particular mine earthquake events differed from ordinary mine earthquakes and that there was a need for further analysis. Moment tensor analysis is a powerful tool for source mechanism studies. Therefore, we analyzed and inverted all the digital waveform recordings of these mine earthquake swarm events to determine the source moment tensor solutions.

2 Data and information

2.1 Distribution of the earthquake swarm and seismic stations

A total of six events with $M_L \geq 2.0$ were recorded in the Boshan earthquake swarm. In addition, one event with M_L 2.7 occurred near the epicenter of the Boshan swarm on September 12, 2010 and, while the swarm was active, one event with M_L 2.7 occurred 16 km to the south of the epicenter on November 25, 2010. As these two events were closely related to the Boshan swarm both tectonically and in time, they were also included in our study. All the broadband digital seismic stations within 150 km of the epicenters of the largest magnitude events in the swarm recorded relatively clear waveform signals. Figure 1 shows

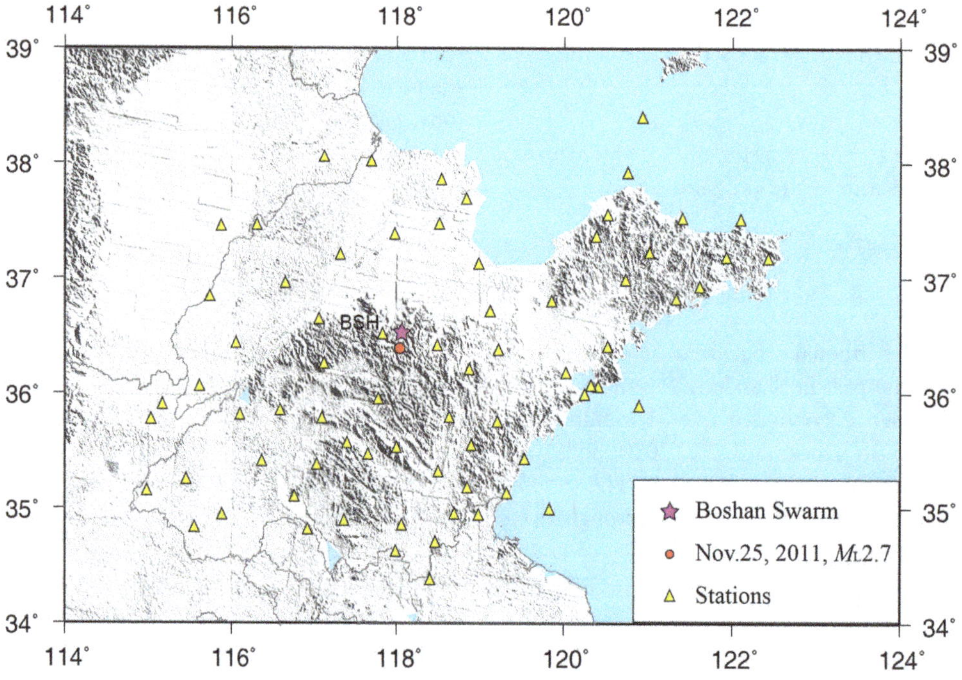

Fig. 1 Distribution of the Boshan swarm and the seismic stations in Shandong Province, China

the distribution of the epicenters of the mine earthquake events and the broadband seismic stations in Shandong area.

After several mine earthquakes in late November 2010, the Earthquake Networks Center of Shandong Province and the Earthquake Administration of Zibo City set up four mobile seismic stations in the vicinity of the epicenter. These four stations plus the BSH station, which is <10 km from the epicenter, formed a temporary seismic network of five stations and recorded several mine earthquake events. Figure 2 shows the relocated epicenters of the Boshan swarm, and the temporary stations at the Boshan area.

The data from the temporary stations show that the S–P value is >0.5 s at the closest station L02; the relocated results show that the focal depths of all these events ranged from 4 to 5 km. The depths of mine earthquake events such as rock bursts and collapses caused by industrial mining activities are usually of the order of hundreds of meters, although they may exceed 2 km in a few deep mine areas [e.g., deep gold mines in South Africa (McGarr et al. 2009)]. However, according to data from the Zibo Mining Group Co. Ltd (Jiao et al. 1990), the Zibo coalfield contains Carboniferous-Permian coal measures with >20 layers of coal, 12 of which are minable in part or in whole; the main coal-bearing strata are 250 m thick and the mining depths are normally in the range of 200–300 m, with a maximum of <400 m. Therefore, the cause of this set of events may not be due purely to mining operations, such as rock bursts. As the epicenter of this earthquake swarm lies in the coal mining area, some researchers have called it a "mine earthquake" swarm; however, we refer to it as an "earthquake swarm" in this study.

Figure 3 shows the velocity model used in our study, which is derived from geophysical exploration in Shandong area (Jiang et al. 2000). The epicenter locations obtained through relocation was used in the inversion of the source

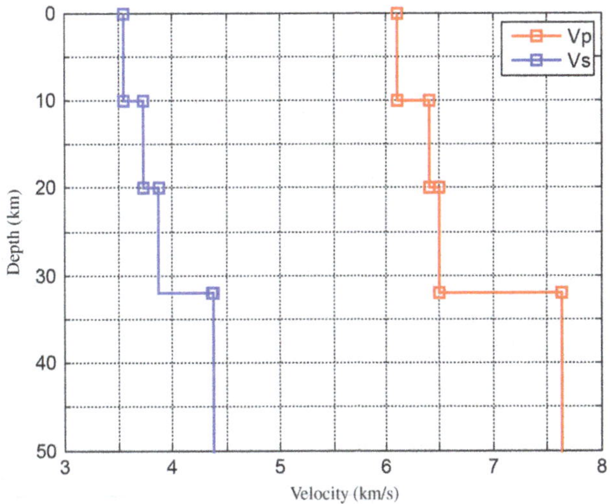

Fig. 3 Velocity model used in our study

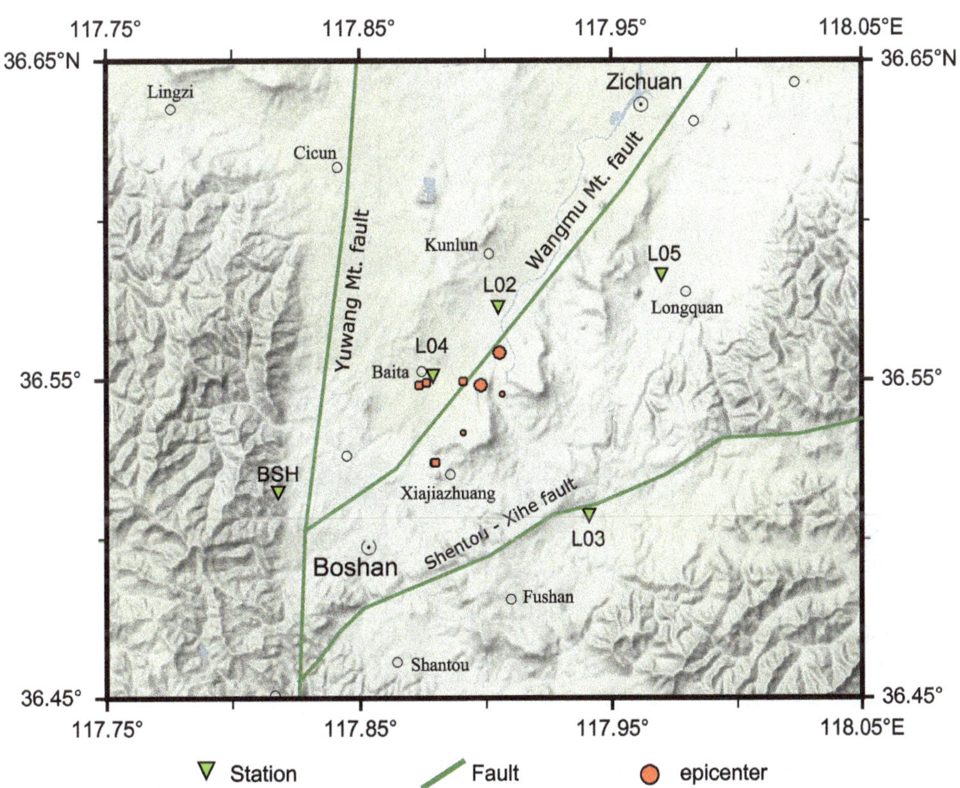

Fig. 2 Distribution of relocated events (*red points*) and stations in the temporary seismic network (*green triangles*)

moment tensor. The search was conducted in the depth direction with a starting depth of 1 km and a step length of 1 km. In total, there were 23 search depths to a maximum of 23 km.

2.2 Waveform time-frequency analysis of Boshan swarm

Zhang et al. (2009) have shown that there are a few clear differences between the waveform time-frequency analysis results for mine and tectonic earthquakes. First, the main frequencies of the waves from tectonic earthquakes are higher than those of mine earthquakes. Second, the frequency components of waves from tectonic earthquakes are richer than those of mine earthquakes and their energy density spectrum is spread over a wider range of frequencies—the bandwidth of tectonic earthquakes is four times as wide as that of mine earthquakes. Third, the peak energy intensity of mine earthquakes over the entire wave train appears much earlier than that of tectonic earthquakes, although the attenuation in energy is more rapid. The time-frequency spectrogram of the Boshan earthquake swarm events shows that the events have distinctive features. The events share some characteristics with general mine earthquake events, such as a low main frequency and rapid attenuation. However, the energy density spectrum of the events is widely spread along the frequency axis and the dominant frequency of the S waves extends from about 7 to 17 Hz (Fig. 4), indicating the rich frequency components of the event.

3 Theory and methods

3.1 Moment tensor decomposition of mine earthquakes

In mathematics, the moment tensor describes the effectiveness of any type of source and can be simply divided into three parts corresponding to different types of sources: an isotropic (ISO) component, a CLVD subcomponent, and a DC component. The DC tensor describes the shear slip in isotropic media along fault planes and the ISO component corresponds to explosive or implosion sources. The ISO and CLVD components are referred to as the non-DC components and are attributable to the opening or closing of faults.

Many studies on the source moment tensor of mine earthquakes have shown that, as a result of the volume change caused by the closure of goafs, the source mechanisms of mine earthquakes (also referred to as "shock bump" or "rock burst") can be decomposed into a combination of two modes: shear slip accompanied by a certain amount of a co-seismic implosion component. McGarr (1994) studied the mechanisms of mine earthquakes and found that the combination of a shear faulting model plus an implosion component could explain the waveforms of similar events observed in deep gold mines in South Africa (McGarr 1992a, b). Increasing numbers of studies are showing that many shallow, naturally occurring mine collapse events have an implosion mechanism as the seismic source (Gibowicz and Kijko 1998; Miller et al. 1998).

Yang et al. (1998) studied explosion-induced mine collapse. The results of source moment tensor inversion showed that the source mechanism of collapse can be expressed by horizontal opening and an implosion. Analysis showed that the unique source characteristics inducing collapse result from the fact that, unlike synchronous collapse, ordinary collapse is initially caused by extension fractures.

Unlike tectonic earthquakes caused by ordinary shear faulting, the fault slip vectors of mine earthquake events are not limited to the fault planes and may deviate from them. For normal seismic sources, the moment tensor can be expressed as a function of a slip vector. Aki and Richards (1980) devised a formula for the point source moment tensor of faults in isotropic media:

$$M_{kl} = \lambda[u_i]n_i\delta_{kl} + \mu([u_k]n_l + [u_l]n_k), \tag{1}$$

where λ and μ are the Lamé constants of the faults, δ_{kl} is the Kronecker symbol, $[\boldsymbol{u}]$ is the slip vector, and $\boldsymbol{n}$ is the normal fault plane. If

$$\boldsymbol{n} = (0,0,1)^{\mathrm{T}},\ [\boldsymbol{u}] = \boldsymbol{u}(\cos\alpha, 0, \sin\alpha)^{\mathrm{T}} \tag{2}$$

then the moment tensor may be in the following form:

$$\boldsymbol{M} = u\begin{bmatrix} \lambda\sin\alpha & 0 & \mu\cos\alpha \\ 0 & \lambda\sin\alpha & 0 \\ \mu\cos\alpha & 0 & (\lambda+2\mu)\sin\alpha \end{bmatrix}, \tag{3}$$

where $\alpha[-\pi/2, \pi/2]$ is the included angle between the slip vector and the fault plane. When $\alpha > 0$, the source is an extension fracture; when $\alpha < 0$, it is a compression fracture; and when $\alpha = 0$, it is pure DC shear faulting.

When using the principal stress axis coordinate system, Eq. (3) can be written in the following diagonal form:

$$\boldsymbol{M} = u\begin{bmatrix} \lambda\sin\alpha - \mu(1-\sin\alpha) & 0 & 0 \\ 0 & \lambda\sin\alpha & 0 \\ 0 & 0 & \lambda\sin\alpha + \mu(1+\sin\alpha) \end{bmatrix}. \tag{4}$$

Then, the matrix trace is as follows:

$$\mathrm{trace}(\boldsymbol{M}) = (3\lambda + 2\mu)u\sin\alpha. \tag{5}$$

Under the constraints that have physical significance, $(\mu > 0, \lambda/\mu > -2/3)$, the trace of matrix $\boldsymbol{M}$ is only related to the included angle α between the slip vector and the fault plane. Therefore, for an extension-type source, the

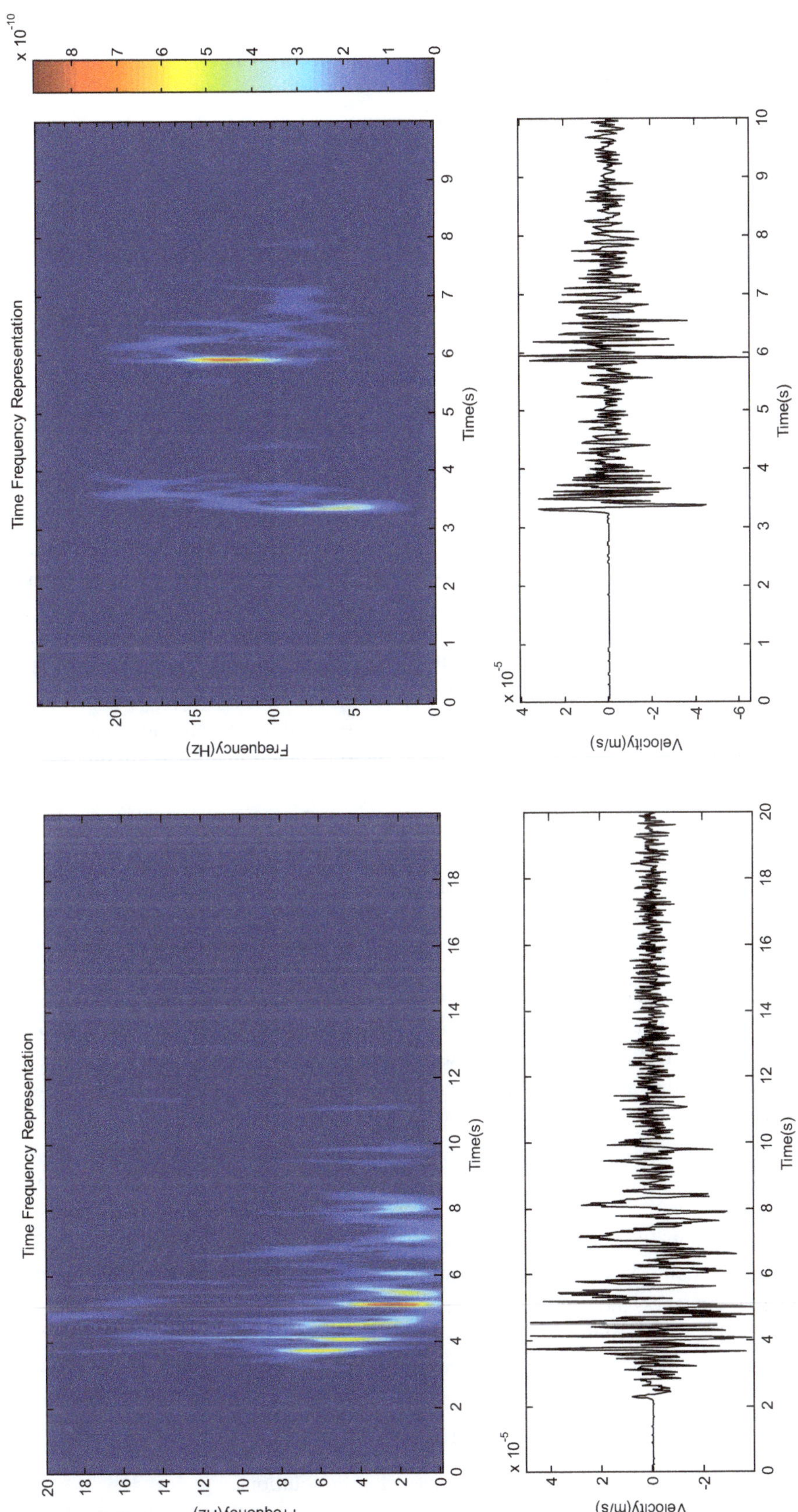

Fig. 4 Energy spectrum of vertical velocity records at BSH station. *Left-hand panels* for Sept. 12 M_L 2.7 event; *right-hand panels* for Nov. 25 M_L 2.4 event

trace of matrix $\boldsymbol{M}$ is positive; otherwise, it is a compression-type source. We designate this by $\lambda/\mu = \kappa$ as the moment tensor can be decomposed into

$$\boldsymbol{M} = \boldsymbol{M}^{\mathrm{ISO}} + \boldsymbol{M}^{\mathrm{CLVD}} + \boldsymbol{M}^{\mathrm{DC}}. \tag{6}$$

We can obtain the various components in the moment tensor in the following proportions:

$$\begin{aligned} c^{\mathrm{ISO}} &= \frac{1}{3}\frac{\mathrm{trace}(\boldsymbol{M})}{|M_{|\max|}|}100\%, \\ c^{\mathrm{CLVD}} &= 2\varepsilon\left(100\,\% - \left|c^{\mathrm{ISO}}\right|\right), \\ c^{\mathrm{DC}} &= 100\,\% - \left|c^{\mathrm{ISO}}\right| - \left|c^{\mathrm{CLVD}}\right|, \end{aligned} \tag{7}$$

where $M_{|\max|}$ is the maximum absolute value of the characteristic value of the moment tensor $\boldsymbol{M}$ and ε is the DC component corresponding to the CLVD component, defined as

$$\varepsilon = -\frac{M^*_{|\min|}}{\left|M^*_{|\max|}\right|}, \tag{8}$$

where $M^*_{|\min|}$ and $M^*_{|\max|}$ are, respectively, the minimum and maximum absolute values of the characteristic value of the deviatoric moment tensor $\boldsymbol{M}^*$.

Based on Eqs. (5), (7), and (8), Vavryčuk (2001) devised the following equation:

$$\begin{aligned} \kappa &= \frac{2}{3}\left[\frac{\mathrm{trace}(\boldsymbol{M})/3}{M^*_{\max} + M^*_{\min}} - 1\right] = \frac{4}{3}\left[\frac{c^{\mathrm{ISO}}}{c^{\mathrm{CLVD}}} - 1\right], \\ \alpha &= \arcsin\left[3\frac{M^*_{\max} + M^*_{\min}}{\left|M^*_{\max}\right| + \left|M^*_{\min}\right|}\right]. \end{aligned} \tag{9}$$

We can infer from the formula that, based on the proportion of different moment tensor components, the specific value of the Lamé coefficient of the fracture fault and the dip of the slip vector of the fault in relation to fault plane can be calculated.

3.2 Method for moment tensor inversion

We used ISOLA package to do the waveform moment tensor inversion in time domain (Sokos and Zahradník 2013, 2008). For small earthquakes, the synthetic seismogram can be approximated with a combination of basic seismograms corresponding to six basic types of source mechanism:

$$S(t) = \mathrm{sum}[a_i \times E_i(t)], \tag{10}$$

where $E_i(t), \quad i = 1, 2, \ldots, 6$, is six basic seismograms; a_i is the coefficient to be solved, which is related to the moment tensor, M_{ij}, in the following geographical coordinates (north, east, and up are positive):

$$\begin{aligned} M_{xx} &= -a_4 + a_6 \\ M_{yy} &= -a_5 + a_6 \\ M_{zz} &= a_4 + a_5 + a_6 \\ M_{xy} &= M_{yx} = a_1 \\ M_{xz} &= M_{zx} = a_2 \\ M_{yz} &= M_{zy} = -a_3 \end{aligned} . \tag{11}$$

The use of the least-squares method to solve the inversion problem minimizes the degree of misfit between the observed waveform and the synthetic waveform. The misfit function is defined as follows:

$$\mathrm{misfit} = \frac{\sum_{i=1}^{n} \frac{\mathrm{obs}_i - \mathrm{syn}_i}{\max(|\mathrm{obs}_i|, |\mathrm{syn}_i|)}}{n}, \tag{12}$$

where $i = 1, 2,\ldots, n$; n = number of stations × number of vector components × number of frequency of use.

Green's functions were calculated using Bouchon's discrete wavenumber method (Bouchon 1981) and the original waveform recordings were calibrated experimentally. The appropriate frequency band was selected for band-pass filtering, e.g., 0.10–0.25 Hz for M_L 3 event. 5 % of cosine taper is used for FFT. A grid search method for a preset trial hypocentral location and time displacement was used to determine the optimum result.

4 Results

We obtained the source moment tensor solutions of eight events in this earthquake swarm (Tables 1, 2, 3). The κ value in Table 2 shows that all the inversion results are in a physically significant range. Figure 5 shows a beach ball representation of the moment tensor solutions.

4.1 December 20, 2010 M_L 3.4 event

As this event was the largest, all the broadband digital seismic stations within 150 km of the epicenter recorded clear waveform signals. We chose the data from six stations (LQU, WLS, YSH, NLA, XIT, and TIA) with epicentral distances within 100 km for the moment tensor studies.

The calculated scalar seismic moment for this event was 1.588×10^{14} N m, the focal depth was 5 km, and the optimized DC source mechanism solutions were nodal plane A strike 302°, dip 38°, rake −49°; and nodal plane B strike 75°, dip 61°, rake −116°; see Fig. 5 for moment tensor solutions.

The correlation coefficient of the optimum synthetic waveform solution was nearly 0.8, showing a good inversion result. The moment tensor solution results showed that

Table 1 Moment tensor components of earthquake events in the Boshan swarm

Event ID	Moment tensor components (10^{10} N m) in geographical coordinate system (NE down)					
	M_{rr}	M_{tt}	M_{pp}	M_{rt}	M_{rp}	M_{tp}
201009121138	−158.17	183.03	171.84	−45.57	68.27	−108.23
201011241356	790.90	−529.93	152.01	196.71	−45.95	30.67
201011252055	−36.164	3.398	32.767	−100.034	−1.624	53.241
201011271942	58.88	218.50	60.14	−217.34	69.71	−89.90
201011291539	−50.78	223.94	102.24	34.03	−16.90	76.76
201012020553	2.99	3.08	1.688	−2.26	4.18	0.60
201012020624	6.91	13.80	1.98	−4.01	10.29	1.22
201012201555	368.5	2047.00	546.60	−416.20	104.75	156.23

Table 2 Source parameters for events in the Boshan swarm

Event ID	Seismic moment (10^{12} N m)	Moment magnitude M_W	Average correlation coefficient	ISO (%)	CLVD (%)	DC (%)	κ
201009121138	2.499	2.2	0.454	21.9	3.3	75.0	8.18
201011241356	7.117	2.6	0.611	16.7	3.3	80.0	6.08
201011252055	1.185	2.0	0.503	0	24.5	75.5	0
201011271942	2.960	2.3	0.373	27.6	43.8	28.6	0.17
201011291539	1.981	2.2	0.339	34.7	15.8	49.4	2.26
201012020553	0.059	1.1	0.532	36.6	16.7	46.7	2.25
201012020624	0.156	1.4	0.551	44.3	37.1	18.6	0.93
201012201555	158.78	3.4	0.667	45.8	36.5	17.7	1.01

Table 3 Parameters of fault planes of events of Boshan swarm

Event ID	Nodal plane *A*			Nodal plane *B*		
	Strike (°)	Dip (°)	Rake (°)	Strike (°)	Dip (°)	Rake (°)
201009121138	309	35	−97	138	55	−85
201011241356	95	53	93	269	36	84
201011252055	185	30	−161	78	80	−61
201011271942	321	14	−55	105	79	−98
201011291539	7	28	−110	210	63	−79
201012020553	186	5	125	331	86	87
201012020624	232	25	177	325	89	65
201012201555	302	38	−49	75	61	−116

the DC component in the seismic source was relatively small, whereas the non-DC components were about 70 % and the ISO and CLVD subcomponents were dramatic, showing that the swarm events were not shear dislocation induced purely by tectonic causes. The source mechanism solutions in Fig. 6 show that the seismic source contains a large uniaxial tensile component.

Figure 7 shows the match between the synthetic and filtered observed waveform at five stations (WLS, YSH, LQU, XIT, TIA). The three-direction average correlation coefficients of these stations are fairly good, supporting the reliability of the moment tensor solutions obtained in this work.

To explain the authenticity of the non-DC components in the moment tensor solutions, inversion was carried out again with the source mechanisms constrained to pure DC solutions; the solutions at different depths are shown in Fig. 8. This figure shows that the degree of waveform correlation of the optimum pure DC solutions at various depths is significantly lower than the moment tensor solutions, explaining the authenticity and reliability of the non-DC components in the moment tensor solutions.

Table 2 shows that, except for the three initial events, the proportion of the DC components in the events was very small (<60 %). Trifu and Shumila (2002) found that large non-DC components appeared in some mine

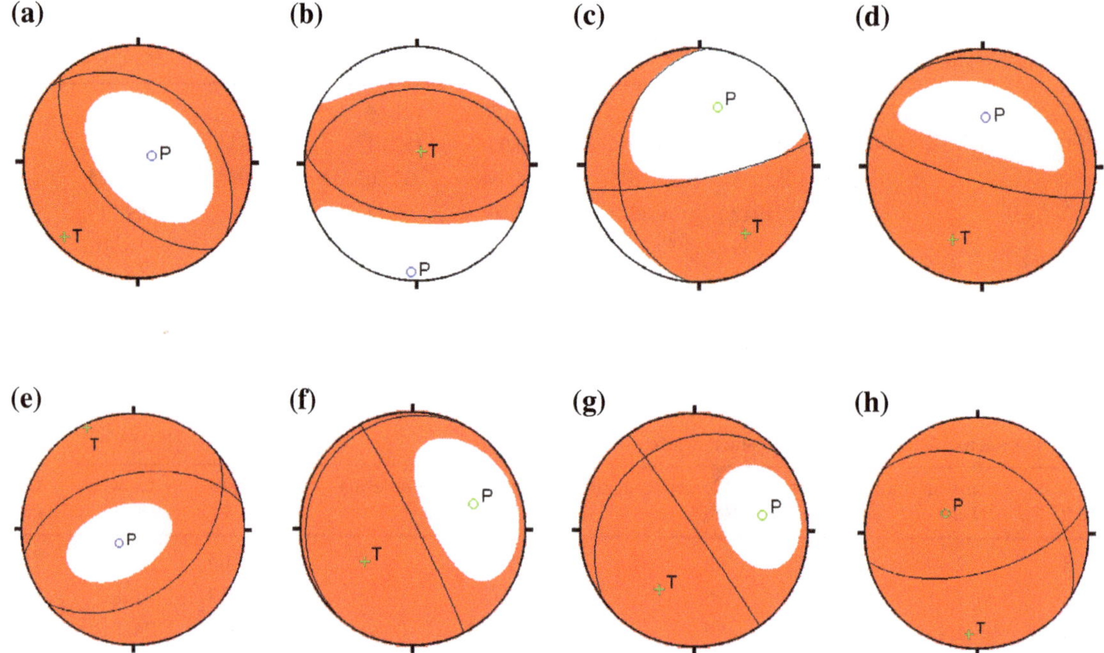

Fig. 5 Source mechanisms of eight events studied in this work. **a** 2009121138, **b** 2011241356, **c** 2011252055, **d** 2011271942, **e** 2011291539, **f** 2012020553, **g** 2012020624, **h** 2012201555

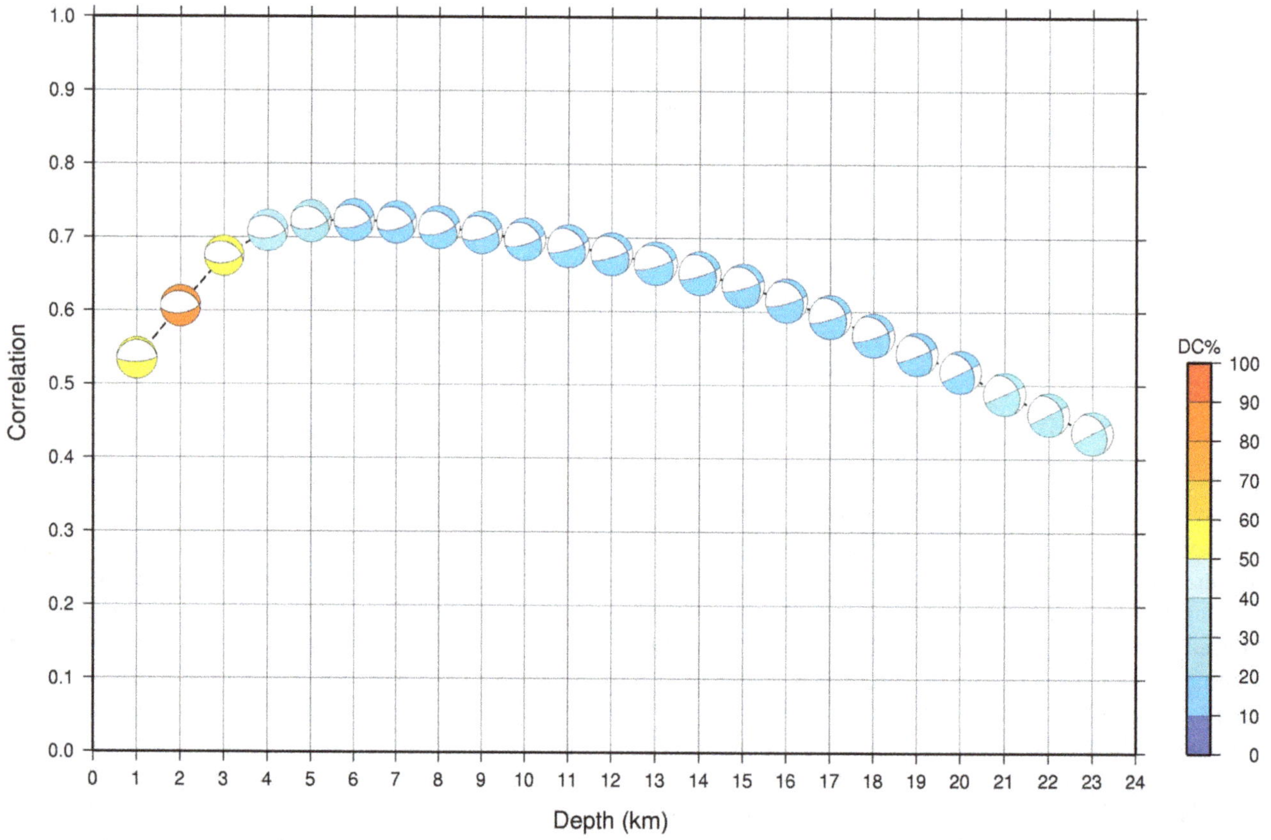

Fig. 6 Waveform misfit function for optimized source mechanisms at different depths

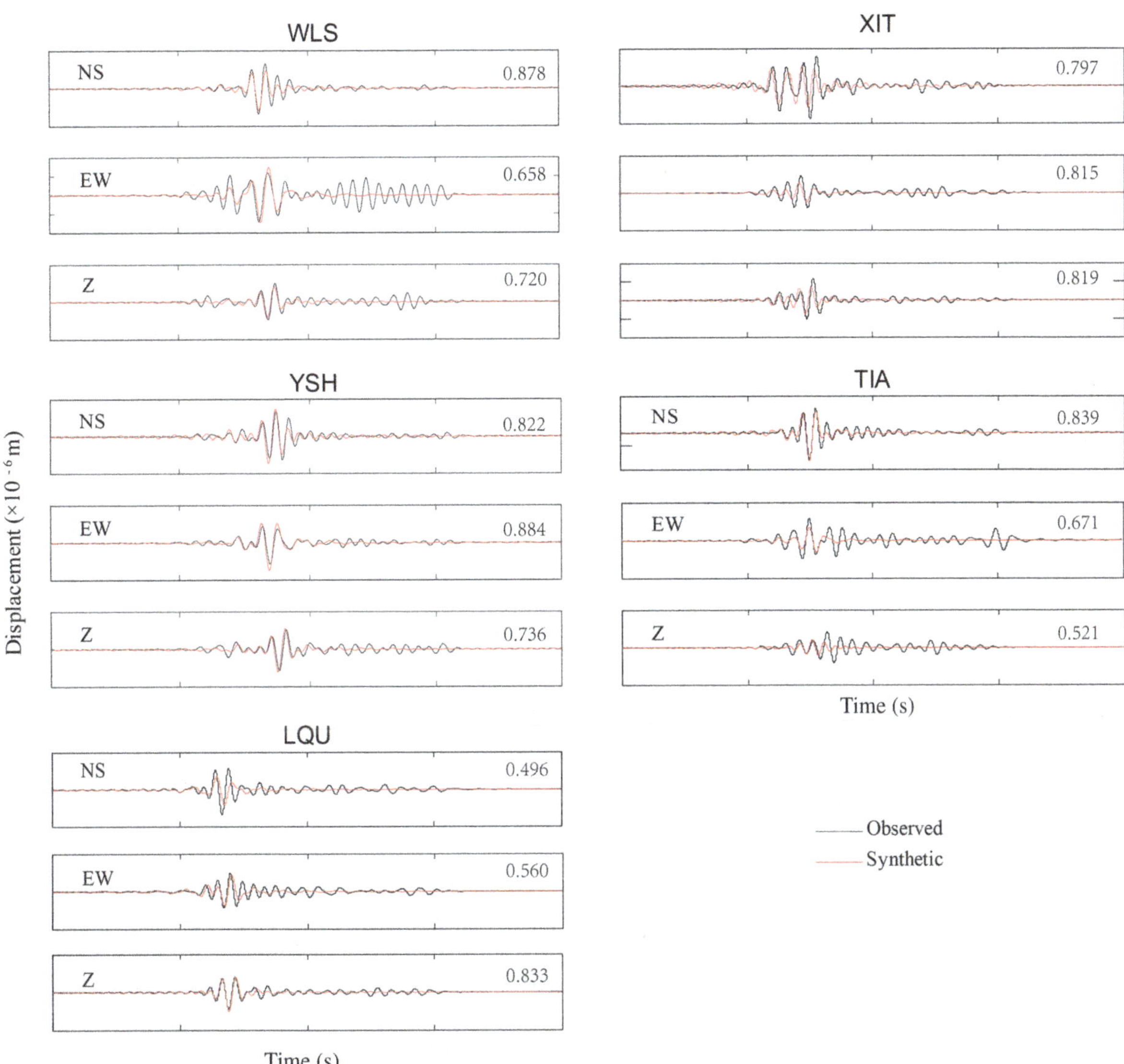

Fig. 7 Fitness of synthetic waveforms and filtered observations, numbers at *right-top* of each component are correlation coefficient between synthetics and observations

earthquake events in which the shear faulting components were 15 %–20 %; their analysis showed that the moment tensor results were reliable. Dahm et al. (2000) studied the accuracy of the seismic source mechanism by comparing the absolute and relative moment tensor solutions; they estimated the error caused by the medium model and inaccurate amplitude data. Dahm et al. (2000) suggested that the significantly high level of similar non-DC components was not caused by inaccurate moment tensor inversion and that a large proportion of these non-DC components reflected the special properties of the seismic sources or media.

Miller et al. (1998) used statistics to show that many mine earthquakes contain non-DC mechanisms and that the first motion is inconsistent with a normal quadrantal distribution. Stickney and Sprenke (1993) studied 21 mine earthquakes (rock burst) in the Coeur d'Alene mining area, Idaho, USA, and found that 90 % of the P-wave first motions observed were downward and that the first motions of 11 of the 21 events were negative. The first motion polarities from 25 stations for the Dec 20, 2010 $M_L = 3.4$ event were obtained from the waveform data recorded by the Shandong Seismic Networks. In contrast to ordinary mine earthquakes, most of the first motions were upward

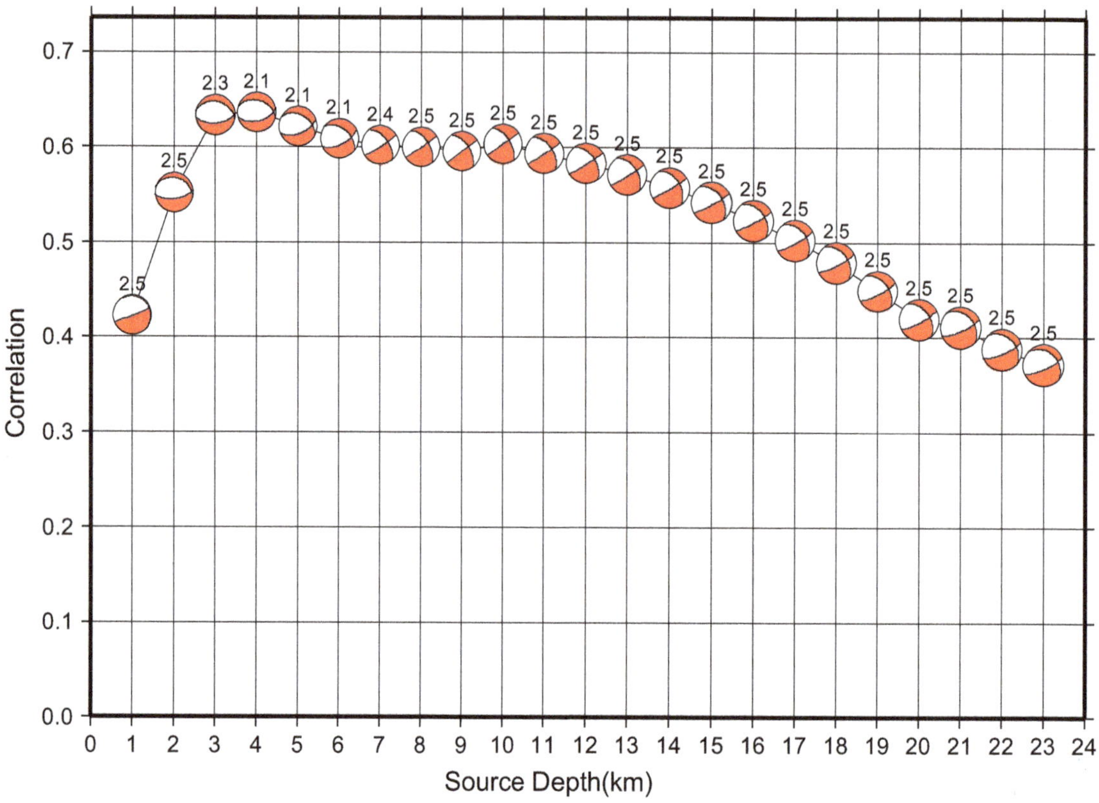

Fig. 8 Waveform misfit function for optimized DC source mechanisms at different depths. The values above the beach balls give the time shift

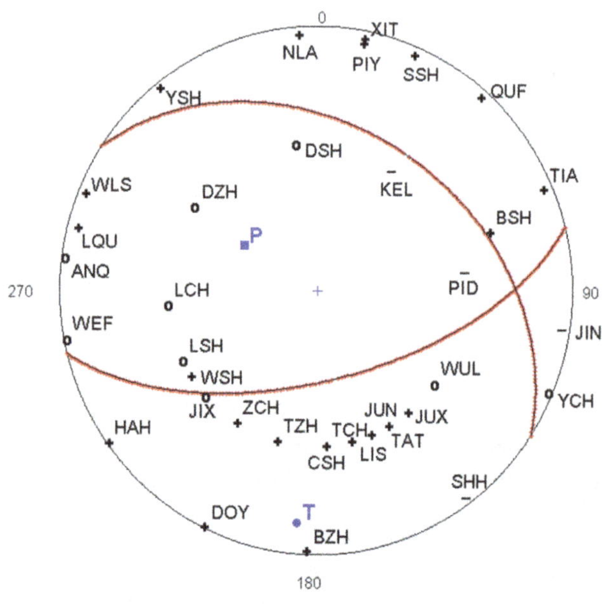

Fig. 9 Source mechanisms and first motion polarities of December 20 event with M_L 3.4; *Symbols* 'o' on the beach ball denote that the first motions are ambiguous at these stations

and even the downward first motions observed by three stations were not obvious; the possibility of misreading cannot be ruled out. Figure 9 shows the correspondence between the source mechanism solutions obtained by MT inversion and the first motion polarities; the results include nine stations with an epicentral distance within 200 km, but which could not determine a clear first motion. Figure 9 shows that because the volume components in this seismic source were large, in the first motion expansion zone in which the first motion should originally be downward, the first motions observed by many stations were blurred. Many positive first motion polarities were observed and this further supports the reliability of solutions from moment tensor inversion.

5 Discussion

For non-DC components in seismic moment tensor solutions, some seismic moment tensor analysis has shown (Miller et al. 1998; Trifu and Shumila 2002; Horálek et al. 2010) that, under certain circumstances (e.g., mine earthquakes due to reservoir filling, water injection in oil production, volcanic activity and geothermal eruption, landslides, collapses, and subsidence), the non-DC components (the CLVD and ISO components) may form a large proportion of the total. Julian et al. (1998) summarized the moment tensor expansion of the equilibrating force system and systematically analyzed the possible physical processes of non-DC earthquakes. They gave the theoretically reasonable explanations that the non-DC rupture process may have its roots in the shear faulting and extension fracture of complex faults, or in shear faulting or polymorphic phase transition in anisotropic media.

The results for the solution of typical mine earthquake focal mechanisms induced by mining operations often show a large implosion component and, as the goafs collapsed, non-DC components in moment tensor solutions. The ISO and CLVD components are negative, indicating the existence of a large amount of uniaxial compression accompanying volume reduction (McGarr 1992a, b; Julià et al. 2009). The computation of this Boshan earthquake swarm events showed that, at the beginning of swarm activity, the DC components formed a large proportion of the source mechanisms (>75 %), whereas the non-DC components may be attributed, to some extent, to inversion error due to constraints in the observed data. However, the moment tensor in the later events showed very large non-DC components and the ISO and CLVD components were both positive, indicating that the seismic sources were tensile cracks.

5.1 Stress state and orientation of *P*, *T* axes

Previous studies of the source mechanisms of typical mine earthquake events have shown (McGarr 1992a, b; Julià et al. 2009) that the maximum stress in mine earthquake sources is compressive stress in a nearly vertical direction, accompanied by a reduction in volume and normal faulting; this is consistent with the closure of goafs as a result of gravity.

Figure 10 shows the orientation of the *P*, *T* axes in eight of the mine earthquake events studied in this work. Except for one tectonic earthquake event (201011241356), the *P* axis in the other seven events is orientated in a nearly vertical direction and the DC components show normal faulting. However, in contrast with the findings of Julià et al. (2009), the maximum stress in these events was tensile stress. The *T* axis was orientated to the southwest in the projection direction of the focal sphere, indicating that the source rupture in these mine earthquake events was largely caused by the action of tensile stress in the southeast in a nearly horizontal direction. The projection of the slip vectors on the focal sphere shows that, apart from the event 16 km south of the earthquake swarm on November 25, 2010, the projection direction of the seismic source slip vectors of the earthquake swarm was relatively stable.

Feignier and Young (1992) created many non-DC mechanism earthquakes by explosions in a 3.5-m-wide mine tunnel cut into complete granite in the Manitoba URL Laboratory, Canada. They calculated the moment tensors of 33 recorded microquakes with moment magnitudes ranging from −2 to −4 and found that tensile events mainly occurred around the caved areas at the top of the mine tunnels and that the implosion events usually occurred in front of active surfaces. Their computation showed that the maximum principal stress in the area was characterized by a positive tensile stress and that the tectonic stress background could explain the existence of a tensile source.

In Boshan District, Zibo City is considered to be a seismic window for Shandong because there are frequent small swarms of earthquakes and microquakes. This area often had an obvious concentrated activity of small earthquakes before the occurrence of moderately strong earthquakes in the zone (within about 500 km). (Su 1997) Boshan District lies around the top of the arc of a horseshoe-shaped structure in west Shandong where stress tends

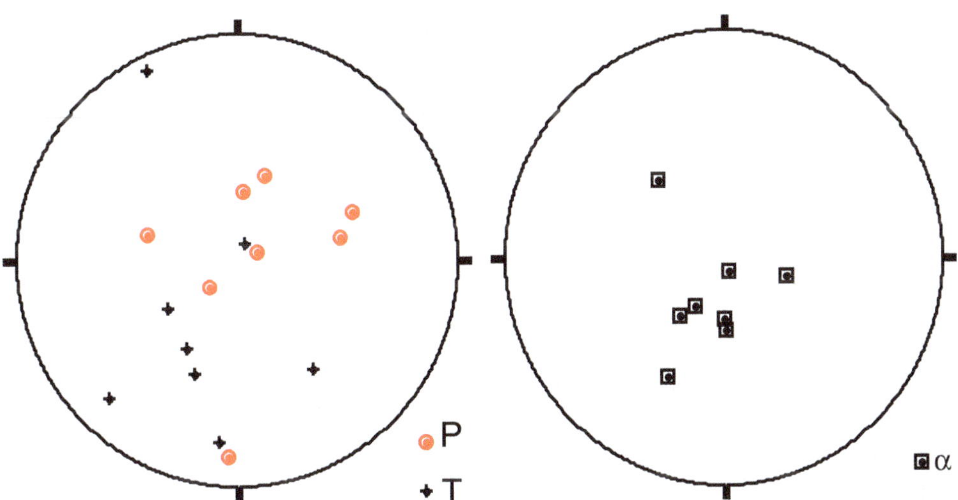

Fig. 10 Positions of *P*, *T* axes (*left*) and slip vectors (α in *right panel*) on equal area lower focal hemisphere projections

to concentrate. Faults are developing in this structure, mostly in the form of high-level normal faults; reverse faults are rare. The data show that (Jiao et al. 1990) the epicenters are mainly in the Zibo syncline; this strikes *c.* 40°–50° NE and trends to NW at an angle of 15°–18°. The direction of tensile stress caused by pure rock gravity therefore does not fit with the computation. The epicenter of this earthquake swarm was mainly the Wangmu Mountain fault in a northeasterly direction. According to Shi et al. (2009), the Wangmu Mountain fault trends west with a dip of about 75° and the activity is mainly in the form of normal faults. Although these fracture properties are close to our moment tensor for tensile cracking, the strike of this fault is inconsistent with the inversion result. The Yuwang Mountain fault lies 6 km to the west of this earthquake swarm and strikes 2°–10° NE, trending approximately east with a dip of 60°–80°, and is mainly characterized by a tensile structure. About 5 km to the south of the epicenter, the Shentou-Xihe fault strikes 40°–60° NE, trending to the SE with a dip of 60°–70°. Therefore, within the small structural unit enclosed by the above three areas of extension tectonics, there is also tension in the main stress direction. The small earthquake moment tensor we obtained through inversion is consistent with this analytical result.

According to data from the Mining Bureau of Zibo (Jiao et al. 1990), there are two groups of secondary faults in research area besides large fault shown in Fig. 2. The 1st group spread approximately in EW direction; the 2nd group arranged in NW direction with normal faulting. As no more accurate data about these secondary faults are available, it is impossible to judge which group of faults these swarm events are related to. However, we are certain that the swarm is caused by fracture tectonic activity related to the Wangmu Mountain fault under the conditions of stress adjustment in this small area of Boshan District.

5.2 Analysis of causes for Boshan earthquake swarm activities

There are many possible sources for non-DC source mechanisms. If multiple shear faulting occurs in conterminous, but different, fault planes, or the shear faulting of non-planar faults is mistakenly considered as an event, then the real DC tensor and moment tensor of the combined events are usually non-DC tensors (Frohlich 1994). In this case, shear faulting may generate a non-DC mechanism; however, there is no isotropic volume component. As the trace of the combined moment tensor is equal to the sum of the traces of its components, the combination of DC tensors cannot generate non-zero ISO components (Horálek et al. 2010).

The heterogeneity or anisotropism of the underground media may also result in the appearance of non-DC components in the moment tensor solutions of shear faulting. The experiments of Kawasaki and Tanimoto (1981) and the studies of Vavryčuk (2002) have shown that the anisotropism and heterogeneity of the underground media are not the main causes of higher non-DC mechanisms. Considering that the data from field mobile stations were used in our studies of the earthquakes that occurred in the period December 2–5, 2010 within a small range with a radius of <10 km, the properties of the underground media can be assumed to be uniform and the influence of lateral heterogeneity is negligible.

Theoretical and experimental studies on the source causes of tensile earthquakes have been reported previously (Walter and Brune 1993; Scholz 1990). Studies have shown that tensile earthquakes usually occur in areas with large amounts of underground fluids, such as in a geothermal field or near volcanoes (Miller et al. 1998; Baisch et al. 2009). Julian et al. (1998) proposed that high-pressure underground fluids are the primary cause of tensile cracking because the existence of fluid pressure may offset rock stresses in the Earth's interior. Fluids under high pressure may rapidly fill the rupture zones when the rocks burst, giving rise to large tensile components (Fig. 11).

Walter et al. (2009) studied the source mechanisms in icequake events caused by the opening of crevasses on the surface layer of the Gornergletscher glacier, Switzerland, using the full waveform inversion method. The moment tensor results showed extensional fractures, indicating the reliability of moment tensor inversion. Walter et al. (2010) showed that the involvement of high-pressure liquids in the rupture process may lead to tensile cracking by studying the moment tensor solutions in icequake events that occurred at the base of glaciers.

In Boshan District, Zibo City has been long known as the Second Spring City of Shandong because there are four groups of springs (Shentou, Qiugu, Zhulong, and Liangzhuang) and more than 40 recorded springs. In addition, a number of geothermal areas and abnormalities in

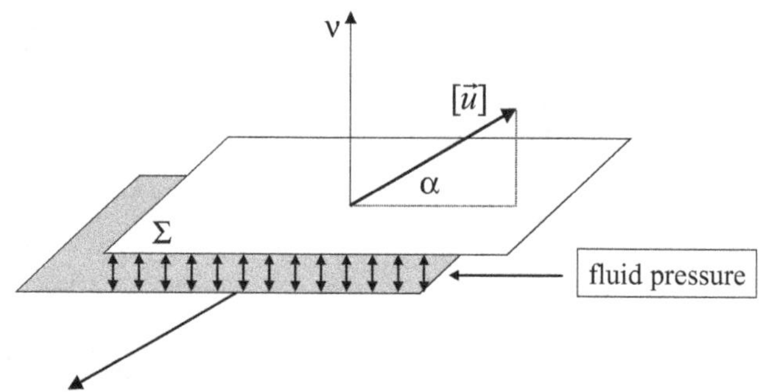

Fig. 11 Model for the tensile source. $\sum$ is the fault plane, v is the normal to the fault, $[\bar{u}]$ is the slip vector, and α is the inclination of the slip vector from the fault plane

hydrochemistry exist in this area (Yang et al. 2005), which matches the observations in earlier studies that tensile earthquakes with non-DC mechanisms mainly occur in areas where there are large underground geothermal fluids or volcanoes. Therefore, the large non-DC components in the moment tensor solutions of the Boshan swarm may be caused by the involvement of underground fluids in the rupture process in the source area.

Moment tensor analysis showed that (except for the three events that occurred on September 12, November 24, and November 25, 2010) the mechanism in the early stages of the swarm was primarily tensile cracking. As massive underground fluids are present in this area, the main cause of the earthquakes was shear faulting at the beginning of swarm activity. After fracturing, underground fluids with a high pore pressure filled the burst area, reducing the friction stress and making a second rupture easier, which resulted in a continuously active earthquake swarm. The high pore pressure also gives rise to the source mechanism of tensile cracking. The results of rock crushing tests support tensile cracking (Byerlee 1990). Špičák and coworkers (2000) and Horálek et al. (2002) studied the West Bohemian region of the Czech Republic and they proposed that high fluid pressures may be a key factor in the frequent swarm activities in this region. This feature is similar to that of the Boshan seismic window. Therefore, we conclude that underground fluids with high pore pressures were involved in the rupture of the hypocentral region and played an important role in the activity of the Boshan earthquake swarm.

It is usually thought that full moment tensor solution retrieved by inverting seismograms may include errors due to incomplete knowledge of the structure along the propagation path. Studies in gold mine in South Africa shows that the full MT solutions inverted from close stations within a few kilometers to the source are believable, even in homogeneous model (Šílený and Milev 2006). In our study, the farthest temporary station to the swarm is no more than 10 km. Despite a few local stations, the final result should still reliable.

6 Conclusions

Using data from the Shandong seismic station networks and waveforms recorded by the temporary station network, we determined the source mechanisms of eight events in a mine earthquake swarm in November and December, 2010 in Boshan District using directional instrumentation and the time domain moment tensor inversion method. We also determined the source mechanism of an earthquake of M_L 2.7 that occurred on September 12, 2010. The following conclusions were drawn.

(1) At the beginning of the swarm activities, the DC components formed a large proportion of the source mechanisms, which may be primarily attributable to shear faulting under tectonic stresses. However, the moment tensors in the following events showed very large non-DC components and the ISO and CLVD components were both positive, indicating that the seismic source was tensile cracking.
(2) The source mechanisms of the swarm events showed that—except for the earthquake event 16 km to the south on November 24, 2010, which showed reverse faulting—the DC components of the events showed normal faulting. The P axis was almost vertical and the maximum stress was tensile; the projection direction of the source slip vector of the swarm events was relatively reliable.
(3) We analyzed the epicentral tectonic features and suggest that this swarm event was caused by fracture tectonic activity related to the Wangmu Mountain fault undergoing stress adjustment in the small area of Boshan District.

Based on analyses of the reliability of the moment tensor solutions, we proposed that underground fluids with high pore pressures were involved in the rupture process in the source area and played an important role in these Boshan swarm activities.

The results obtained by this study, constrained by the original data and velocity model, showed that the average waveform correlation coefficient of an individual event (e.g., 11291539) was not high; however, the results are still acceptable for microquakes of M_L 2 (Fig. 1).

Although these findings showed that the source type at the beginning of the Boshan earthquake swarm activities was mainly fault rupture under tectonic stress, these activities occurred in a mining area where coal goafs have altered the stress distribution inside the underground formations. However, we can make no clear judgment about whether mining activities have caused these swarms. The moment tensor results for this swarm show that the source activities can be divided into two stages. Rupture in the earlier stage may have an influence on the activity in the next stage, although the physical mechanism of their interactions still unclear and requires further study.

Acknowledgments We thank the Earthquake Networks Center of Shandong Province for their hard work and the reviewers of this paper for their valuable comments. This study was partly supported by the program of Science for Earthquake Resilience (XH15026) provided by China Earthquake Administration, and the Science and Technology Development Plan Project of Shandong Province, China (GRANT-2014GSF120007).

References

Aki K, Richards PG (1980) Quantitative Seismology: Theory and Methods. Translated into Chinese by Li QZ, Zou QJ, et al. W. H. Freeman, New York, pp 55–60 **(in Chinese)**

Baisch S, Vörös R, Weidler R, Wyborn D (2009) Investigation of fault mechanisms during geothermal reservoir stimulation experiments in the Cooper Basin, Australia. Bull Seismol Soc Am 99(1):148–158

Bouchon M (1981) A simple method to calculate Green's functions for elastic layered media. Bull Seismol Soc Am 71(4):959–971

Byerlee J (1990) Friction, over pressure and fault normal compression. Geophys Res Lett 17(12):2109–2112

Chen JH, Gan JS, Wu HB, Zhang LF, Ding YW, Wei GC, Shen XL (2009) Analysis of micro-earthquake swarms in Luoquanhuang area of Zigui County. J Geodesy Geodyn 29(6):45–48 **(in Chinese with English abstract)**

Dahm T, Horálek J, Šílený J (2000) Comparison of absolute and relative moment tensor solutions for the January 1997 West Bohemia earthquake swarm. Stud Geophys Geod 44(2):233–250

Feignier B, Young RP (1992) Moment tensor inversion of induced micro-seismic events: evidence of non-shear failures in the $-4 < M < -2$ moment magnitude range. Geophys Res Lett 19(14):1503–1506

Fletcher JB, McGarr A (2005) Moment tensor inversion of ground motion from mining-induced earthquakes, Trail Mountain, Utah. Bull Seismol Soc Am 95(1):48–57

Frohlich C (1994) Earthquakes with non-double-couple mechanisms. Science 264(5160):804–809

Gibowicz SJ (1995) Scaling relations for seismic events induced by mining. Pure appl Geophys 144(2):191–209

Gibowicz SJ, Kijko A (1998) In: An introduction to mining seismology. Translated into Chinese by Xiu JG, Xu P, Yang XP. Seismological Press, Beijing pp 232–238

He XS, Li SY, Shen P, Feng QX (2006) A wavelet packet approach to wave classification of earthquakes and mining shocks. Earthq Res China 22(4):425–434 **(in Chinese with English abstract)**

He X, Li S, Pan K, Zhang TZ, Wang LY, Xu ZH, Jiang XQ, Song XY, Lu QH, He SY (2007) Mining seismicity, gas outburst and the significance of their relationship in the study of physics of earthquake sources. Acta Seismol Sinca 29(3):314–327 **(in Chinese with English abstract)**

Horálek J, Šílený J, Fischer T (2002) Moment tensors of the January 1997 earthquake swarm in NW Bohemia (Czech Republic): double-couple vs. non-double-couple events. Tectonophysics 356(1–3):65–85

Horálek J, Jechumtálová Z, Dorbath L, Šílený J (2010) Source mechanisms of micro-earthquakes induced in a fluid injection experiment at the HDR site Soultz-sous-Forêts (Alsace) in 2003 and their temporal and spatial variations. Geophys J Int 181(3):1547–1565

Jiang WW, Hao TY, Jiao CM, Song HB (2000) The characters of gravity and magnetic fields and crustal structure from Qingzhou to Muping, Shandong Province. Prog Geophys 15(4):18–26 **(in Chinese with English abstract)**

Jiao RF, Yang CY, Liu BL et al (1990) Annals of Boshan. People's Press of Shandong Province, Jinan, pp 1–682 **(in Chinese)**

Julià J, Nyblade AA, Durrheim R, Linzer L, Gök R, Dirks P, Walter W (2009) Source mechanisms of mine-related seismicity, Savuka Mine, South Africa. Bull Seismol Soc Am 99(5):2801–2814

Julian BR, Miller AD, Foulger GR (1998) Non-double-couple earthquakes 1: theory. Rev Geophys 36(4):525–549

Kawasaki I, Tanimoto T (1981) Radiation patterns of body waves due to the seismic dislocation occurring in an anisotropic source medium. Bull Seismol Soc Am 71(1):37–50

Li T, Cai M, Zuo Y, Liu YQ (2005) Features of focal mechanisms of mining-induced earthquakes: a case study of the Fushun Laohutai coal mine, Liaoning Province. Geol Bull China 24(2):136–144 **(in Chinese with English abstract)**

Lin HC, Wang BP, Liu HR, Jiang JK (1990) Comparative study of tectonic and collapse earthquakes. Acta Seismol Sinca 12(4):448–455 **(in Chinese with English abstract)**

Liu XQ, Du YH, Xu B, Li H, Shen P, Zhang P (2005) The mode identification method and its application to regional mine and natural earthquakes. Earthq Res China 21(1):50–60 **(in Chinese with English abstract)**

McGarr A (1992a) An implosive component in the seismic moment tensor of a mining-induced tremor. Geophys Res Lett 19(15):1579–1582

McGarr A (1992b) Moment tensors of ten Witwatersrand mine tremors. Pure appl Geophys 139(3–4):781–800

McGarr A (1994) Earthquake source mechanics and fracture mechanics: theory and observation: some comparisons between mining-induced and laboratory earthquakes. Pure appl Geophys 142(3–4):467–489

McGarr A, Boettcher M, Fletcher JB, Sell R, Johnston MJS, Durrheim R, Spottiswoode S, Milev A (2009) Broadband records of earthquakes in deep gold mines and a comparison with results from SAFOD, California. Bull Seismol Soc Am 99(5):2815–2824

Miller A, Foulger G, Julian B (1998) Non-double-couple earthquakes 2: observations. Rev Geophys 36(4):551–568

Oye V, Bungum H, Roth M (2005) Source parameters and scaling relations for mining-related seismicity within the Pyhäsalmi Ore Mine, Finland. Bull Seismol Soc Am 95(3):1011–1026

Phillips WS, Pearson DC, Yang X, Stump BW (1999) Aftershocks of an explosively induced mine collapse at White Pine, Michigan. Bull Seismol Soc Am 89(6):1575–1590

Scholz CH (1990) The mechanics of earthquakes and faulting. Cambridge University Press, New York, pp 66–100

Shen P, Yang XH, Mao ZY, Gao MQ, Wu SY, Ye TL (2006) Research on nucleation of earthquake in km-size experiment. Prog Geophys 21(3):717–721 **(in Chinese with English abstract)**

Shi QM, Wang HS, Wang T (2009) Geophysical exploration of Wangmushan fault and the activities of the times. Eng Des Ground 11:96–98 **(in Chinese with English abstract)**

Šílený J, Milev A (2006) Seismic moment tensor resolution on a local scale: simulated rockburst and mine-induced seismic events in the Kopanang gold mine, South Africa. Pure appl Geophys 163(8):1495–1513

Sokos E, Zahradník J (2008) ISOLA a Fortran code and a Matlab GUI to perform multiple-point source inversion of seismic data. Comput Geosci 34:967–977

Sokos E, Zahradník J (2013) Evaluating centroid moment tensor uncertainty in new version of ISOLA software. Seismol Res Lett 84:656–665

Špičák A (2000) Earthquake swarm and accompanying phenomena in intraplate regions: a review. Stud Geophys Geod 44(2):89–106

Stec K (2007) Characteristics of seismic activity of the Upper Silesian Coal Basin in Poland. Geophys J Int 168(2):757–768

Stickney M, Sprenke K (1993) Seismic events with implosional focal mechanisms in the Coeur d'Alene Mining District, Northern Idaho. J Geophys Res 98(B4):6523–6528

Su LS (1997) A study of prediction effectiveness of Boshan seismic situation window. North China Earthq Sci 15(3):40–44 **(in Chinese with English abstract)**

Trifu C, Shumila V (2002) Reliability of seismic moment tensor inversions for induced micro seismicity at Kidd Mine, Ontario. Pure appl Geophys 159(1–3):145–164

Vavryčuk V (2001) Inversion for parameters of tensile earthquakes. J Geophys Res 106(B8):16339–16358

Vavryčuk V (2002) Non-double-couple earthquake of 1997 January in West Bohemia, Czech Republic: evidence of tensile faulting. Geophys J Int 149(2):364–373

Walter WR, Brune JN (1993) Spectra of seismic radiation from tensile crack. J Geophys Res 98(B3):4449–4459

Walter F, Clinton JF, Deichmann N, Dreger DS, Minson SE, Funk M (2009) Moment tensor inversions of icequakes on Gornergletscher, Switzerland. Bull Seismol Soc Am 99(2A):852–870

Walter F, Dreger DS, Clinton JF, Deichmann N, Funk M (2010) Evidence for near-horizontal tensile faulting at the base of Gornergletscher, a Swiss Alpine glacier. Bull Seismol Soc Am 100(2):458–472

Yang X, Stump BW, Phillips WS (1998) Source mechanism of an explosively induced mine collapse. Bull Seismol Soc Am 88(3):843–854

Yang S, Zhang HD, Li WZ, Wu L (2005) Genesis of geothermal anomaly in Southern Zhangdian District of Zibo City. Hydrogeol Eng Geol 32(3):59–62 **(in Chinese with English abstract)**

Zhang SQ, Zhang C, Xiu JG, Guan J, Liu YC, Zhang ZP, Dong JP, Xiao LX, Chen G, Li WP (1993) Review of mines seismicity. Progress in Geophysics 8(3):69–85 **(in Chinese with English abstract)**

Zhang LF, Liao WL, Zeng XS, Zhong YY (2009) Analysis of time-frequency characteristics of wave spectrum between tectonic earthquake and mine earthquake. Seismol Geol 31(4):699–706 **(in Chinese with English abstract)**

11

Shear-wave velocity structure of the crust and uppermost mantle in the Shanxi rift zone

Meiqing Song · Yong Zheng · Chun Liu · Li Li · Xia Wang

Abstract The Shanxi rift zone is one of the largest and active Cenozoic grabens in the world, studying the velocity structure of the crust and upper mantle in this region may help us to understand the mechanisms of rift processes and the seismogenic environment of active seismicity in continental rifts. In this work, using the broadband seismic data of Shanxi, Hebei, Henan, Shaanxi provinces, and the Inner Mongolia Autonomous Region from February 2009 to November 2011, we have picked out 350 high-quality phase velocity dispersion curves of fundamental mode Rayleigh waves at periods from 8 to 75 s, and Rayleigh wave phase velocity maps have been constructed from 8 to 75 s period with horizontal resolution ranging from 40 to 50 km by two-station surface-wave tomography. Then, using a genetic algorithm, a 3D shear-wave speed model of the crust and uppermost mantle have been derived from these maps with a spatial resolution of 0.4° × 0.4°. Four characteristics can be outlined from the results: (1) Except in the Datong volcanic zone, in the depth range of 11–30 km, the location of a transition zone between the high- and low-velocity regions is in agreement with the seismicity pattern in the study region, and the earthquakes are mostly concentrated near this transition zone; (2) In the depth range of 31–40 km, shear-wave velocities are higher to the south of the Taiyuan Basin and lower to the north, which is similar to the distribution pattern of Moho depth variations in the Shanxi region; (3) The shear-wave velocity pattern of higher velocities to the south of 38°N and lower velocities to the north is found to be consistent with that from the upper crustal levels to depth of 70 km. At the deeper depths, the spatial scale of the low-velocity anomalies zone in the north is gradually shrinking with depth increasing, the low-velocity anomalies are gradually disappearing beneath the Datong volcanic zone at the depth of 151–200 km. We proposed that the root of the Datong volcano may reach to a depth around 150 km; (4) Along the N–S vertical profile at 112.8°E, the 38°N latitude is the boundary between high and low velocities, arguing the tectonic difference between the Shanxi rift zone and its flanks, in the rift zone the seismic velocity is dominated by low-velocity anomalies while in the flanks it is high.

Keywords Shanxi rift zone · Two-station method · Surface wave · Shear-wave velocity · Tomography

1 Introduction

The Shanxi rift zone is one of the second-order active tectonic zones in China (Zhang et al. 2003a, 2004), which absorbs the convergence rates between many tectonic blocks, such as the Ordos Block, the North China Block, and the South China Block (Liu 2008). Because of this kind of tectonic environment, a lot of active tectonic features

M. Song
Earthquake Prediction Center, Earthquake Administration of Shanxi Province, Taiyuan 030021, China
e-mail: smq28@126.com

Y. Zheng (✉)
Institute of Geodesy and Geophysics, Chinese Academy of Sciences, Wuhan 430077, China
e-mail: zhengyong@whigg.ac.cn

C. Liu
Earthquake Administration of Shaanxi Province, Xi'an 710068, China

L. Li · X. Wang
Earthquake Administration of Shanxi Province, Taiyuan 030021, China

can be seen in this area. In the inner part of this rift zone, a series of Cenozoic rift basins distribute from north to south, including the Datong, the Xingding, the Taiyuan, the Linfen, the Yuncheng, and the Weihe rift basins (EAC 1988). Historical records show that there have been eight earthquakes with magnitudes equal to or larger than *M* 7.0 occurred in the Shanxi rift zone, which account for one quarter of the large-scale earthquakes that occurred in North China in the historical records (Li 1981). Due to the intensity and frequent occurence of the strong earthquakes in this region, the Shanxi rift is presently identified as one of the most tectonically active Cenozoic rift zones in the world (Xing et al. 2005). The strong and unbalanced rifting may be closely related to geodynamic processes and deep structures within the rift as well as the surrounding tectonic environment. Thus, the Shanxi rift zone not only presents tectonic characteristics that are common to seismic environments but also exhibits uniquely tectonic processes (Yang et al. 1995). Moreover, the active tectonic environment also caused complex tectonic deformation in this area, and the Shanxi rift is spreading until now. In the north end of the Shanxi rift, there is Datong volcanic zone, which is a significant Quaternary volcanic zone in the North China Craton, and the tectonic evolution of the volcanic zone may play an important role for the rejuvenation of the North China Craton (Deng et al. 2007). Therefore, the Shanxi rift and the surrounding region are ideal regions for studying continental rift dynamics. A deep-going and systematic study of the velocity structure of the crust and upper mantle in this region would be of great importance. It would help to raise the location precision of Shanxi digital seismic network to determine the focal mechanisms of local earthquakes, to study the deep tectonic setting of the genesis of rift earthquakes and to reveal dynamic rift processes.

In the past two decades, a large number of geophysical studies have been done to image the velocity structure in Shanxi rift, including deep seismic sounding (DSS) (Zhu et al. 1994, 1999; Liu et al. 2000; Zhao et al. 1999, 2006; Zhang et al. 1997), waveform modeling (Zhang et al. 2003b), teleseismic receiver function (Ge et al. 2011; Wu et al. 2011; Tang et al. 2010), body-wave tomography (Wang 2005; Xu et al. 1997; Zhang et al. 2011), ambient seismic noise and two-station analysis (Zheng et al. 2011; Fang 2010; Tang et al. 2011), surface-wave tomography (Li et al. 2010, 2012), and magnetotelluric sounding (Wei et al. 2006; Zhao et al. 1997). These results have provided us the overall information of the crust and upper mantle in this region. However, due to the constraints from the datasets and the methods, the previous studies cannot provide detailed structure for the whole region of the Shanxi rift. For example, although DSS method can resolve the crustal structure well, only six unevenly distributed sections have crossed this region. In addition, the magnetotelluric sounding work in this region only allows us to acquire the media properties under the Datong basin near Ying Xian. The resolution and reliability of the results from body-wave tomography (Wang 2005; Xu et al. 1997; Zhang et al. 2011), ambient noise tomography (Zheng et al. 2011; Fang 2010; Tang et al. 2011), and receiver function analysis (Ge et al. 2011; Wu et al. 2011; Tang et al. 2010) were not enough to comprehensively study the deformation mechanisms within the Shanxi rift zone. On the other hand, earthquake surface-wave tomography can reveal the velocity structure from the surface to a depth of 150 km, or even deeper, thus is free from the limitation of the ANT method that can only resolve the structure in shallow depth. Using this method, Li et al. (2012) obtained the Rayleigh-wave phase velocity structure as well as 3D shear-wave velocity images with period ranges from 12 to 125 s in the Ordos Block and its neighboring rift basin. However, due to the limitation of station coverage and azimuthal distribution of the earthquakes, the spatial resolution of these previous studies are worse than 100 km, which is unsatisfied for the requirement of deciphering the detailed tectonic features in the Shanxi rift zone whose spatial scale is only around 100 km.

In general, shear-wave velocity structure can provide better rheological images than that of the P-wave tomography, and surface-wave phase velocity is sensitive to the variation of shear-wave velocity, thus, surface-wave tomography is one of the most important tools to understand the rheological and tectonic structures in the crust and the upper mantle (Li and Burke 2006; Jia and Zhang 2008; Ding et al. 2008; Jobert et al. 1985; Toksoz and Anderson 1966; Kanamori 1970; Forsyth 1975; Zhang and Lay 1996; Zhang and Ma 1997). A number of studies have resolved the crustal and upper-mantle velocity structures using long-period surface-wave data (Zhu et al. 2007; Yanovskaya and Kozhevnikov 2003; Huang et al. 2003; He et al. 2002, 2009). However, due to the sparse station distribution, most of the previous studies used the single-station method to determine group velocity along mixed paths and further to obtain the crustal and upper-mantle velocity structures. Compared with traditional single-station surface-wave tomography (SWT), the precision of the phase dispersion curves measured by two-station method can be improved significantly because it is nearly free from the errors of epicenter location and the original time of the earthquake (Yanovskaya and Kozhevnikov 2003). With increasing Digital Seismic Network development in recent years, many scholars have been able to obtain very good local and regional crustal shear velocity structures with higher resolution by the use of two-station phase velocity dispersion (He et al. 2002, 2009; Prindle and Tanimoto 2006; Xu et al. 2000, 2007). For example, Yi et al. (2008)

obtained phase velocity images in mainland China and its adjacent regions using the Rayleigh-wave average phase velocity at periods from 20 to 120 s with 102 stations and 538 two-station paths.

Due to the characteristics of the two-station method, the two stations should be approximately on the same great-circle path along with the seismic source. Before 2008, the Shanxi Digital Seismic Network only contained 21 stations with uneven azimuthal distribution, and some of them were short-period seismographs, so the previous seismic network did not fulfill the requirements of the two-station method. Fortunately, with the development of the Shanxi Digital Seismic Network, the original 21 stations have been replaced with digital broadband seismometers, and 12 new broadband seismic stations were added into the network system. With the accumulation of data from the upgraded network, it is possible to study the high-resolution 3D fine structure of the crust and upper mantle in local regions by the use of the surface waves. In this paper, we try to study the fine structures of complex media in the crust and uppermost mantle by surface-wave tomographic images with the accumulated data from the upgraded seismic network.

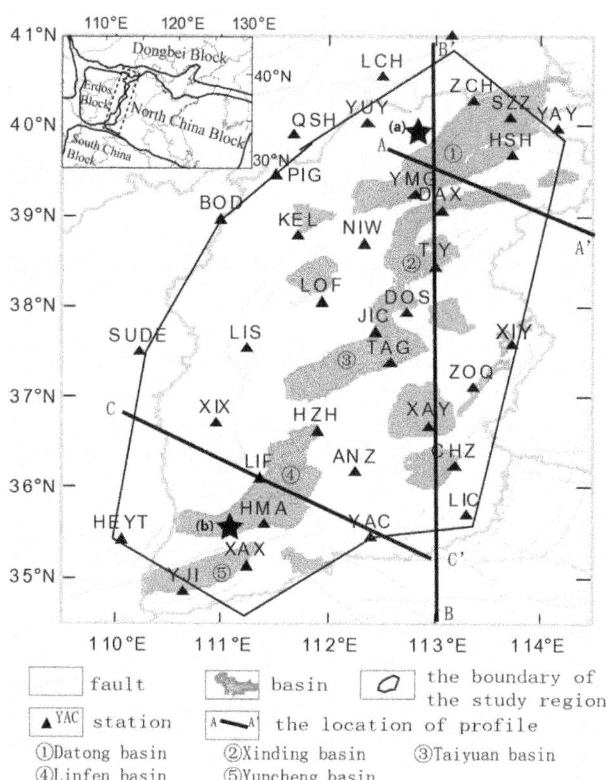

Fig. 1 The tectonic sketch map and distribution of stations and cross sections in the study region. The *solid triangles* are the broadband seismic stations used in this work, and the *solid lines* delineate the locations of the profiles shown in Fig. 9. The *stars* (**a**) and (**b**) denote the locations referred to in Fig. 6. The *rectangle* in the *inset* outlines the area of study region in the large-scale map

2 Data

In this work, we collected the surface-wave data from 31 stations in Shanxi Digital Seismic Network. In order to increase the ray density at the boundaries of the studied area to ensure the reliability of the inversion, we also collected data from another six stations in adjacent provinces, including the Hebei, Henan, Shaanxi provinces, and the Inner Mongolia Autonomous Region (Fig. 1). Finally, we obtained seismic data with evenly distributed path coverage in the study area.

The data are selected to meet the following requirements: (1) the selected earthquake should be shallow events with focal depth less than 100 km, and the epicentral distance is between 30° and 90° and the magnitude is greater than 5.0 in order to obtain the fundamental mode Rayleigh waves; (2) the deviation angles between the path of the two stations to seismic source should be less than 3°; (3) the distance between the two stations in a path should be larger than 100 km; and (4) the vertical waveform records should have high signal-to-noise ratios (SNR) without obvious disturbed signals, so that clear arrival times for wave-packet energy can be observed through the frequency time analysis (FTAN).

The seismic data were recorded from February 2009 to November 2011, and the number of earthquakes that satisfied the first requirement was greater than 100. Because there were five types of seismometers in the seismic network, we first removed the instrument responses for each station before choosing the pairs of the stations. Then, we manually removed the low SNR data and chose fundamental mode Rayleigh-wave phases. After these procedures, long-period vertical surface-wave data from 37 earthquakes that satisfied conditions (2), (3), and (4) were selected (Fig. 2). Finally, we obtained 350 high-quality phase velocity dispersion curves of fundamental mode Rayleigh waves from all of the paths.

3 Method

In this work we first measured the surface-wave dispersion by two-station method, and built the phase velocity maps. Then, based on the dispersion curves, we further built the shear-wave velocity maps for the studied area by genetic algorithm, the methodologies are simply described in the follow sections:

3.1 Measurement of the dispersion curves

Picking Rayleigh-wave dispersion curves from the waveform data is critically important in surface-wave tomography. We measured the dispersion curves using the two-

Fig. 2 Epicenter distribution of the earthquakes used in this study. The *star* denotes the center of the Shanxi Digital Seismic Network, and *black dots* denote locations of earthquake epicenters

station method, which needs a large number of earthquakes with uniform distribution. Figure 1 shows the distribution of stations in the Shanxi region. We found that the azimuthal distribution is quite uniform. The narrow-band filtering cross-correlation method (Yao et al. 2004; Ditmar and Yanovskaya 1987) is used in this paper; the detailed description of the phase dispersion picking method can be found in Song et al. (2013).

Figure 3 shows an example of measuring the phase velocity with narrow-band filtering cross-correlation method. Figure 3a shows high-quality vertical-component waveforms recorded by the XAX and YUY stations from the Sumatra earthquake that occurred on August 22, 2010. The waveforms have high signal-to-noise ratios, and contain abundant surface-wave frequency components. By tracking the maximum correlation coefficient value, we can identify the Rayleigh speed for the corresponding period by FTAN method, the maximum of the correlation coefficient can be easily tracked from 5 to 75 s periods (Fig. 3b). In general, the phase velocity errors are small and increase with period, for long-period range (e.g., from 65 to 75 s) the errors are relatively larger.

In this work, we obtained 350 phase velocity paths for all of the selected station pairs through FTAN (Fig. 4a), these paths have covered Shanxi region evenly at period range between 11 and 75 s. However, the paths are relatively poorer on the two sides of the rift region because the station spacing in the peripheral region is relatively sparse. At most periods, the numbers of the paths are near or greater than 300, except at some short periods ($\leq$10 s, shown in Fig. 4b).

3.2 Inversion for the phase velocity maps

Using the methods of Ditmar and Yanovskaya (1987) and Yanovskaya and Ditmar (1990), we obtained phase velocity dispersion maps at 37 central periods from 5 to 75 s. Because the model basis function uses the integral form of the group arrival times, it does not require initial

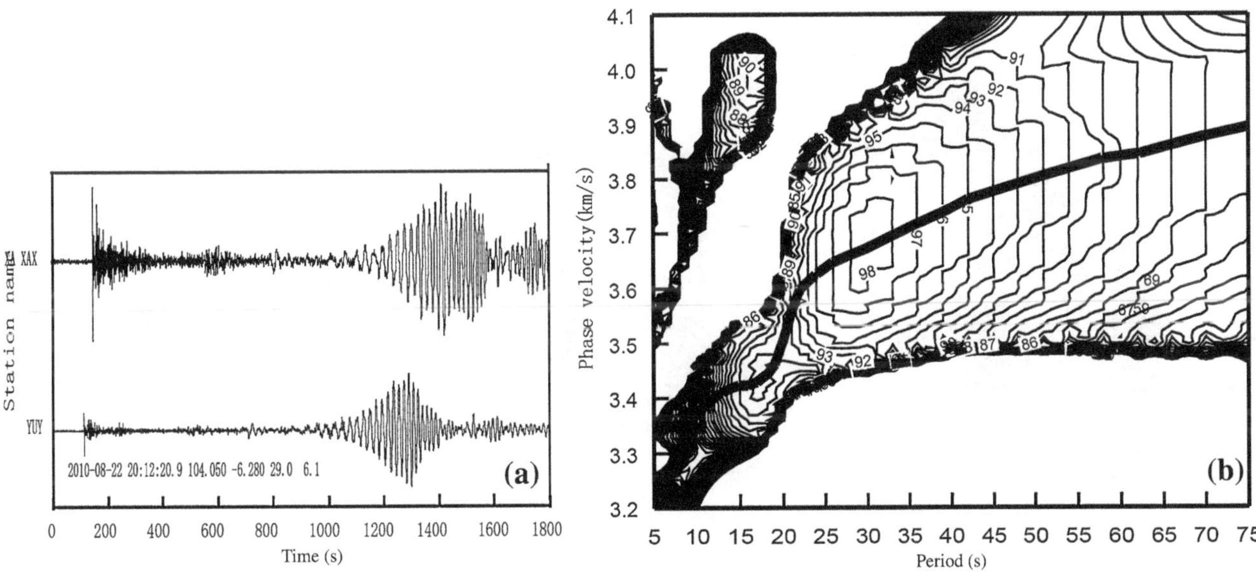

Fig. 3 An example of phase velocity on the same path between two stations: **a** recorded waveforms and **b** phase velocity versus period. The *bold line* outlines the trajectory of the maximum value of cross correlation, it allows us to identify the phase velocity at the corresponding period

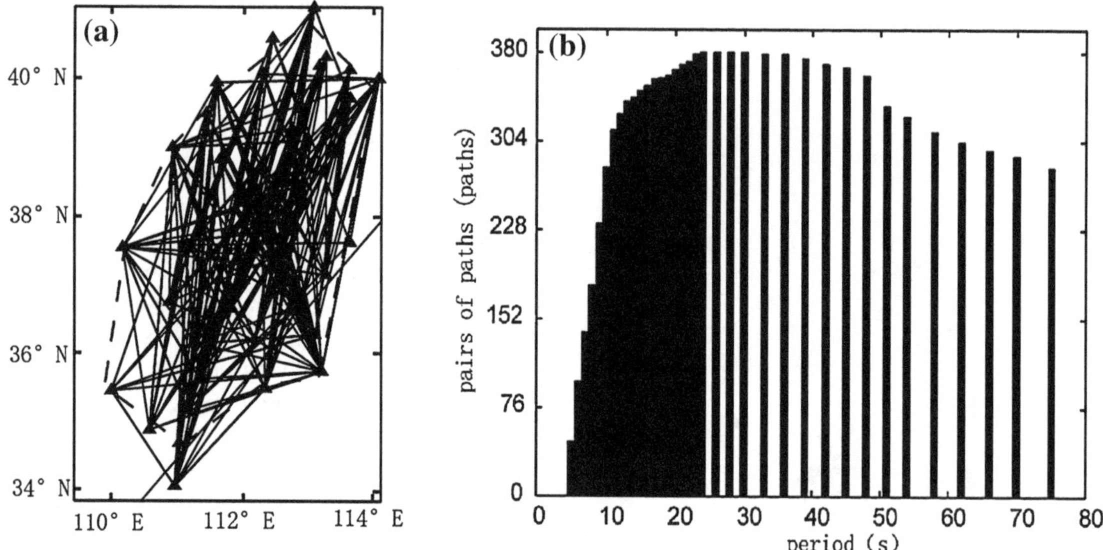

Fig. 4 **a** Distribution of the paths for surface waves, **b** the measured number of raypaths at different periods

parameters or constraint conditions. The detailed method can be found in the literature (Xu et al. 1997). The fundamental concept is: if we input a phase velocity corresponding to the period T_n in the mixed L paths, we can obtain the phase velocity $[C(\phi, \lambda)]$ and the horizontal distribution of the resolution $[R(\phi, \lambda)]$ at this period. Here, ϕ is latitude, and λ is longitude at each node. The phase velocity contour outlines the lateral variation of the phase velocity distribution at a certain depth. Based on this method, we obtained phase velocity dispersion maps at 37 periods, ranging from 5 to 75 s with grid size of $0.2° \times 0.2°$. Figure 5 shows three path-density grade-resolution tests at periods 10, 33, and 62 s. The path numbers are 288, 346, and 276, respectively. The resolution ranges from 40 to 50 km and reaches 70 km near the border of the study region.

3.3 Inversion for the SV structure

In this work, the genetic algorithm performed the inversion (Wu et al. 1997). Since the genetic algorithm is a kind of method of non-linear inversion, it requires a large amount of computation to find the suitable inversion result. In order to minimize the search space, improve the inversion efficiency, and diminish the non-uniqueness of the inversion, we constructed the reference model using a two-step

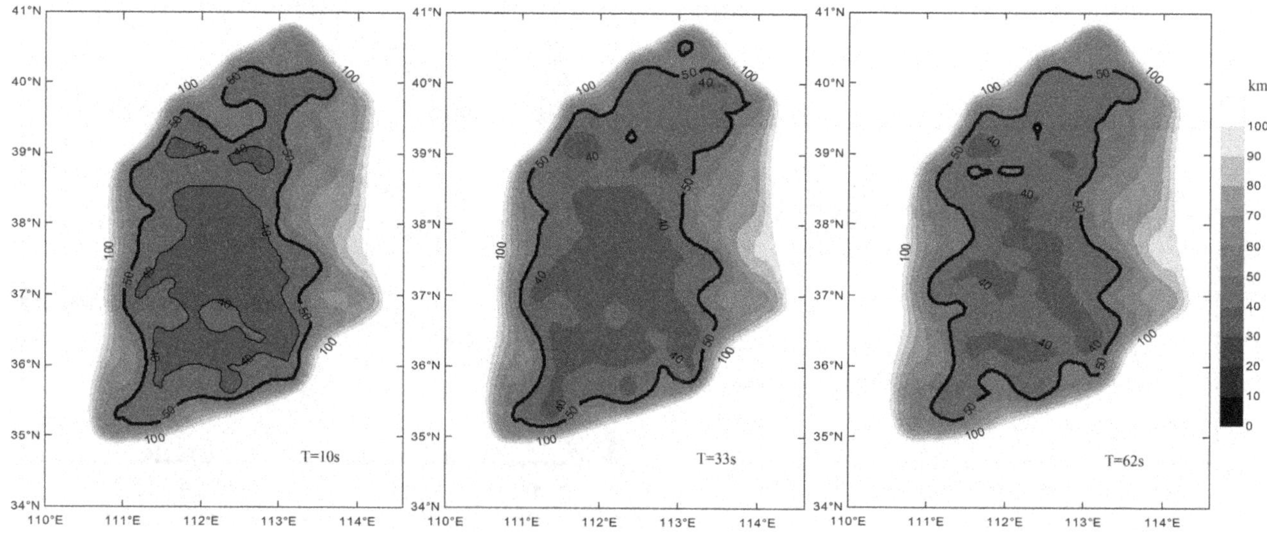

Fig. 5 Maps of phase velocity resolution in the studied area. The *heavy lines* are the 50 km marginal distribution resulting from the inversion of surface-wave phase velocities

procedure. In the first step, at all points we collected DSS results (Zhu et al. 1994, 1999; Liu et al. 2000; Zhao et al. 1999, 2006; Zhang et al. 1997), referred to the previous work on the Moho and lithosphere (Xing et al. 2002), the AK135 model (http://rses.anu.edu.au/seismology/ak135/ak135f.html) and the Moho thickness (Xing et al. 2002) and P/S velocity ratio determined by H–K stacking on teleseismic receiver functions (Tang et al. 2010). Considering that the longest period in the phase velocity dispersion curve is 75 s, the depth in the inversion was set from the Earth's surface to a depth of 200 km. In the second step, smoothing constraints were applied to the model parameterization, each layer thickness was fixed, and the shear-wave velocities were inverted in each horizontal layer. The smoothing coefficient was chosen based on the thickness of the layer. If the layer thickness is less than 3 km, the smoothing coefficient should be set between 0.2 and 0.4, and when the layer thickness is larger than 5 km, the smoothing coefficient is usually set in the range between 0.5 and 0.7. In shallow parts of the crust, the smoothing coefficient is about 0.6, and near the Moho discontinuity, the coefficient is set as 1.0.

Using these constraints, we first invert the shear-wave velocities on the grids which are close to the DSS profiles (the location shown in Fig. 1), and then we use these grid results as initial constraints to further invert the shear-wave velocities on the remaining grids. Overall, we have obtained the shear-wave velocities at each node with grid size of 0.4° × 0.4° (165 grids) from 0 to 200 km, then, by assembling all the 1D v_s models for each grid point, we form the 3D model using the kriging gridding method (Jorge 2007). To clarify, because only Rayleigh waves are used in the inversion, the data are mainly sensitivity to S_V velocity (v_{SV}). Figure 6 shows an example of the v_{SV} inversion from phase velocity dispersion curves at locations (a) and (b) marked in Fig. 1. With the picked out dispersion curve with uncertainty (shown in the right column), the shear-wave velocity can be inverted within the upper and lower bounds.

4 Results and discussion

4.1 Phase velocity maps

Based on the horizontal distribution maps of phase velocities at six representative periods (Fig. 7), we analyzed the lateral heterogeneity of the crustal and upper-mantle velocity structures in the Shanxi region. Considering the shear velocity sensitivity kernels of Rayleigh-wave phase velocities (He et al. 2009; Zheng et al. 2010), the phase velocity maps can be used to analyze the shear-wave structure for different depths.

The distribution map of phase velocity at period 10 s is as shown in Fig. 7a. It mainly deciphers the features of velocity structure of the shallow crust (about 10 km). The distribution of low-velocity and high-velocity zones is closely related to the thickness of sedimentary layers and regional geological structures (He et al. 2004). The bedrock mountain areas and transverse uplift (Lingshi uplift) along the two sides of the Shanxi rift zone are high-velocity areas, while the basins in the central part of the Shanxi rift and the shallow depression in the two sides of the mountain area are low-velocity areas. These low-velocity areas include: Datong, Xinding, Taiyuan, Linfen basins, and Yuxian basin in the Taihang mountain uplift. Shanxi geological data (Xing and Ye 1991) and the results of DSS (Zhu et al. 1994, 1999; Liu et al. 2000; Zhao et al. 1999, 2006; Zhang et al. 1997) indicated that the thickness of sedimentary layers in these areas is between 1.4 and 5.1 km. The thicknesses of the Taiyuan basin and Datong basin are between 3.5 and 3.9 km, and the thickest sediment is located in the Yuncheng basin. The central areas of the subsided fault basins (e.g., Datong, Daixian, Qingxu, Hongtong, and Wuxiang) (Xu et al. 2007) show low-phase velocity anomalies that may be related to the thickness of sedimentary cover in the basins. Magnetotelluric sounding profiles from Yanggao in Datong to Rongcheng in the Shandong province show that there is a low-resistivity belt at the surface near Yanggao in Datong, suggesting that there is a Cenozoic sedimentary cover in this region (Zhao et al. 1997). In general, the Rayleigh-wave phase velocity map at 10 s clearly outlines the lateral variations of seismic-wave speed in the upper crust here, whereas depression basins display low-velocity anomalies, and uplifted mountains or bedrock outcrops show high-velocity anomalies.

The distribution map of phase velocities at period 15 s is as shown in Fig. 7b. This period range expresses the features of velocity structure in a range from the ground surface to about 20 km depth where the speed is still slightly affected by the presence of sedimentary. The low-velocity anomalies shown in the 10-s map also exist in the 15-s map, but their areas become smaller. This is due to an improvement in speed with increasing depth. In the northern Daixian and Datong region, there is an obvious low-velocity anomaly distributed in the middle-upper crust. The results of DSS and crustal electrical structure in the Datong and Yingxian–Fuping areas show low-speed anomalies (crustal high-conductive strata) near Yingxian at a depth range of 17–22 km (Zhang et al. 1997; Zhao et al. 1997; Fang et al. 1995; Liu et al. 1991), which is in good agreement with the phase velocity map pattern at 15 s in these areas.

The distribution map of phase velocities at periods 20 and 26 s are as shown in Fig. 7c, d. The influenced depth

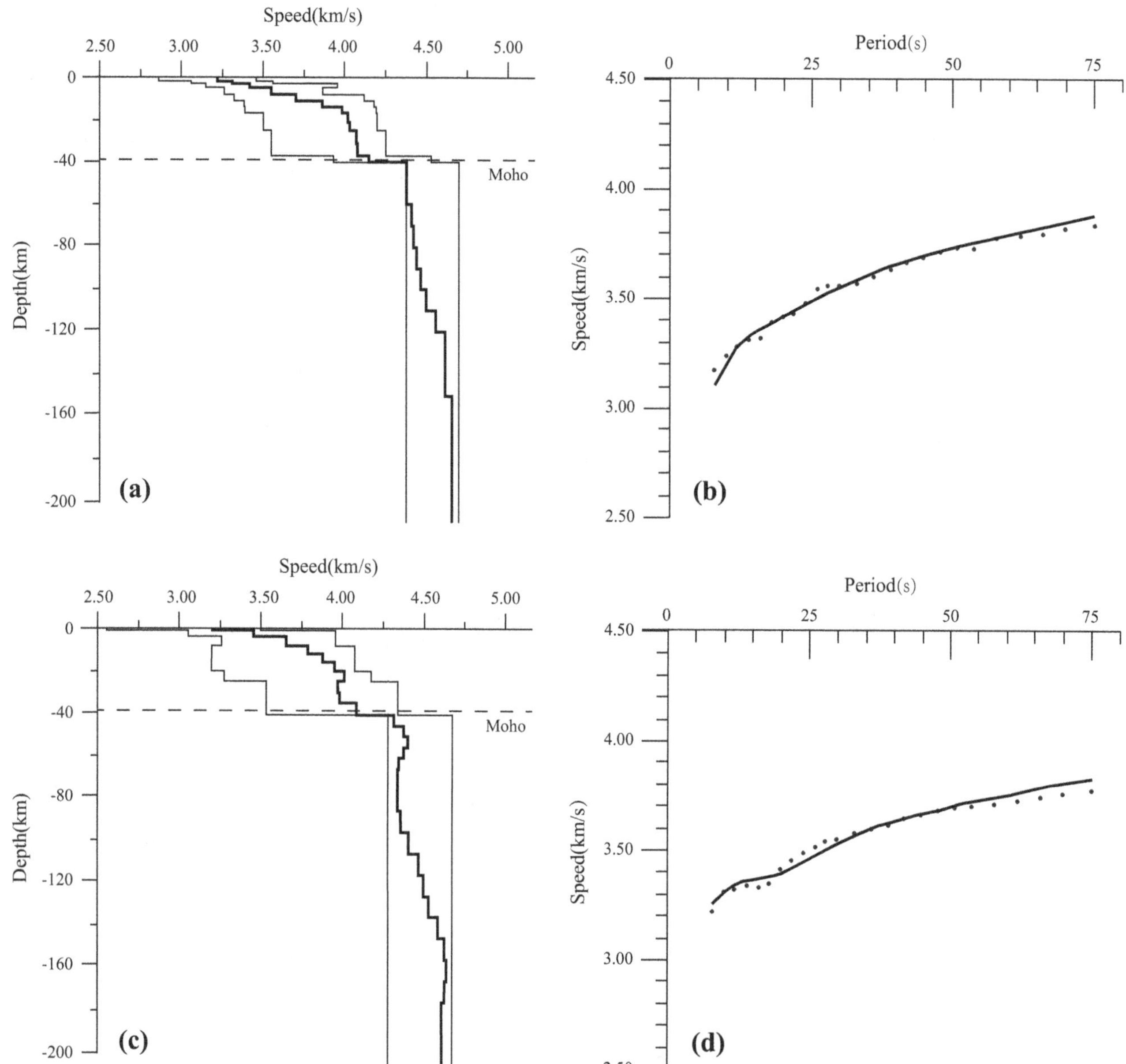

Fig. 6 Examples of the v_{sv} inversion from phase velocity dispersion curves at locations (**a**) and (**b**) marked in Fig. 1 as *stars*, respectively. **a** and **c** The shear-wave velocity profile along the depth. The *light gray curves* represent the upper and lower bounds of the shear-wave velocity in the studied area, the *black thick curve* is the final inverted shear velocity profile by the genetic algorithm inversion. **b** and **d** The observed and computed Rayleigh-wave phase velocity curve at locations (**a**) and (**b**), the *dots* are observed as dispersion curve, and the *line* is fitted dispersion curve

range is about 30–40 km, this depth range is also close to the average depth of the Moho discontinuity in Shanxi. These maps reflect mainly the variation features of the upper-mantle and lowermost crust. Because the shear-wave velocity on the top of the upper mantle is usually about 4.5 km·s^{-1}, which is much higher than that of the lowermost crust (about 3.9 km·s^{-1}), thus the variation of Moho depth may have a large effect on the phase velocity (Tang et al. 2010). In Fig. 7c, d, phase velocities to the south of 38°N are higher than those to the north. Considering the effect of the Moho variation, the high-velocity areas may be correspondent to shallower Moho depths, while the low-velocity area may reflect deeper Moho discontinuities. Figure 7e, f, are the phase velocity maps of Rayleigh waves at 36 and 54 s, which show the velocity structures at the top of the upper mantle. The phase velocity to the south of 38°N is higher than to the north, which is consistent with the surface-wave tomography results of Li et al. (2012). The scale and the range of low-velocity areas vary with period; the low-velocity area at 54 s is smaller than that at

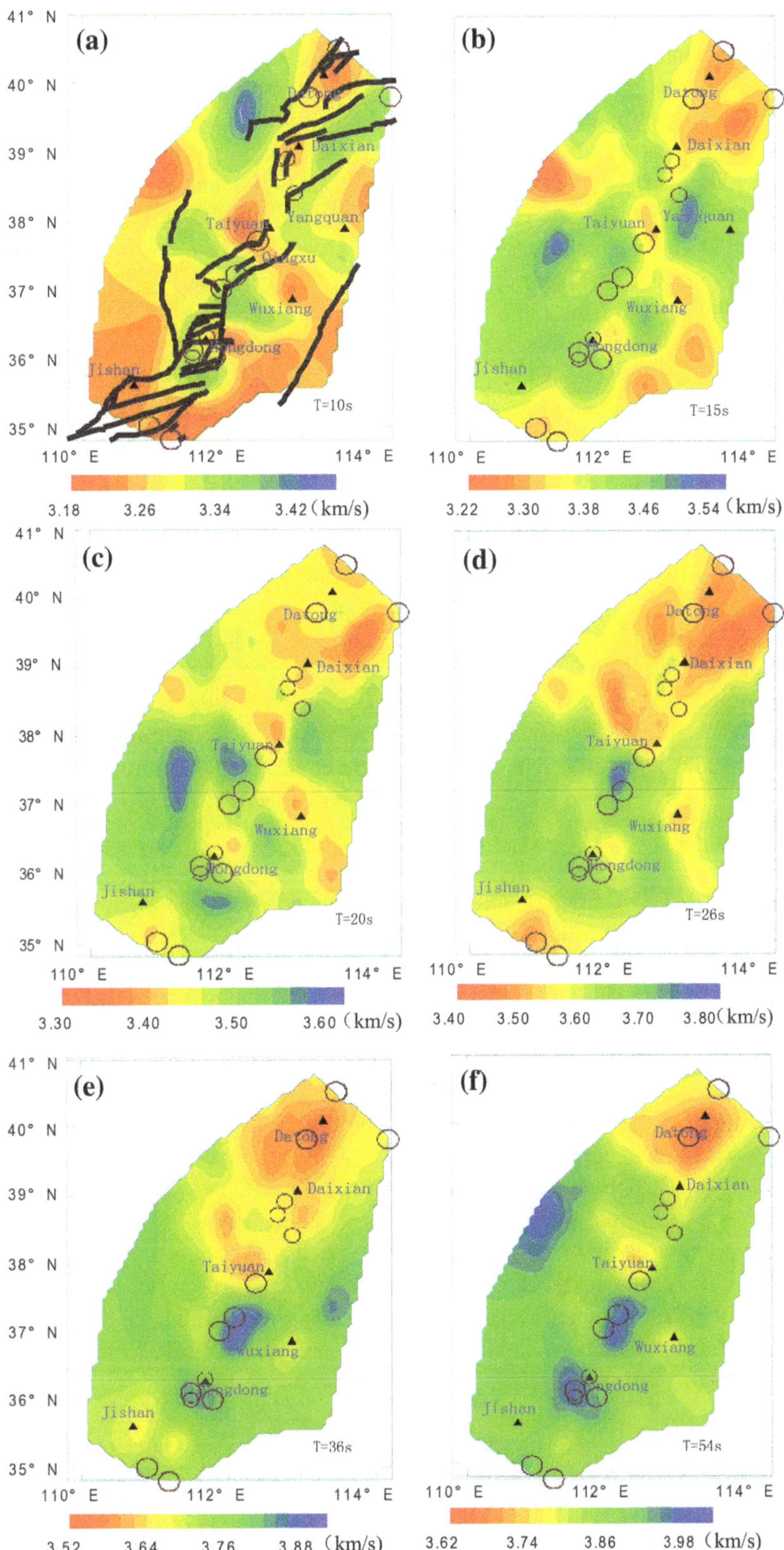

Fig. 7 Phase velocity images from the Shanxi region: **a–f** are phase velocity maps at periods of 10, 15, 20, 26, 36, and 54 s, respectively. *Solid black lines* delineate the active faults in (**a**). *Brown circles* mark the locations of historical earthquakes $M_S \geq 6.0$. The *black triangles* are the locations of the cities in the region

36 s and is divided into two areas: one is near Taiyuan, and the other is in the Datong volcanic area.

Near Taiyuan, there is a persistent low-velocity anomaly from 8 to 75 s periods. Anomaly region extends from the Xishan fault of the Taiyuan basin to the west mountain area. There are many small earthquakes at the same region, however, no similar anomalies can be observed from studies of the geologic structure, magnetotelluric structure, or gravity field. Field surveys indicate that there are some hot springs distributed along the Xishan fault in this region (Liang et al. 2000). Since shear-wave velocities are more sensitive to the fluid content than P-wave velocities, fluids may be one cause of the low-velocity anomaly at the short-period interval. Another possible cause could be the existence of upwelling, asthenospheric material beneath the rift. It is needed to obtain the high-resolution velocity structure at greater depths to reveal this speed anomaly.

4.2 v_{SV} structure

The distribution pattern of the shear-wave velocities delineates the crustal and upper-mantle S-wave velocity structures quite well in Fig. 8. The shear-wave velocity varies with areas and depth, and is described in the following sections.

The upper crust model velocity structure depends on the quality of the phase dispersion curves at 5–10 s periods, however, the curve paths were poor in short periods in our work. Thus, we greatly focus on the deeper structures hereinafter.

In the crust, lateral heterogeneities can be observed in the horizontal S-wave velocity structures, and S-wave velocity ranges from 3.78 to 4.38 km s^{-1}. Shear velocity anomalies typically are similar with depth, except beneath Taiyuan, Wuxiang, Hongdong, as well as Jishan near basins. The shear-wave velocities in these areas exhibit significant low anomalies in the upper crust, which are mainly caused by the slow velocity in the sediments. The boundaries of basins cannot be clearly identified from velocity gradient, which is likely due to the horizontal span of the basins being smaller than the spatial resolution of this work. The most significant velocity variation feature in the middle to lower crust (10–40 km) (Fig. 8a–c) is that in general, high velocities zone mainly cover southern areas of Shanxi region, including Taiyuan, Yuncheng and Linfen basins, while low velocities areas concentrate in north areas of Shanxi region. These may be associated with the distribution of the Moho. As we know, the Moho is the discontinuity between the crust and the mantle; if the depth is shallower than the Moho discontinuity, the rocks still belong to crustal materials so that the velocity should be quite different from that in the mantle. Previous studies (Xing et al. 2005; Xing and Ye 1991) have shown that the Moho depth is about 38 km beneath Taiyuan basin and about from 35 to 37 km under the southern region (including the Yuncheng basin and the Linfen basin) and about 40–41 km below the northern region (including the Xinding basin and the Datong basin); at 40 km under Datong to north Taiyuan basin the rocks may still be the crustal materials, thus their velocity is shown in low anomalies. Furthermore, modest low velocities are observed throughout the Datong basin in the crust, maybe also caused by the elevated crust temperatures resulting from relatively thin lithosphere and young magmatism and extension.

The relocation of small to moderate earthquakes (Song et al. 2012) mostly occured in the velocity anomalies gradient transition zone except in the Datong volcanic zone (Fig. 8a–c). This phenomenon is in agreement with the previous studies. For example, studies of traveltime tomography in North China (Wang 2005; Zhu et al. 1990; Yu et al. 2003) show that distribution of the strongest earthquakes also has similar pattern. During the process of tectonic movement, a large amount of strain energy can be accumulated in the relatively weak zones (near anomalies gradient areas), thereby potentially causing earthquakes (He et al. 2009). Therefore, this work may be the key to a better understanding of relative rift motion and patterns of seismicity.

One noteworthy observation is that the latitude of 38°N is one boundary that separates the Shanxi region into two different blocks. The northern block is relatively ‘soft’ with lower velocities, and the southern block is just the opposite. At the same time, there is a NW-trending velocity gradient belt along Lanxian–Qingxu–Wuxiang, which crosses the 38°N boundary in the northern tip of the Taiyuan basin (Figs. 7b, c, 8a, b). The gradient belt can also be found in the Bouguer gravity anomaly map of the Shanxi province (1:500,000 scale; Liu 2008). In addition, Wei et al. (2007) showed that there was a NNW P-wave velocity transitional areas at the depth of 10–20 km in the southwestern portion of the Shanxi province. The relocation of small to moderate earthquakes in the same area revealed one belt of sparse seismicity, suggesting that energy is able to easily accumulate and release suddenly under the structure strength long-term function.

In the uppermost mantle (Fig. 8d), the image revealed the shear-wave velocities ranges from 4.23 to 4.53 km s^{-1} and the speeds are mostly larger than 4.45 km s^{-1}, except in the Datong region, which is probably due to being partially molten in the lower crust.

Velocity anomalies in the upper mantle (Fig. 8e, f) are distinct from those observed in the overlying crust and uppermost mantle. The shear wave increases with depth, on the other hand, the lateral variation gradually decreases with depth. With increasing depth, the spatial scale of the

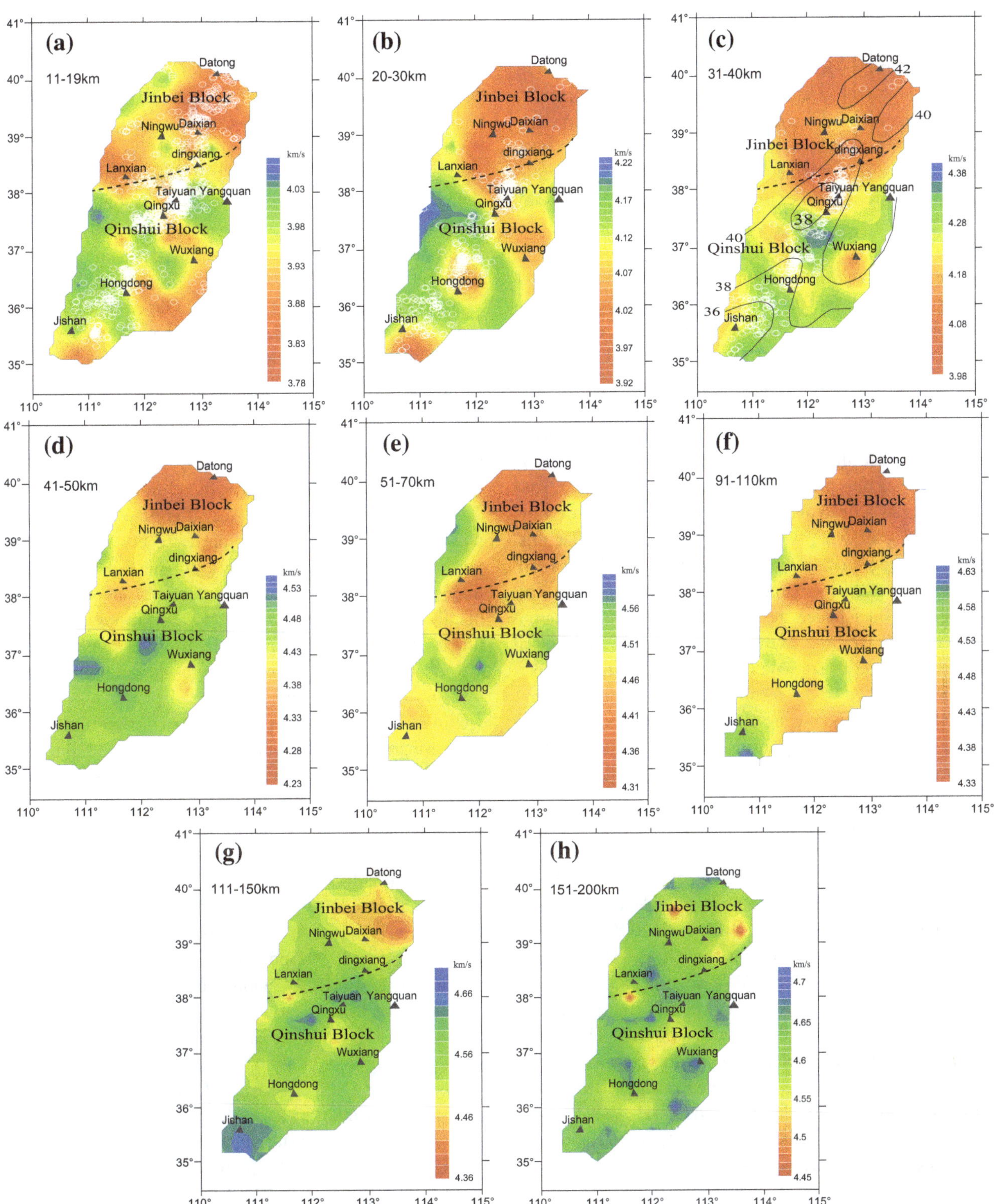

Fig. 8 v_{vs} maps at depths of 11–19 km (**a**), 20–30 km (**b**), 31–km (**c**), 41–50 km (**d**), 51–70 km (**e**), 91–110 km (**f**), 111–150 km (**g**), and 151–200 km (**h**); the *white spots* indicate the small to moderate earthquakes versus depth ($M_L \geq 2.0$). The *dashed lines* denote the boundary between the Jinbei and Qinshui ancient block (the details described are given from Sun et al. 1992). The *solid lines* mark the thickness of Moho discontinuity in (**c**)

low-velocity anomalies in the north gradually shrank and finally concentrated in the Datong volcanic zone. Regional geological surveys (Xu et al. 2005; Zhang 1986) show that there are large-scale Cenozoic mantle-originated magmas, and the low-velocity anomalies are probably due to the elevated temperatures in the lower crust and uppermost mantle resulting from the upwelling of the hot mantle materials. Moreover, the low-velocity anomalies disappeared beneath the Datong volcanic zone at the depth of 151–200 km as shown in Fig. 8h. We propose that the root of the Datong volcano may reach a depth of around 150 km. Previous studies have indicated a large area of Cenozoic mantle-rooted magmatic rock covered in this area (Xu et al. 2005; Sun et al. 1992). Thus the Datong area is an active Cenozoic volcanic zone, which needs a relatively large and deep root magma chamber under the Datong volcanic are, the deep rooted low-velocity anomalies under the Datong revealed by this work is a good evidence for previous studies. As the Datong volcanic zone has a deep and broad root, it may be caused by the upwelling of hot asthenospheric materials under this region. As suggested by previous studies, the rejuvenation of the North China Craton may be caused by the upwelling materials under the central North China Craton (Deng et al. 2007); the Datong volcanic zone could be a result of this kind of tectonic movement. In order to make sure the reliability of the result, we also compare our result with that of Zheng et al. (2011), and both results exhibit deep root low-velocity anomalies under the Datong volcanic zone.

4.3 The maps along vertical profiles

Theoretical study indicates that the energy of Rayleigh waves is mainly concentrated towards the depth within the range of about half a wavelength (He et al. 2009). Thus, we can consider approximately that the phase velocity at a certain period is the average of S-wave velocities over a depth range of half a wavelength (in fact, the phase velocity is a little less than the S-wave velocity). Therefore, the variation of phase velocity when the period changes from short to long can be considered resulting from variation of materials from shallow to deep (He et al. 2009). Similarly, we can also find lateral material variation by comparing the phase velocities of different areas. In this work, we constructed the shear-wave velocity map for each grid in our study areas, and picked out three shear-wave vertical profiles (*A–A′*, *B–B′*, *C–C′*; locations shown in Fig. 1). Furthermore, in order to compare the velocity distribution between the surface-wave and shear-wave velocities, we also built phase velocity profiles along the same locations. The comparison is shown in Fig. 9.

The *A–A′* profile traverses the Datong basin, Xinding basin, and Wutaishan uplift to the east, which belongs to the Yingxian–Fuping segment of Zibo–Yingxian DSS profile (Zhao et al. 2006). The phase- and shear-wave velocity variations are similar to the velocity structure revealed by the DSS profile. (Figure 9a, b, profile *A–A′*), and the contours of 3.8 km s^{-1} of phase velocity and 4.45 km s^{-1} of shear-wave velocity become shallower from west to east. This was consistent with the previous results about Moho discontinuity at the same locations. (Zhao et al. 2006).

The profile *B–B′* that extends along longitude 113°E crosses the Qinshui block, Zhongtiaoshan block, and Jinbei block from south to north; it can express the velocity distribution features in N–S direction.

In the vicinity of latitude 38°N, there was a transition zone where the phase velocity changed rapidly. The phase velocity from 18 to 25 s varies gently on the south side of the transition zone, while sharply below the Xinding and Taiyuan basins on the north side. It is noteworthy that the phase velocity in the south is much higher than that in the north between 36 and 75 s periods (these periods reflect mainly the features of velocity structure of the top of upper mantle) (Fig. 9c). Sun et al. (1992) proposed that the Shanxi block could be divided into two different tectonic blocks at a boundary located at 38°N. In the northern part, the Jinbei block is 'soft' for asthenospheric material upwelling to the depths of ~80 km; the Qinshui block in the southern portion is relatively 'hard', where the depth to the lithosphere is between 120 and 150 km. The phase velocities shown in Fig. 9c are very similar to the geological observations.

In Fig. 9d, the 38°N latitude is also the boundary between high and low velocities at depths between 40 and 100 km, and to the south of the 38°N latitude, shear-wave velocity is higher than to the north. Moreover, in Jinbei block between 39°N and 40°N, we can clearly observe prominent low-velocity anomalies, which provided the evidence that there is an upwelling asthenosphere channel in this region and that the Shanxi rift may be generated by this upwelling flow.

The profile *C–C′* crosses the Lvliangshan uplift in the west and the Linfen basin in the central rift and Taihangshan uplift in the east, which belong to the Daning–Jincheng segment of DDS profile from Yinchuan to Zhengzhou (Zhu et al. 1994). The phase velocity structure revealed by this profile is very similar to the result of DDS. In Fig. 9e, due to the upwelling of materials from great depths below the Linfen Basin, the Moho interface and interfaces within the crust are uplifted, and the high-conductivity layer in the upper mantle shows a strong escalating trend (Zhu et al. 1994, Fig. 9f).

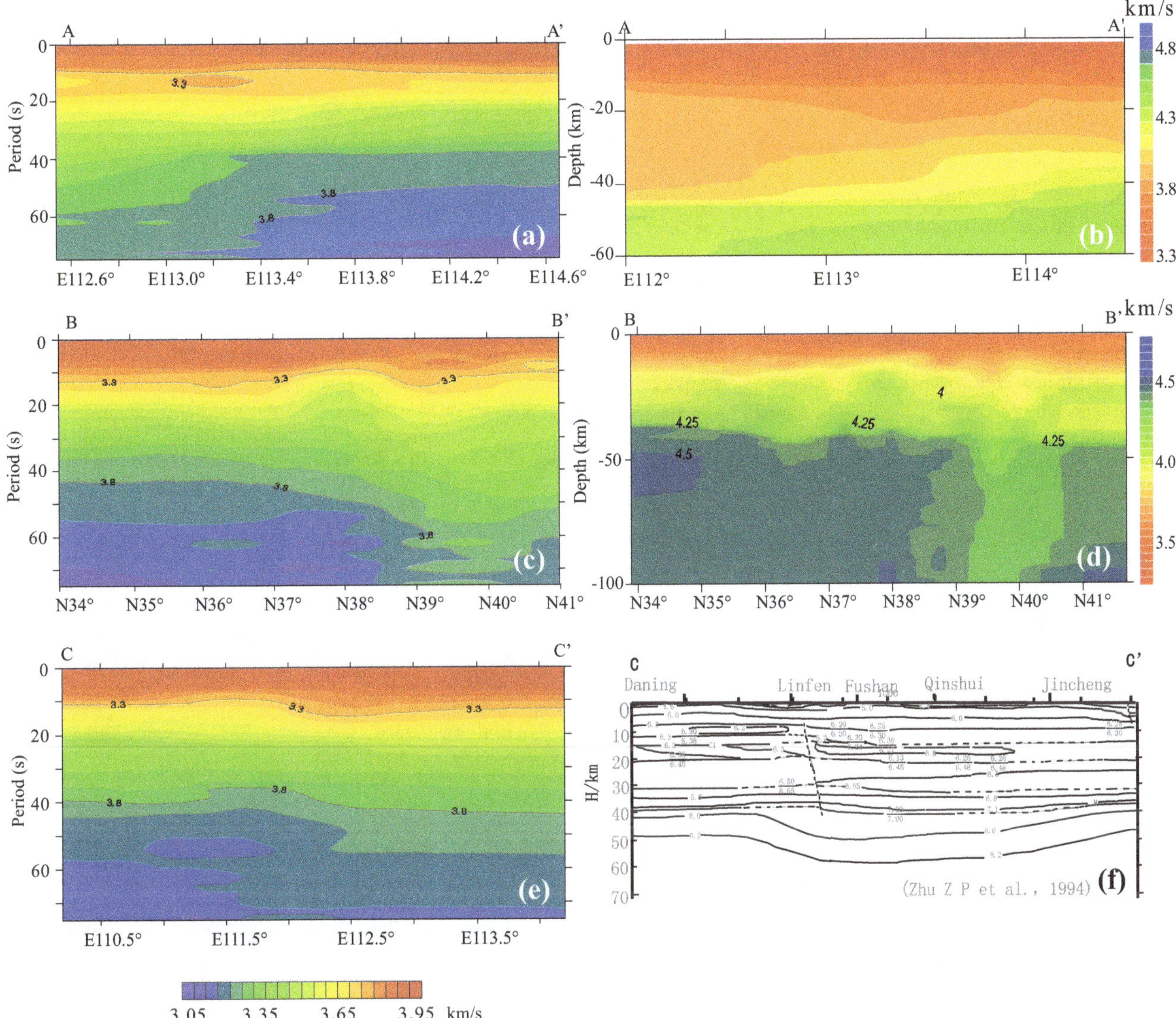

Fig. 9 Vertical cross sections of phase velocity and v_{sv} along the A–A', B–B', and C–C' profiles identified in Fig. 1. **a**, **c**, and **e** represent the phase speed versus depth. **b** and **d** display the shear-wave speed at corresponding depths. **f** The P-wave velocity section results taken from the DDS experiment by Zhu et al. (1994) at the same location along the C–C' profile

5 Conclusions

Based on teleseismic waveforms recorded by the Shanxi seismic network and networks of surrounding provinces, and removing instrument responses and applying the two-station (TS) method, we have obtained phase velocity distribution images from the inversion of phase velocity dispersion of Rayleigh waves. The horizontal resolution is between 40 and 50 km. Furthermore, the 1D S-wave velocity was inverted from the Earth's surface to 200 km beneath 165 nodes in the Shanxi region using a genetic algorithm, then, by assembling all the 1D *vs* models for each grid point, we form the 3D model using kriging gridding method (Jorge 2007). The main characteristics were as follows:

In the crust, the most significant difference between velocity anomalies in the middle/lower crusts (10–40 km (Moho)) (Fig. 8a–c) is that there are high velocities from Taiyuan basin to the Yuncheng basin in the southern parts of Shanxi region, while there are low speeds from the north of Taiyuan basin to the Datong basin. These may be associated with the distribution of the Moho discontinuity. Especially modest low velocities are observed throughout the Datong basin in the crust, probably due to the elevated crust temperatures resulting from relatively thin lithosphere and young magmatism and extension.

In the uppermost mantle, the image revealed that the shear-wave velocities are mostly larger than 4.45 $km{\cdot}s^{-1}$, except in the Datong region, which is probably due to the partial melting in the lower crust. Moreover, in the upper mantle (Fig. 8e–h), shear-wave velocity increased with depth and the lateral heterogeneity gradually decreased. With increasing depth, the spatial scale of the low-velocity anomalies in the north gradually shrank, and the low-velocity anomalies disappeared beneath the Datong volcanic zone at the depth of 151–200 km. We proposed that the root of the Datong volcano may be a depth of around 150 km.

Except in the Datong volcanic zone, most of the relocated events on the north of the Taiyuan basin were located in the transition zone between high and low velocities at depths between 11 and 19 km, while the ones on the south were closer to the velocity gradient belt at depths between 11 and 30 km (Fig. 8a–b). The findings are in agreement with the results of some researchers who showed that the earthquakes mainly concentrated in the velocity anomalies gradient transition zone, therefore the seismicity is mainly controlled by the upper crustal structures.

The latitude of 38°N is one boundary that separates the Shanxi region into two different blocks. The northern block is relatively 'soft' with lower velocities, and the southern block is just the opposite. Same pattern can be observed in the depth from 40 to 100 km that 38°N is the boundary of high velocity and low velocity and the shear-wave velocity increases from north to south. Moreover, in Jinbei block between 39°N and 40°N, we can clearly observe prominent low-velocity anomalies, which provided the evidence that there is an upwelling asthenosphere channel in this region.

Although our study has characterized the crust–mantle velocity structure of the Shanxi region, better seismic velocity structure can be expected in the future. In our work, we already collected seismic data in 45 stations, which are more than any previous studies in this region, however, there are 56 stations running in the Shanxi rift zone, and in recent days, additional 12 stations have been deployed, so that the ray paths in the studied area are much more evenly distributed. So, in the near future, we will collect more data and apply new methods other than only two-station methods, such as ambient noise tomography (zheng et al. 2011) and two-plane wave tomography (Li et al. 2012) to build better resolution crustal and uppermost mantle shear-wave velocity structure model for the Shanxi rift zone, so that we can understand more clearly about the tectonic settings in this region.

Acknowledgments The authors thank three anonymous reviewers and editors for constructive critical reviews that improved this manuscript. This work was financially supported by Open Grant from State key Laboratory of Geodesy and Earth's Dynamics (Grant No. SKLGED2014-4-4-E), and Office of Science and Technology in Shanxi province based on research Projects (2012011029), and Scientific and Technological Research Projects in Shanxi province (20100311129-2, 20090311084), and the China Earthquake Administration spark Project (XH15007).

References

Deng JF, Su SG, Niu YL, Liu C, Zhao GC, Zhao XG, Zhou S, Wu ZX (2007) A possible model for the lithospheric thinning of North China Craton: evidence from the Yanshanian (Jura-Cretaceous) magmatism and tectonism. Lithos 96:22–35

Ding ZF, He ZQ, Sun WG (2008) 3-D crust and upper mantle velocity structure in eastern Tibetan plateau and it surrounding areas. Chin J Geophys 51(5):1431–1443 (in Chinese with English Abstract)

Ditmar PG, Yanovskaya TB (1987) Generalization of Backus–Gilbert method for estimation lateral variations of surface wave velocities. Phys Solid Earth Izvestia Acad Sci USSR 23(61):470–477

Fang LH (2010) Rayleigh surface wave ambient noise tomographic imaging of the North China. Natl Earthq Dyn 4:40–42 (in Chinese with English Abstract)

Fang SM, Yu QF, Lou H, Xing JS (1995) Study of gravity anomaly in crustal low velocity zone in the WUTAI region. Seismol Geol 17(2):109–113 (in Chinese with English Abstract)

Forsyth DW (1975) The early structural evolution and anisotropy of the oceanic upper marital. Geophys J R Astron Soc 43:103–162

Ge C, Zheng Y, Xiong X (2011) Study of crustal thickness and Poisson ratio of the North China Craton. Chin J Geophys 54(10):2538–2548 (in Chinese with English Abstract)

He ZQ, Ding ZF, Ye TL, Sun WG, Zhang NL (2002) Group velocity distribution of Rayleigh waves and crustal and upper mantle velocity structure of the Chinese mainland and its vicinity. Acta Seismol Sin 24(3):252–259 (in Chinese with English Abstract)

He ZQ, Su W, Ye TL (2004) Seismic tomography of Yunnan region using short-period surface wave phase velocity. Acta Seismol Sin 26(6):583–590 (in Chinese with English Abstract)

He ZQ, Yie DL, Ding ZF (2009) Surface wave tomography for the phase velocity in the northeastern part of North China. Chin J Geophys 52(5):1233–1242 (in Chinese with English Abstract)

Huang ZX, Su W, Peng YJ, Zheng YJ, Li HY (2003) Rayleigh wave tomography of China and adjacent regions. J Geophys Res 108(B2):2073 (in Chinese with English Abstract)

Jia SX, Zhang XK (2008) Study on the crust phases of deep seismic sounding experiments and fine crust structures in the northeast margin of Tibetan plateau. Chin J Geophys 51(5):1431–1443 (in Chinese with English Abstract)

Jobert N, Joumet B, Him A, Zhong SK (1985) Deep structure of southern Tibet inferred from the dispersion of Rayleigh waves through a long-period seismic network. Nature 313:386–388

Jorge KY (2007) On unbiased backtransform of lognormal kriging estimates. Computat Geosci 11(3):219–234

Kanamori H (1970) Velocity and Q of mantle waves. Phys Earth Planet Inter 2:259–275

Li SB (1981) Earthquake Research in China. Seismological Press, Beijing, pp 120–121 (in Chinese with English Abstract)

Li A, Burke K (2006) Upper mantle structure of southern Africa from Rayleigh wave tomography. J Geophys Res 111:10303–10318

Li P, Zhou SY, Chen YS, Feng YG, Jiang MM, Tang YC (2010) 3D velocity structure in Shanxi graben and Ordos from two plane waves method. CT Theory Appl 19(3):47–60 (in Chinese with English Abstract) (54(5):1233–1242)

Li D, Zhou SY, Chen YS, Feng YG, Li P (2012) 3-D lithospheric structure of upper mantle beneath Ordos region from Rayleigh wave tomography. Chin J Geophys 55(5):1613–1623 (in Chinese with English Abstract)

Liang WB, Li HJ, Yang CY (2000) Edge geomorphology and neotectonic movement in Taiyuan basin and their meanings. J Shanxi Univ (Nat Sci Ed) 23(2):178–181 (in Chinese with English Abstract)

Liu JL (2008) Recognite division tectonic by using the new technology of gravity and magnetic potential field in Shanxi fault basin. Changan University, Xian

Liu CQ, Qia SX, Du GH (1991) Result of seismic refraction sounding along the transect from Xiangshui Jiangsu, to Mondula, Nei Monggol. Seismol Geol 13(3):193–203 (in Chinese with English Abstract)

Liu BF, Zhang XK, Zhang CK, Song SY, Zhou XS (2000) Geological interpretation of S wave data along the Wen'an-Yuxian-Cahay-ouzhongqi profile in north China. Seismol Geol 22(1):81–88 (in Chinese with English Abstract)

Prindle K, Tanimoto T (2006) Teleseismic surface wave study for S-wave velocity structure under an array: Southern California. Geophys J Int 166:601–621

Song MQ, Zheng Y, Ge C, Li B (2012) Relocation of small to moderate earthquakes in Shanxi Province and its relation to the seismogenic structures. Chin J Geophys 55(2):513–525 (in Chinese with English Abstract)

Song MQ, He ZQ, Zheng Y, Lv F, Liu C, Liang XJ, Su Y, Li L (2013) Rayleigh-wave phase velocity distributions in Shanxi region. Prog Geophys 28(4):1836–1848 (in Chinese with English Abstract)

Sun JY, Xing JS, Ye ZG, Chen CW (1992) The research of intraplate tectonics and deep process in North China. Geol Sci Technol Inf 11(1):1–12 (in Chinese with English Abstract)

Tang YC, Feng YG, Chen YS, Zhou SY, Ning JY, Wei SQ, Li P, Yu CQ, Fan WY, Wang HY (2010) Receiver function analysis at Shanxi Rift. Chin J Geophys 53(9):2102–2109 (in Chinese with English Abstract)

Tang YC, Chen YS, Yang YJ, Ding ZF, Liu RF, Feng YG, Li P, Yu CQ, Wei SQ, Fan WY, Wang HY, Zhou SY, Ning JY (2011) Ambient noise tomography in north China craton. Chin J Geophys 54(8):2011–2022 (in Chinese with English Abstract)

The active fault system in Erdos block and its margins subject group (1988) The active fault system in Erdos block and its margins. Seismological Press, Beijing

Toksoz MN, Anderson DL (1966) Phase velocity of long-period surface waves and structure of the upper Mantle. J Geophys Res 71:1649–1658

Wang ZS (2005) Tomographic imaging of the 3-D Crust and upper mantle velocity structures beneath North China and its vicinity. Institute of Geophysics, China Earthquake Administration, Beijing

Wei WB, Jin S, Ye GF, Deng M, Jing JE (2006) MT sounding and lithosphere thickness in North China. Geol China 33(4):762–772 (in Chinese with English Abstract)

Wei WB, Ye GF, Jin S, Deng M, Jing JE (2007) Three dimensional p-wave velocity structure of the crust of North China. Earth Sci J China Univ Geosci 32(4):441–452 (in Chinese with English Abstract)

Wu JP, Ming YH, Tang Y (1997) The application in uppermost mantle velocity structure study using the genetic algorithm. Seismol Geomagn Obs Res 18(2):11–25 (in Chinese with English Abstract)

Wu Y, Ding ZF, Zhu LP (2011) Crustal structure of the North China Craton from teleseismic receiver function by the Common Conversion Point stacking method. Chin J Geophys 54(10):2528–2537 (in Chinese with English Abstract)

Xing JS, Ye ZG (1991) Preliminary discussions on interpolate structural features and their evolution in Shanxi Province. Shanxi Geol 6(1):3–15 (in Chinese with English Abstract)

Xing JS, Liu JH, Zhao JQ (2002) Deep-seated Tectonics in the North China intraplate. Earthq Res Shanxi 4:3–12 (in Chinese with English Abstract)

Xing ZY, Zhao B, Tu MY, Xing JS (2005) The information of the Fenwei rift valley. Earth Sci Front 12(2):247–262 (in Chinese with English Abstract)

Xu Y, Tian Y, Chuo YQ, Wang J, Wu L (1997) The P wave velocity structure of cruster in Datong-Yanggao 6.1 seismic area and its surrounding regions. Earthq Res Shanxi 1–2:24–29

Xu GM, Li GP, Wang SE, Chen H, Zhou HS (2000) The 3-D structure of shear waves in the crust and mantle of east continental China inverted by Rayleigh wave data. Chin J Geophys 43(3):366–376 (in Chinese with English Abstract)

Xu YG, Ma JL, Frey FA, Feigenson MD, Liu JF (2005) Role of lithosphere-asthenosphere interaction in the genesis of Quaternary alkali and tholeiitic basalts from Datong, western North China Craton. Chem Geol 224(4):247–271

Xu GM, Yao HJ, Zhu LB, Shen YS (2007) Shear wave velocity structure of the crust and upper mantle in western China and its adjacent area. Chin J Geophys 50(1):193–208 (in Chinese with English Abstract)

Yang WR, Sun JY, Ji KC, Xing JS (1995) The contrast of continental rift—the analysis between the Fen_Wei rift and Baikal rift. China University of Geoscience Press, Wuhan

Yanovskaya TB, Ditmar PG (1990) Smoothness criteria in surface wave tomography. Geophys J Int 102(1):63–72

Yanovskaya TB, Kozhevnikov VM (2003) 3D S-wave velocity pattern in the upper mantle beneath the continent of Asia from Rayleigh wave data. Phys Earth Planet Inter 138:263–278

Yao HJ, Xu GM, Xiao X, Zhu LB (2004) A quick tracing method based determination of dual stations surface wave on image analysis technique for the phase velocities dispersion curve of surface wave. Seismol Geomagn Obs Res 25(1):1–8 (in Chinese with English Abstract)

Yi GX, Yao HJ, Zhu JS, van der Hilst RD (2008) Rayleigh-wave phase velocity distribution in China continent and adjacent regions. Chin J Geophys 51(2):402–411 (in Chinese with English Abstract)

Yu XW, Chen YT, Wang PD (2003) Three dimensional P wave velocity structure in Beijing Tianjin–Tangshan area. Acta Seismol Sin 25(1):1–14 (in Chinese with English Abstract)

Zhang YB (1986) Characteristics of petrology of the Datong Basalts. Seismol Geol 8(1):59–67 (in Chinese with English Abstract)

Zhang YS, Lay T (1996) GlobM surface wave phase velocity variations. J Geophys Res 101:8415–8436

Zhang YS, Ma SZ (1997) Global surface wave phase velocity variations and their tectonic implications. Chin J Geophys 40(2):181–192 (in Chinese with English Abstract)

Zhang JS, Zhu ZP, Zhang XK, Zhang CK, Gai YJ, Nie WY (1997) The seismic velocity structure of cruster and upper mantle and deep structure feature in north Shanxi plateau. Seismol Geol 19(2):220–226 (in Chinese with English Abstract)

Zhang PZ, Deng QD, Zhang GM, Ma J, Gan WJ, Min W, Mao FY, Wang Q (2003a) The strong earthquakes mainland and activity blocks. Sci China Earth Sci (Ser D) 33:12–19 (in Chinese with English Abstract)

Zhang XM, Shu PY, Diao GL (2003b) Study on S wave velocity structure under part stations in Shanxi province. Acta Seismol Sin 25(4):341–350 (in Chinese with English Abstract)

Zhang GM, Ma HS, Wang H, Li L (2004) Relations between the active blocks mainland and the strong seismic activity. Sci China Earth Sci (Ser D) 34(7):591–599 (in Chinese with English Abstract)

Zhang FX, Li YH, Wu QJ, Ding ZF (2011) The P wave velocity structure of upper mantle beneath the North China and surrounding regions from FMTT. Chin J Geophys 54:1233–1242 (in Chinese with English Abstract)

Zhao GZ, Liu TS, Jiang Z, Tang J, Xu CF, Zhan Y, Bai DH, Liu GD (1997) Investigation on MT data along Yanggao-Rongcheng profile by two-dimensional inversion. Chin J Geophys 40(1):38–46 (in Chinese with English Abstract)

Zhao JR, Zhang XK, Zhang CK, Ren QF, Zhu ZP, Wu T (1999) The structure features of deep crustal structure in seismic area of Linxian, Henan Province. Earthq Res China 15(3):229–236 (in Chinese with English Abstract)

Zhao JR, Zhang XK, Zhang CK, Zhang JS, Liu BF, Pan SZ (2006) Features of deep crustal structure beneath the Wutai mountain area of Shanxi province. Chin J Geophys 49(1):123–129 (in Chinese with English Abstract)

Zheng Y, Yang YJ, Ritzwoller MH, Zheng XF, Xiong X, Li ZN (2010) Crustal structure of the northeastern Tibetan plateau, the Ordos block and the Sichuan basin from ambient noise tomography. Earthq Sci 23:465–476

Zheng Y, Shen WS, Zhou LQ, Yang YJ, Xie ZJ, Michael H (2011) Crust and upper most mantle beneath the North China Craton, northeastern China, and the Sea of Japan from ambient noise tomography. J Geophys Res 116:B12312. doi:10.1029/2011JB008637

Zhu LP, Zeng RS, Liu FT (1990) Three-dimensional P-wave velocity structure under the Beijing network area. Chin J Geophys 30(3):267–277 (in Chinese with English Abstract)

Zhu ZP, Zhang JS, Zhou XS, Zhu WX (1994) Study on the structure of crust and upper mantle in Linfen earthquake region in Shanxi. Earthq Sci North China 12(1):77–84 (in Chinese with English Abstract)

Zhu ZP, Zhang JS, Zhang CK, Zhao JR, Liu MQ, Tang ZQ, Gai YJ, Ren QF, Nie WY, Yang Q (1999) Study of crust mantle structure in central and southern Shanxi. Acta Seismol Sin 21(1):42–49 (in Chinese with English Abstract)

Zhu JS, Cao JM, Yan ZQ (2007) High-resolution Rayleigh surface wave tomographic imaging of China and adjacent regions and its geodynamic implications. Geol China 34(5):759–767 (in Chinese with English Abstract)

12

Dynamic soil-tunnel interaction in layered half-space for incident P- and SV-waves

Jia Fu · Jianwen Liang · Lin Qin

Abstract The dynamic soil-tunnel interaction is studied by the model of a rigid tunnel embedded in layered half-space, which is simplified as a single soil layer on elastic bedrock to the excitation of P- and SV-waves. The indirect boundary element method is used, combined with the Green's function of distributed loads acting on inclined lines. It is shown that the dynamic characteristics of soil-tunnel interaction in layered half-space are different much from that in homogeneous half-space, and that the mechanism of soil-tunnel interaction is also different much from that of soil-foundation-superstructure interaction. For oblique incidence, the tunnel response for in-plane incident SV-waves is completely different from that for incident SH-waves, while the tunnel response for vertically incident SV-wave is very similar to that of vertically incident SH-wave.

Keywords Underground tunnel · Layered half-space · P-wave and SV-wave · Indirect boundary element method · Soil-tunnel interaction · Site dynamic characteristics

1 Introduction

Dynamic soil-structure interaction (SSI) is an interdisciplinary field involving in the knowledge of soil and structural dynamics, earthquake engineering, and geophysics. Most studies on this problem mainly focus on soil-superstructure interaction, using a model of a rigid foundation with or without a building on it. For example, the classic solutions of a semi-circular rigid foundation with a shear wall on it were obtained by a kind of analytical method (Luco 1969; Trifunac 1972). More recently, the influences of site dynamic characteristics on SSI were studied separately using the same model in elastic layered half-space (Liang et al. 2013a, b) by indirect boundary element method.

The scholars have already obtained the solutions of dynamic responses of underground tunnel by analytical methods (Lee and Trifunac 1979) or numerical methods (Luco and De Barros 1994; De Barros and Luco 1994) for several decades. However, the interaction between soil and underground structure, although as an important part of soil-structure interaction, are rarely studied up to now, only Hatzigeorgiou and Beskos (2010) compared damage evolution between lined tunnel and soil cavity to study the interaction between an underground runnel and the surrounding soil. Parvanova et al. (2014) analyzed the surface displacement and stress distribution along tunnel circumference to study respectively the interaction between one tunnel or twin tunnels and local topography.

In a companion paper (Fu J, Liang J and Qin L (2015). Dynamic soil-tunnel interaction in layered half-space for incident plane SH waves. In review, cited as "(Fu et al. 2015)" in the following for convenience), the influence of site dynamic characteristics on soil-tunnel interaction is already studied by the model of a rigid tunnel embedded in layered half-space to the excitation of SH-waves, and the main milestones and methods on dynamic responses of underground tunnel are reviewed in the introduction. In this paper, the problem is continuously discussed using the same model to the excitation of P- and SV-waves, by indirect boundary element method combined with Green's functions

J. Fu · J. Liang (✉) · L. Qin
Department of Civil Engineering, Tianjin University, Tianjin 300072, China
e-mail: liang@tju.edu.cn

J. Liang
Tianjin Key Laboratory of Civil Engineering Structures & New Materials, Tianjin 300072, China

of distributed loads acting on inclined lines (Wolf 1985). The further study on soil-tunnel interaction to incident surface waves is our work in the future.

2 Methodology

2.1 Model

In Fig. 1, an underground lined tunnel is completely rigid and infinitely long, with outer radius of a, the inner radius of b, the mass of M_0, and mass density of ρ_0, also its embedded depth from ground surface to the center is d. It is bonded tightly to the surrounding soil at interface Γ without slippage. The layered half-space is simplified to a single soil layer with thickness D over bedrock. Both the soil layer and bedrock are elastic, homogeneous, and isotropic medium. The material parameters of the bedrock are characterized by shear-wave velocity β_R, mass density ρ_R, Poisson ratio ν_R, and damping ratio ξ_R, while the material parameters of the soil layer are characterized by shear-wave velocity β_L, mass density ρ_L, Poisson ratio ν_L, and damping ratio ξ_L. The plane P-waves or SV-waves are incident from depth D' with horizontal incident angle θ, circular frequency ω, and unit amplitude.

In order to use indirect boundary element method (IBEM) to solve the problem, the layered half-space should be divided into sub-layers, with the boundary Γ of the tunnel divided into N elements of straight lines meanwhile, and it is better to make all elements the same length in order that IBEM can perform best. Also, as the Green's functions used in this paper are distributed line loads in horizontally layered half-space, the elements should be symmetrical about z-axis.

2.2 Impedance function

In order to apply IBEM, a set of fictitious horizontal loads $q_j e^{i\omega t}$ and vertical loads $r_j e^{i\omega t}$ ($j = 1, 2, \ldots, N$) which compose the fictitious load vector

$$\boldsymbol{P} = [q_1, q_2, \ldots, q_N, r_1, r_2, \ldots, r_N]^{\mathrm{T}} \tag{1}$$

is imposed onto every element as Fig. 2, with time factor $e^{i\omega t}$ is omitted hereafter. The values of these loads are all unknowns which should be determined by boundary condition that the tunnel produces the rigid displacement to the excitation of these loads.

For in-plane excitation, the rigid displacement of the tunnel is $\Delta = [\Delta_x, a\varphi, \Delta_z]^{\mathrm{T}}$, with Δ_x and Δ_z being the horizontal and vertical displacements, and φ being the rotational angle about its center, respectively. So on boundary Γ, the no slippage assumption gives

$$\boldsymbol{U}(x,z) = \begin{bmatrix} 1 & -z/a & 0 \\ 0 & x/a & 1 \end{bmatrix} \begin{bmatrix} \Delta_x \\ a\varphi \\ \Delta_z \end{bmatrix} = \Omega(x,z)\Delta \quad (x, z) \in \Gamma \tag{2}$$

Symbol $\boldsymbol{U}(x, z)$ is a two-dimensional vector whose elements represent the horizontal and vertical displacement at the point (x, z), respectively.

Under the excitation of fictitious loads, the displacements at the point (x, z) belonging to lth element Γ^l can also be represented by

$$\boldsymbol{U}(x,z) = \boldsymbol{g}^{\mathrm{h}}(x,z)\boldsymbol{P} \qquad (x, z) \in \Gamma^l \tag{3}$$

in which $\boldsymbol{g}^{\mathrm{h}}(x,z)$ is a $2 \times 2N$ matrix of displacement Green's functions

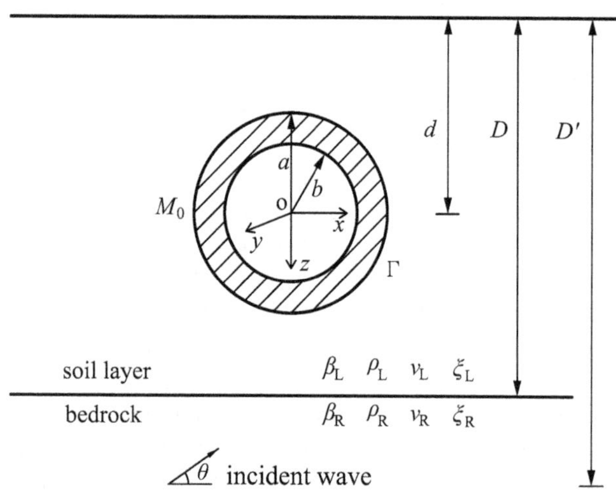

Fig. 1 Cross-section of an infinitely long tunnel (with rigid lining) embedded in layered half-space simplified as a single soil layer on elastic bedrock

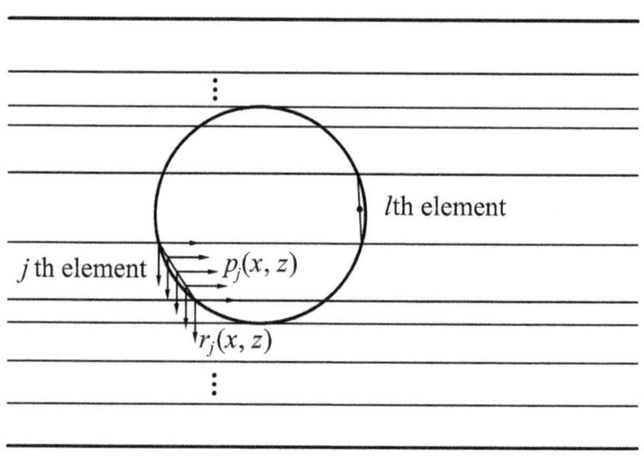

Fig. 2 Green's functions of horizontally and vertically loads distributed on an inclined line

$$\boldsymbol{g}^{\mathrm{h}}(x,z)=\begin{bmatrix} g^{\mathrm{h}}_{l1}(p_1) & g^{\mathrm{h}}_{l2}(p_2) & \dots & g^{\mathrm{h}}_{lN}(p_N) & g^{\mathrm{h}}_{l1}(r_1) & g^{\mathrm{h}}_{l2}(r_2) & \dots & g^{\mathrm{h}}_{lN}(r_N) \\ g^{\mathrm{v}}_{l1}(p_1) & g^{\mathrm{v}}_{l2}(p_2) & \dots & g^{\mathrm{v}}_{lN}(p_N) & g^{\mathrm{v}}_{l1}(r_1) & g^{\mathrm{v}}_{l2}(r_2) & \dots & g^{\mathrm{v}}_{lN}(r_N) \end{bmatrix} \tag{4}$$

with g^{h}_{lj} and g^{v}_{lj} are the horizontal and vertical displacements at the point (x, z) of lth element when a unit distributed load p_j or r_j is imposed onto jth element (Wolf 1985).

If it is assumed that

$$\boldsymbol{P}=\boldsymbol{\Lambda}\boldsymbol{\Delta} \tag{5}$$

the symbol Λ is a $2N \times 3$ matrix, with its three columns being the values of fictitious loads when the tunnel moves with unit horizontal displacement Δ_x, unit rotation arc-length $a\varphi$, and unit vertical displacement Δ_z, respectively. Introducing Eqs. (3) and (5) into (2) gives

$$\boldsymbol{g}^{\mathrm{u}}(x,z)\boldsymbol{\Lambda}=\boldsymbol{\Omega}(x,z) \qquad (x,\ z)\in\Gamma^{l}. \tag{6}$$

For each column of Λ and $\boldsymbol{\Omega}(x,\ z)$, every point on boundary Γ determines a set of $2 \times 2N$ equations like Eq. (6), and if N target points on boundary Γ are chosen (usually one target point from one element in order that IBEM can perform best), there comes a set of $2N \times 2N$ equations, from which this column of Λ can be solved.

Then, the traction at the point (x, z) of lth element is

$$\boldsymbol{T}(x,z)=\boldsymbol{g}^{\mathrm{t}}(x,z)\boldsymbol{\Lambda}\boldsymbol{\Delta} \qquad (x,\ z)\in\Gamma^{l} \tag{7}$$

in which $\boldsymbol{T}(x,z)$ is a two-dimensional vector whose elements represent horizontal and vertical tractions at the point (x, z), respectively, and $\boldsymbol{g}^{\mathrm{t}}$ is a $2 \times 2N$ matrix of traction Green's functions

$$\boldsymbol{g}^{\mathrm{t}}=\begin{bmatrix} \Pi_{l1}(p_1) & \Pi_{l2}(p_2) & \dots & \Pi_{lN}(p_N) & \Pi_{l1}(r_1) & \Pi_{l2}(r_2) & \dots & \Pi_{lN}(r_N) \\ \Theta_{l1}(p_1) & \Theta_{l2}(p_2) & \dots & \Theta_{lN}(p_N) & \Theta_{l1}(r_1) & \Theta_{l2}(r_2) & \dots & \Theta_{lN}(r_N) \end{bmatrix}$$
$$\Pi_{lj}=e_{xl}(g\sigma x_{lj})+e_{zl}(g\tau_{lj})$$
$$\Theta_{lj}=e_{zl}(g\sigma z_{lj})+e_{xl}(g\tau_{lj}) \tag{8}$$

with Π_{lj} and Θ_{lj} are the horizontal and vertical traction at the point (x, z) when a unit load p_j or r_j is imposed onto jth element (Wolf 1985), and e_{xl} and e_{zl} are the unit normal vector in x-direction and z-direction of lth element.

Finally, the force vector $\boldsymbol{F}=[F_x,\ M/a,\ F_z]^{\mathrm{T}}$, with F_x, M, and F_z being the total horizontal force, rotational moment, and vertical force imposed on the tunnel, is obtained by integral with respect to the tractions along Γ

$$\boldsymbol{F}=\int_{\Gamma}\boldsymbol{\Omega}(x,z)^{\mathrm{T}}\boldsymbol{T}\mathrm{d}S. \tag{9}$$

Introducing Eq. (7) into (9) gives the desired relationship between tunnel displacement and the force imposed on it

$$\boldsymbol{F}=\boldsymbol{K}\boldsymbol{\Delta} \tag{10}$$

So impedance function matrix of the tunnel is

$$\boldsymbol{K}=\int_{\Gamma}\boldsymbol{\Omega}(x,z)^{\mathrm{T}}\boldsymbol{g}^{\mathrm{t}}\boldsymbol{\Lambda}\mathrm{d}S \tag{11}$$

and its form is as follows

$$\boldsymbol{K}=\beta_{\mathrm{L}}^{2}\rho_{\mathrm{L}}\begin{bmatrix} K_{\mathrm{HH}} & K_{\mathrm{HM}} & 0 \\ K_{\mathrm{MH}} & K_{\mathrm{MM}} & 0 \\ 0 & 0 & K_{\mathrm{VV}} \end{bmatrix} \tag{12}$$

in which K_{HH}, K_{MM}, K_{VV}, K_{MH}, and K_{HM} being the horizontal, rotational, vertical, and two-coupling impedance functions, respectively, with $K_{\mathrm{MH}}=K_{\mathrm{HM}}$. Taking K_{HH} for example, it is convenient to write the impedance function as

$$K_{\mathrm{HH}}=k_{\mathrm{HH}}+\mathrm{i}\frac{\omega a}{\beta_{\mathrm{L}}}c_{\mathrm{HH}} \tag{13}$$

with $\mathrm{i}=\sqrt{-1}$.

2.3 Tunnel response

Effective input motion Δ is the tunnel displacement under harmonic-wave excitation, and it can be decomposed into two parts (Luco and Wong 1987)

$$\boldsymbol{\Delta}=\boldsymbol{\Delta}_{1}+\boldsymbol{\Delta}_{2} \tag{14}$$

in which $\boldsymbol{\Delta}_1$ corresponds to the tunnel displacement when its mass M_0 is not taken into account (Luco 1986)

$$\boldsymbol{\Delta}_{1}=\boldsymbol{K}^{-1}\int_{\Gamma}\left[\boldsymbol{g}^{\mathrm{t}}\boldsymbol{\Lambda}^{\mathrm{T}}\ \boldsymbol{U}_{f}(x,z)-\boldsymbol{\Omega}(x,z)^{\mathrm{T}}\ \boldsymbol{T}_{f}(x,z)\right]\mathrm{d}S \tag{15}$$

with $\boldsymbol{U}_f(x,z)$ and $\boldsymbol{T}_f(x,z)$ are two-dimensional vectors, corresponding to the displacements and tractions in two directions of free-field response, respectively. Symbol $\boldsymbol{\Delta}_2$ is the additional displacement caused by inertia force $\boldsymbol{F}=[F_x,\ M/a,\ F_z]^{\mathrm{T}}$, with $F_{\mathrm{T}x}$, M_{T}, and $F_{\mathrm{T}z}$ being the horizontal force, rotational moment, and vertical force caused by tunnel mass, respectively, and based on the

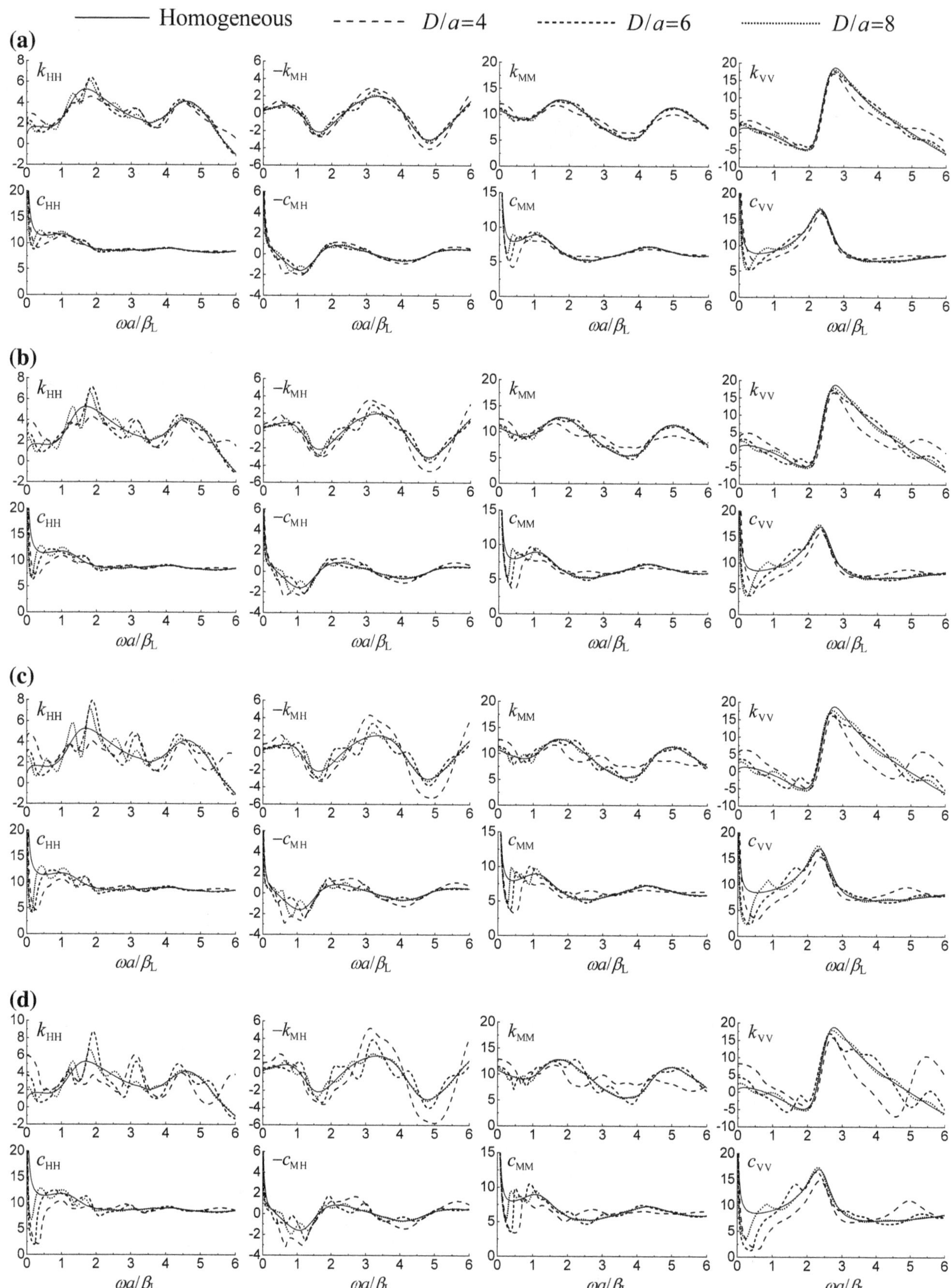

Homogeneous
D/a=4
D/a=6
D/a=8
(a)
(b)
(c)
(d)
k_{HH}
$-k_{MH}$
k_{MM}
k_{VV}
c_{HH}
$-c_{MH}$
c_{MM}
c_{VV}
$\omega a/\beta_L$

◂**Fig. 3** Spectrum of tunnel impedance functions ($d/a = 2$, $\rho_R = \rho_L = 2000$ kg/m^3, $\nu_R = \nu_L = 0.25$, damping ratio $\xi_R = 0.05$ and $\xi_L = 0.02$ for layered half-space, and $\xi_R = \xi_L = 0.05$ for homogeneous half-space). **a** $\beta_R/\beta_L = 2$. **b** $\beta_R/\beta_L = 3$. **c** $\beta_R/\beta_L = 5$. **d** $\beta_R/\beta_L = \infty$

concept of impedance functions, the additional displacement is solved by

$$\Delta_2 = \boldsymbol{K}^{-1}\boldsymbol{F}_T. \tag{16}$$

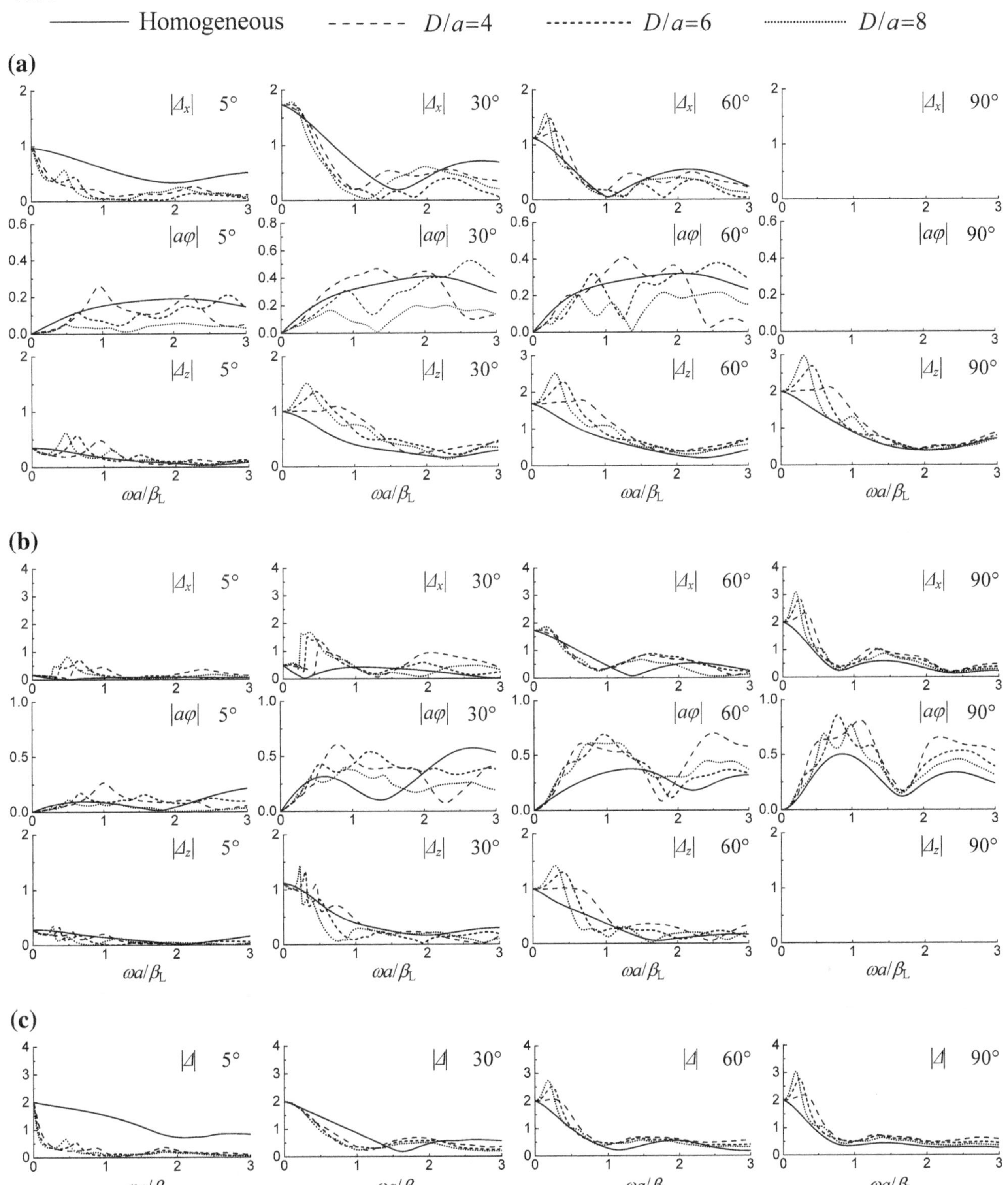

Fig. 4 Spectrum of tunnel displacements in homogeneous half-space and layered half-space with bedrock stiffness $\beta_R/\beta_L = 2$ ($d/a = 2$, $\rho_R = \rho_L = 2000$ kg/m^3, $\nu_R = \nu_L = 0.25$, $\rho_0 = 2500$ kg/m^3, $M_0/M_s = 1/4$, $D'/a = 8$, damping ratio $\xi_R = 0.05$ and $\xi_L = 0.02$ for layered half-space, $\xi_R = \xi_L = 0.05$ for homogeneous half-space). **a** P-wave. **b** SV-wave. **c** SH-wave (Fu et al. 2015)

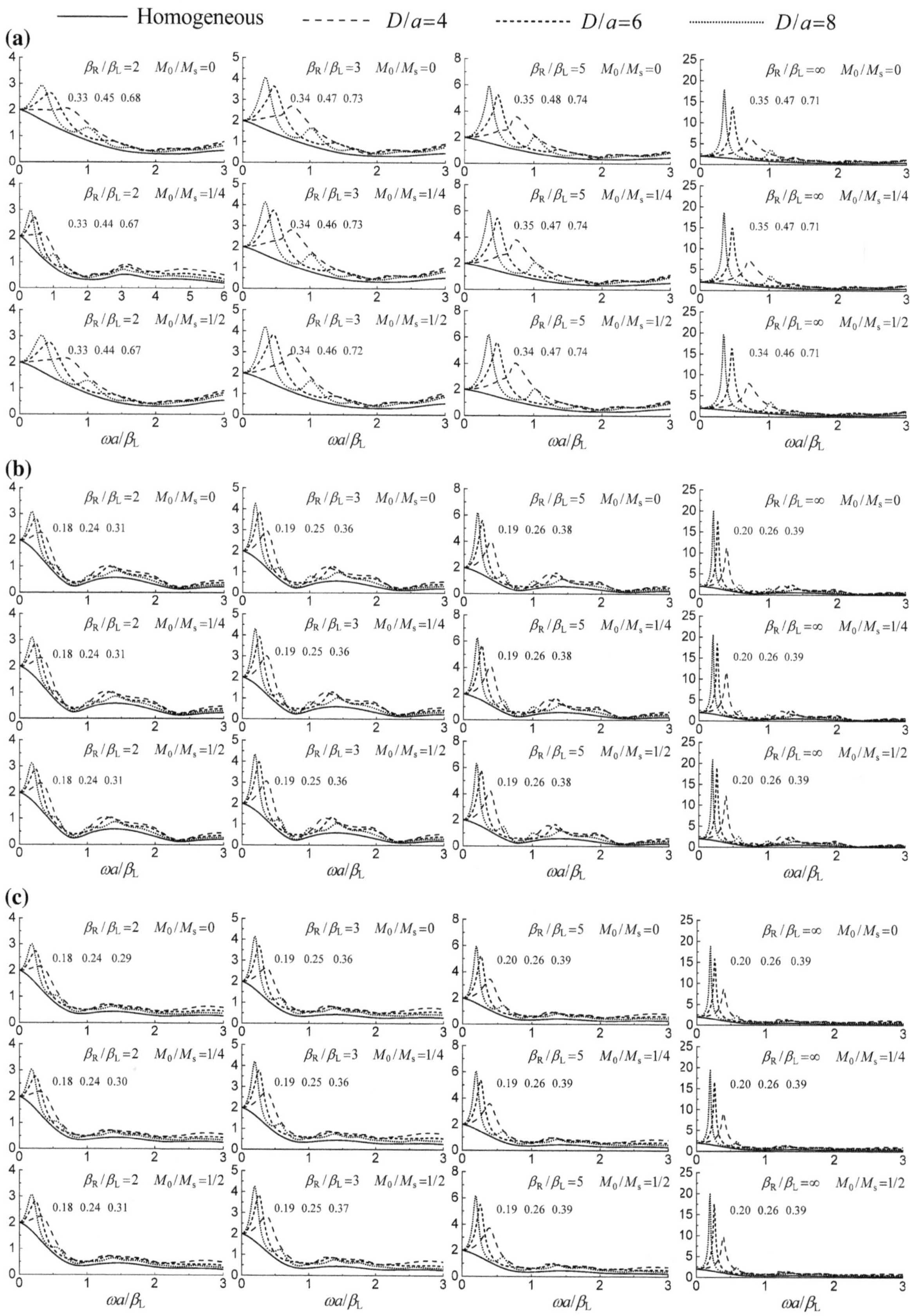

Homogeneous
$D/a=4$
$D/a=6$
$D/a=8$
(a)
(b)
(c)
$\omega a/\beta_L$

◀**Fig. 5** Spectrum of tunnel displacement in homogeneous half-space and layered half-space for vertical incidence ($\theta = 5°$, $d/a = 2$, $\rho_R = \rho_L = 2000$ kg/m^3, $\nu_R = \nu_L = 0.25$, $\rho_0 = 2500$ kg/m^3, $D'/a = 8$, damping ratio $\xi_R = 0.05$ and $\xi_L = 0.02$ for layered half-space, $\xi_R = \xi_L = 0.05$ for homogeneous half-space). **a** Tunnel displacement $|\Delta_z|$ for vertically incident P-wave. **b** Tunnel displacement $|\Delta_x|$ for vertically incident SV-wave. **c** Tunnel out-of-plane displacement for vertically incident SH-wave (Fu et al. 2015)

For a rigid body

$$\boldsymbol{F_T} = \omega^2 \boldsymbol{M_0 \Delta} = \omega^2 \begin{bmatrix} M_0 & 0 & 0 \\ 0 & I_0/a^2 & 0 \\ 0 & 0 & M_0 \end{bmatrix} \boldsymbol{\Delta} \tag{17}$$

with I_0 being the rotational inertia with respect to tunnel center. Introducing Eqs. (16) and (17) into (14) gives the final solution of tunnel displacement

$$\boldsymbol{\Delta} = \left(\boldsymbol{I} - \omega^2 \boldsymbol{K}^{-1} \boldsymbol{M_0}\right)^{-1} \boldsymbol{\Delta_1} \tag{18}$$

As the amplitude of incident wave is assumed to be unit 1, the tunnel displacements Δ_x, $a\varphi$, and Δ_z are all dimensionless, in fact they represent the amplification factor of incident excitation.

3 Numerical results and analysis

3.1 Impedance function

Figure 3 is the impedance function of tunnel in frequency domain of homogeneous half-space and layered half-space. The embedded depth of the tunnel is $d/a = 2$. The parameters of homogeneous half-space are $\rho_R = \rho_L = 2000$ kg/m^3, $\nu_R = \nu_L = 0.25$, and $\xi_R = \xi_L = 0.05$. While the parameters of layered half-space are $\rho_R = \rho_L = 2000$ kg/m^3, $\nu_R = \nu_L = 0.25$, $\xi_R = 0.05$, $\xi_L = 0.02$, with the shear-wave velocity ratio of the soil layer to the bedrock ("bedrock stiffness" for short) varying with four values $\beta_R/\beta_L = 2, 3, 5$, and ∞, and the ratio of the soil-layer thickness to the tunnel radius ("soil-layer thickness" for short in the following) varying with three values $D/a = 4$, 6, and 8, so the tunnel is completely embedded within the soil layer in order to be convenient to analyze the site effect on tunnel response.

The impedance functions of homogeneous half-space are vibrating functions and the impedance functions of layered half-space vibrate around that of homogeneous half-space, because the layered half-space involves the dynamic characteristics of site while homogeneous half-space cannot reflect these characteristics. When the bedrock stiffness β_R/β_L increases, the influence of the site dynamic characteristics also increase in the way that the vibrating period of impedance functions keeps invariable, and the shape of the curves is such as that the curves are multiplied by a factor in y-axis.

3.2 Tunnel response

Figure 4a and b are the spectrum of tunnel horizontal displacement Δ_x, rotational-arc $a\varphi$, and vertical displacement Δ_z in homogeneous half-space and layered half-space with bedrock stiffness $\beta_R/\beta_L = 2$ for incident P-wave and SV-wave, respectively. The mass density of the tunnel is $\rho_0 = 2500$ kg/m^3, and the dimensionless tunnel mass is $M_0/M_s = 1/4$ with M_s being the mass of soil replaced by the tunnel. The incident P- and SV-wave comes from $D'/a = 8$ for all sites with four incident angle $\theta = 5°$, 30°, 60°, and 90°, and the parameters of the half-space are the same as that in Fig. 3. The tunnel displacement spectrum for incident SH-wave is also plotted in Fig. 4c (Fu et al. 2015) for comparison, there is only out-of-plane translational displacement $|\Delta|$ in this condition.

For oblique incidence, the spectrum of in-plane displacements is more complicated than that of incident SH-wave, especially for small incident angle ($\theta = 5°$ and 30°). Also, the displacements in layered half-space increase with incident angle increasing for incident SH-waves, while this is not the fact for in-plane excitation. For vertical incidence ($\theta = 90°$), the symmetry of the tunnel gives that $\Delta_x^{\mathrm{P}} = a\varphi^{\mathrm{P}} = \Delta_z^{\mathrm{SV}} = 0$, it is also noticed that the tunnel displacements in layered half-space are larger than those in homogeneous half-space.

The tunnel displacement spectrum of homogeneous half-space is much smoother than that of layered half-space; also, there is an evident peak for layered half-space on the spectrum of translational displacements Δ_x and Δ_z for large incident angle ($\theta = 60°$ and 90°), while the peak does not exist for homogeneous half-space. This is because the site dynamic characteristics introduce much influence on tunnel response of layered half-space, while the homogeneous half-space does not involve these characteristics. It is noticed that there is also a peak on the spectrum of Δ_x for incident SV-wave with small angle ($\theta = 5°$ and 30°), but there is no interest in this peak which does not reflect the site dynamic characteristics.

Figure 5a and b are the spectra of tunnel translational displacement in homogeneous half-space and layered half-space for vertically incident P-wave (Δ_z) and SV-wave (Δ_x), respectively. The dimensionless tunnel mass is $M_0/M_s = 0$, 1/4, 1/2, with other parameters the same as those in Figs. 3 and 4. For comparison, the tunnel displacement spectrum for vertical incident SH-wave is also plotted in Fig. 5c (Fu et al. 2015). It is noticed that for both incident P-wave and SV-wave, the peak value becomes larger with soil-layer thickness increasing, which is similar to incident

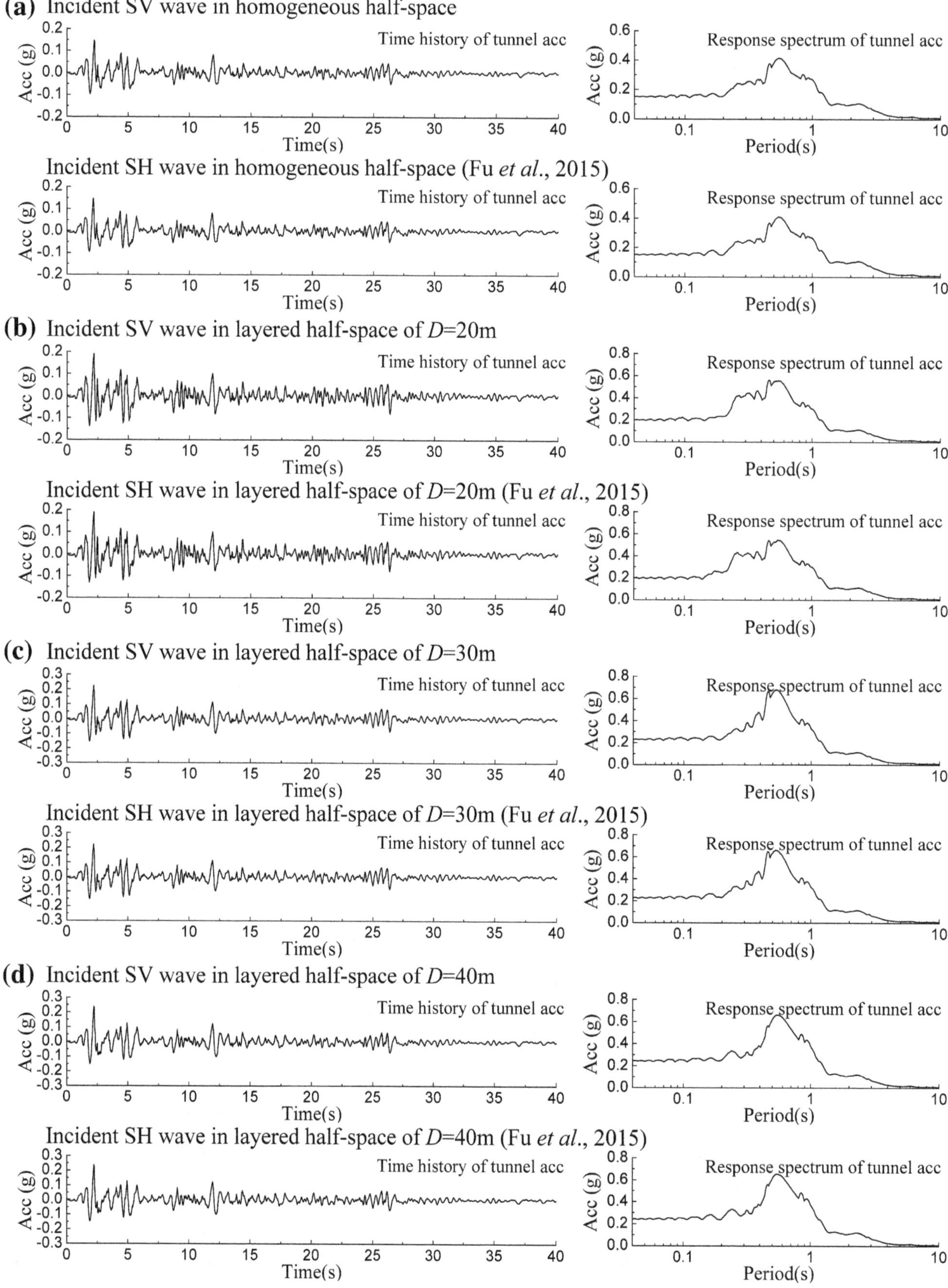
(a) Incident SV wave in homogeneous half-space
Time history of tunnel acc
Response spectrum of tunnel acc
Acc (g)
Time(s)
Period(s)
Incident SH wave in homogeneous half-space (Fu et al., 2015)
Time history of tunnel acc
Response spectrum of tunnel acc
(b) Incident SV wave in layered half-space of D=20m
Time history of tunnel acc
Response spectrum of tunnel acc
Incident SH wave in layered half-space of D=20m (Fu et al., 2015)
Time history of tunnel acc
Response spectrum of tunnel acc
(c) Incident SV wave in layered half-space of D=30m
Time history of tunnel acc
Response spectrum of tunnel acc
Incident SH wave in layered half-space of D=30m (Fu et al., 2015)
Time history of tunnel acc
Response spectrum of tunnel acc
(d) Incident SV wave in layered half-space of D=40m
Time history of tunnel acc
Response spectrum of tunnel acc
Incident SH wave in layered half-space of D=40m (Fu et al., 2015)
Time history of tunnel acc
Response spectrum of tunnel acc

◀**Fig. 6** Time history (*left*) and response spectrum (*right*) of tunnel acceleration for vertically incident El Centro wave with peak acceleration of 0.1 g in homogeneous half-space (**a**) and in layered half-space of $D = 20$ m (**b**), $D = 30$ m (**c**), and $D = 40$ m (**d**) (parameters: $a = 5$ m, $b = 4$ m, $d = 10$ m, $D' = 40$ m, $\rho_0 = 2500$ kg/m^3; for homogeneous half-space $\rho_R = \rho_L = 2000$ kg/m^3, $\nu_R = \nu_L = 0.25$, $\beta_R = \beta_L = 250$ m/s, for layered half-space $\rho_R = \rho_L = 2000$ kg/m^3, $\nu_R = \nu_L = 0.25$, $\xi_R = 0.05$, $\xi_L = 0.02$, $\beta_L = 250$ m/s, $\beta_R = 500$ m/s)

SH-waves because the path the incident wave propagates and amplifies is longer in thicker soil layer (Fu et al. 2015).

The tunnel mass have little influence on tunnel displacement spectrum for both incident P-wave and SV-wave as the condition of incident SH-wave because the tunnel mass itself is small. It can be concluded that the kinematic interaction also overwhelmingly dominates for in-plane excitation, and the inertia interaction can hardly have influence on soil-tunnel interaction.

For free-field ground motion to vertically incident P-wave, the frequencies for which interference produces maximum response of the soil layer are ("resonant frequencies" for short)

$$\omega_\alpha = \frac{(2j-1)\pi\alpha_L}{2D} \qquad (j = 1, 2, 3\ldots). \tag{19}$$

So $D/a = 4$ corresponds to $\omega_\alpha a/\beta_L = 0.68, 2.04, 3.40, \ldots$, $D/a = 6$ corresponds to $\omega_\alpha a/\beta_L = 0.45, 1.36, 2.27, \ldots$, $D/a = 8$ corresponds to $\omega_\alpha a/\beta_L = 0.34, 1.02, 1.70, \ldots$ and so on. It is observed that the peak frequency of tunnel displacement evidently becomes lower with soil-layer thickness increasing, and it is lower than the first resonant frequency ω_α of free-field response for $\beta_R/\beta_L = 2$, while higher than ω_α for $\beta_R/\beta_L = 3$, 5 and ∞. Nevertheless, the difference between the peak frequency of tunnel displacement and ω_α is not large. While for free-field ground motion to incident SV-wave, the resonant frequencies of the soil layer are

$$\omega_\beta = \frac{(2j-1)\pi\beta_L}{2D} \qquad (j = 1, 2, 3\ldots). \tag{20}$$

So $D/a = 4$ corresponds to $\omega_\beta a/\beta_L = 0.39, 1.18, 1.96, \ldots$, $D/a = 6$ corresponds to $\omega_\beta a/\beta_L = 0.26, 0.79, 1.31, \ldots$, $D/a = 8$ corresponds to $\omega_\beta a/\beta_L = 0.20, 0.59, 0.98, \ldots$ and so on. The peak frequency of tunnel displacement also becomes lower with soil-layer thickness increasing as incident P-wave, but it is lower than the first resonant frequency ω_β of free-field response for all bedrock stiffness, and the difference between the peak frequency of tunnel response and ω_β is not large either. Moreover, it is noticed that the peak frequency of tunnel response for both incident P-wave and SV-wave becomes lower with bedrock stiffness decreasing (there exists abnormal case for $\beta_R/\beta_L = \infty$ to incident P-wave), but this phenomenon is not evident.

While in the papers by Liang et al. (2013a, b) studying the soil-foundation-superstructure interaction, although the foundation is also assumed to be completely rigid, the difference between the peak frequency of foundation displacement spectrum and the resonant frequency of free-field response is much larger, and the variation of bedrock stiffness can have more evident influence on the peak frequency of foundation displacement. This is because the soil-tunnel interaction is dominated by kinematic interaction, which can be influenced only by site dynamic characteristics, so the dynamic characteristics of tunnel response are similar to site dynamic characteristics; while the dynamic characteristics of foundation response are also influenced strongly by superstructure dynamic characteristics, and the system mass is large with the inertia interaction also introducing much influence on foundation response, so the dynamic characteristics of foundation response are different much from site dynamic characteristics.

It is also noticed that although the spectra shapes of oblique incident SV-wave differ much from that of oblique incident SH-wave in Fig. 4, the two spectra of vertical incidence is very similar to each other, especially for peak value and peak frequency. This is because the dynamic characteristics of underground tunnel can be influenced only by the site characteristics. As the free-field response for vertically incident SV-wave is exactly identical to that of vertically incident SH-wave, the two spectra of tunnel displacement for vertical incidence are similar to each other; while as the free-field ground motion for obliquely incident SV-wave and SH-wave is essentially different, the two spectra of tunnel displacement for oblique incidence are also different much just as the spectrum of free-field response. Nevertheless, in Liang et al. (2013a, b), even for vertically incident SV-wave and SH-wave, the displacement spectrum of rigid foundation still holds much difference. This is because the foundation displacement spectrum can be influenced by both the site characteristics and the superstructure characteristics. As the dynamic characteristics of superstructure of in-plane direction are different from that of out-of-plane direction, the two spectra of foundation displacement holds little similarities although the site characteristics for vertically incident SV- and SH-wave are identical. In conclusion, the mechanism of soil-tunnel interaction which is a rigid system, is totally different from that of soil-foundation-superstructure interaction which is a flexible system.

3.3 Analysis in time domain

The similarity of tunnel response to vertically incident SV-wave and SH-wave can further be justified in time domain. Figure 6 is the tunnel response in time domain for vertically incident El Centro wave with peak ground acceleration of 0.1 g as SV-wave and SH-wave (Fu et al. 2015). The left part of each sub-figure is the time history of tunnel acceleration with x-axis being the time history by interval 0.02 s and y-axis being the acceleration of 1 g; the right part is the response spectrum of tunnel acceleration with x-axis being the period and y-axis being the maximum acceleration of 1 g. In this section, the outer radius of the tunnel is $a = 5$ m, the inner radius is $b = 4$ m, the embedded depth is $d = 10$ m, and the mass density is $\rho_0 = 2500$ kg/m^3. The vertical incident SV-wave or SH-wave all come from depth $D' = 40$ m. For homogeneous half-space (a), the shear-wave velocity is 250 m/s, the mass density is 2000 kg/m^3, and the damping ratio is 0.05. For layered half-space, the soil-layer thickness is $D = 20$ m (b), 30 m (c), and 40 m (d), which corresponds to $D/a = 4$, 6, and 8, respectively. The soil layer is of shear-wave velocity $\beta_L = 250$ m/s, mass density $\rho_L = 2000$ kg/m^3, and damping ratio $\xi_L = 0.02$; the bedrock is of $\beta_R = 500$ m/s, $\rho_R = 2000$ kg/m^3, and $\xi_R = 0.05$. It is observed that in time domain, the tunnel responses for vertically incident SV-wave and SH-wave are more similar than that in frequency domain—they are nearly identical.

4 Conclusions

The spectrum of tunnel impedance function of layered half-space vibrates around that of homogeneous half-space; the mechanism of dynamic soil-tunnel interaction in layered half-space is different much from that in homogeneous half-space, and the former is larger than the latter. This is because the layered half-space involves the site dynamic characteristics while the homogeneous half-space cannot reflect these characteristics.

The mechanism of dynamic soil-tunnel interaction is different much from that of dynamic soil-foundation-superstructure interaction, because the soil-tunnel interaction is dominated by kinematic interaction, so the dynamic characteristics of tunnel response are similar to site dynamic characteristics, while for soil-foundation-superstructure interaction, the foundation response can be influenced by both site dynamic characteristics, and superstructure dynamic characteristics which can be represented by inertia interaction, so the difference of the peak frequency of foundation response to the resonant frequency of the free-field response is much larger than that of tunnel response to the resonant frequency of the free-field response in soil-foundation-superstructure interaction.

For oblique incidence, the tunnel response for in-plane incident waves is completely different from that for incident SH-waves, especially for small incident angle; while the tunnel response for vertically incident SV-wave is very similar to that of vertically incident SH-wave, because the tunnel response is influenced strongly by the site dynamic characteristics which are identical for vertically incident SV-wave and SH-wave, while differ much for oblique incidence.

Acknowledgments This study is supported by the National Natural Science Foundation of China (No. 51378384) and the Key Project of Natural Science Foundation of Tianjin Municipality (No. 12JCZDJC29000).

References

De Barros FCP, Luco JE (1994) Seismic response of a cylindrical shell embedded in a layered viscoelastic half-space. II: validation and numerical results. Earthq Eng Struct Dyn 23:569–580

Hatzigeorgiou GD, Beskos DE (2010) Soil-structure interaction effects on seismic inelastic analysis of 3-D tunnels. Soil Dyn Earthq Eng 30:851–861

Lee VW, Trifunac MD (1979) Response of tunnels to incident SH-waves. J Eng Mech Div ASCE 105:643–659

Liang J, Fu J, Todorovska MI, Trifunac MD (2013a) Effects of the site dynamic characteristics on soil-structure interaction (I): incident SH waves. Soil Dyn Earthq Eng 44:27–37

Liang J, Fu J, Todorovska MI, Trifunac MD (2013b) Effects of the site dynamic characteristics on soil-structure interaction (II): incident P and SV waves. Soil Dyn Earthq Eng 51:58–76

Luco JE (1969) Dynamic interaction of a shear wall with the soil. J Eng Mech Div ASCE 95:333–346

Luco JE (1986) On the relation between radiation and scattering problems for foundations embedded in an elastic half-space. Soil Dyn Earthq Eng 5(2):97–101

Luco JE, De Barros FCP (1994) Seismic response of a cylindrical shell embedded in a layered viscoelastic half-space. I: formulation. Earthq Eng Struct Dyn 23:553–567

Luco JE, Wong HL (1987) Seismic response of foundations embedded in a layered half-space. Earthq Eng Struct Dyn 15(2):233–247

Parvanova SL, Dineva PS, Manolis GD, Wuttke F (2014) Seismic response of lined tunnels in the half-plane with surface topography. Bull Earthq Eng 12:981–1005

Trifunac MD (1972) Interaction of a shear wall with the soil for incident plane SH waves. Bull Seismol Soc Am 62:63–83

Wolf JP (1985) Dynamic soil-structure interaction. Prentice-Hall, Englewood Cliffs, pp 114–178

13

Crustal attenuation characteristics of S-waves beneath the Eastern Tohoku region, Japan

Muhammad Adeel Arshad

Abstract An inversion method was applied to crustal earthquakes dataset to find S-wave attenuation characteristics beneath the Eastern Tohoku region of Japan. Accelerograms from 85 shallow crustal earthquakes up to 25 km depth and magnitude range between 3.5 and 5.5 were analyzed to estimate the seismic quality factor Q_S. A homogeneous attenuation model Q_s for the wave propagation path was evaluated from spectral amplitudes, at 24 different frequencies between 0.5 and 20 Hz by using generalized inversion technique. To do this, non-parametric attenuation functions were calculated to observe spectral amplitude decay with hypocentral distance. Then, these functions were parameterized to estimate Q_S. It was found that in Eastern Tohoku region, the Q_s frequency dependence can be approximated with the function 33 $f^{1.22}$ within a frequency range between 0.5 and 20 Hz. However, the frequency dependence of Q_s in the frequency range between 0.5 and 6 Hz is best approximated by $Q_s(f) = 36 f^{0.94}$ showing relatively weaker frequency dependence as compared to the relation $Q_s(f) = 6 f^{2.09}$ for the frequency range between 6 and 15 Hz. These results could be used to estimate source and site parameters for seismic hazard assessment in the region.

Keywords Tohoku · Generalized inversion technique · Non-parametric attenuation function · Seismic quality factor · S-wave

M. A. Arshad (✉)
Department of Civil Engineering, University of Engineering & Technology, Peshawar, Peshawar 25120, KPK, Pakistan
e-mail: ceadeel@uetpeshawar.edu.pk

1 Introduction

The study region is confined within 38°N to 39.5°N and 140.5°E to 142.5°E, and contains eastern part of Tohoku region in Japan's Honshu Island as shown in Fig. 1. Eastern Tohoku mainly comprises Iwate and Miyagi prefectures. The region is situated over one of the most active tectonic regimes of the world where Pacific plate is subducting under the Eurasian plate, generating numerous earthquakes in this area.

There is a close connection between the topography of the Tohoku region and the distribution of active faults. The topography of the Tohoku region is characterized by alternate mountain ranges and lowlands running in a north–south direction. The primary active faults of the Tohoku region lie at the boundaries of the mountains and the lowlands.

The crustal depth beneath Tohoku is estimated to be 25–30 km (Kaminuma and Aki 1963; Zhao et al. 1992) and many of the large destructive inland earthquakes are known to have occurred at these shallower depths. The shallower crustal earthquakes tend to result in a greater intensity of surface shaking and often cause the greatest loss of life and damage to property. In engineering applications like seismic hazard analysis, S-wave phases of ground motions are of most interest, as most of the seismic energy from the earthquake source is radiated in the form of S-waves. Therefore, quantification of the S-wave attenuation characteristics is necessary.

The records of strong earthquakes can be inverted to obtain the characteristics of attenuation, radiation and propagation pattern of seismic waves and to construct models of ground behaviour at various stations (Pavlenko and Wen 2008). In this study, linear inversion analysis following a non-parametric approach has been performed

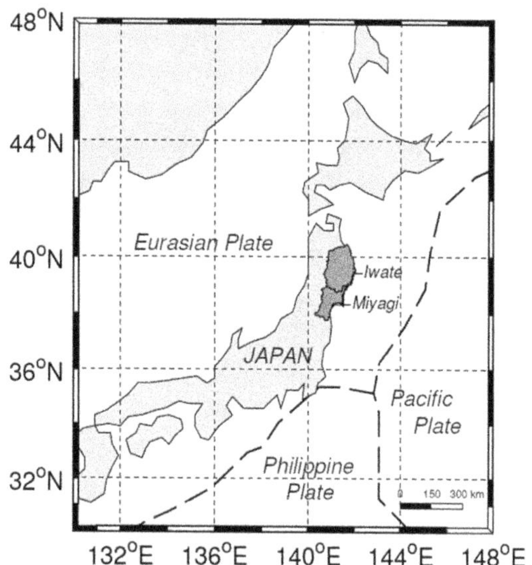

Fig. 1 Map of Japan showing Iwate and Miyagi prefectures which constitute Eastern Tohoku

to investigate the frequency dependence of attenuation of S-waves (Q_s).

2 Dataset

The dataset comprises 315 three-component earthquake records, from 85 earthquakes recorded at 6 K-NET (Kyoshin network) stations installed in the Eastern Tohoku region of Japan in the framework of National Research Institute for Earth Science and Disaster Prevention (NIED). The earthquake records were obtained over a period of nine years from Dec-2003 to May-2012 with magnitudes M_w ranging between 3.5 and 5.5 and hypocentral distances between 35 and 125 km. The data have been sampled at 100 Hz. The data represent no distance dependence in magnitude as shown in Fig. 2.

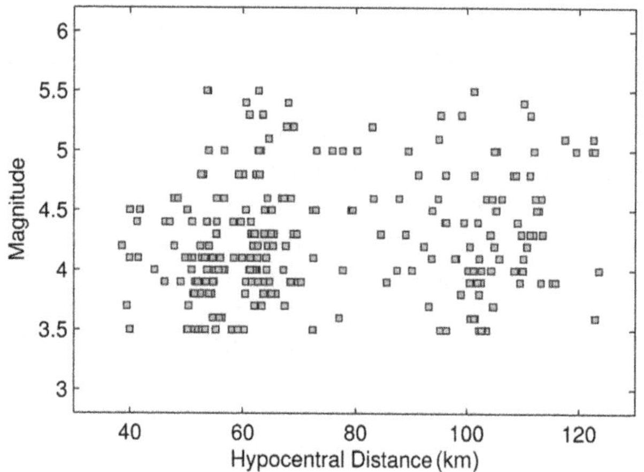

Fig. 2 Distribution of magnitudes with hypocentral distances

The site details of the K-NET stations used in this study are listed in Table 1. These stations were initially installed with K-NET95-type accelerometers which were later replaced with K-NET02-type accelerometers in 2004 (Okada et al. 2004). Since, the crustal thickness of Eastern Tohoku region ranges in between 25 to 30 km, the hypocentral depth of the selected events was restricted not to exceed 25 km. Moreover, only those earthquakes were selected which were recorded by at least three stations. The details of epicentral coordinates of the events used and the hypocentral distances to the recording stations are provided in Appendix Table 2. The ray paths between different source-station pairs in the region are shown in Fig. 3. There are many crossing ray paths between different source station pairs ensuring reliable estimates of homogeneous attenuation model.

3 Data processing and analysis

The records were corrected for instrumental response and baseline. Instrument correction was only applied on the records from K-NET95-type accelerometers for frequencies larger than ~15 Hz, as the response of these accelerometers slightly decrease for frequencies larger than this limit. The baseline was corrected by subtracting the average of all the points of the record. S-wave portion of both the horizontal components of the recorded accelerograms were analyzed. For each horizontal component of the records, the S-wave window was selected such that it starts 1 s before the S-wave onset and ends when 80 % of the total energy of the record is reached as shown in Fig. 4. Typical window lengths range between 5 and 12 s. If in any case the window's length increases over 20 s, it was fixed to a maximum duration of 20 s to avoid having too much coda energy in the analyzed time window (Oth et al. 2011). The beginning and the end of S-wave window were tapered with a 5 % cosine taper.

A plot showing the build-up of Arias intensity with time is known as a Husid plot (Fig. 4) and it serves to identify the interval with the arrival of majority of the energy.

The Fourier amplitude spectra (FAS) were computed for each window and smoothed around 24 frequency points equidistant on logarithmic scale between 0.5 and 20 Hz using the running mean filter. This smoothing technique is optimal for reducing random noise while keeping the sharpest step response (Smith 1999). The smoothing bandwidth is determined by trial-and-error. Having tested various smoothing bandwidths, the following criteria were adopted: for each frequency point, a bandwidth equal to 0.3 of an octave was considered; if this bandwidth was smaller than 0.5 Hz, then 0.5 Hz was used as the smoothing

Table 1 Location of the seismic stations used in this study

Site code	Site name	Lat. (°)	Long. (°)	Elevation (m)	Station
IWT007	KAMAISHI	39.270	141.856	11	K-NET
MYG002	UTATSU	38.726	141.511	79	K-NET
MYG003	TOHWA	38.735	141.311	28	K-NET
MYG004	TSUKIDATE	38.729	141.022	40	K-NET
MYG005	NARUKO	38.796	140.654	300	K-NET
MYG011	OSHIKA	38.297	141.504	13	K-NET

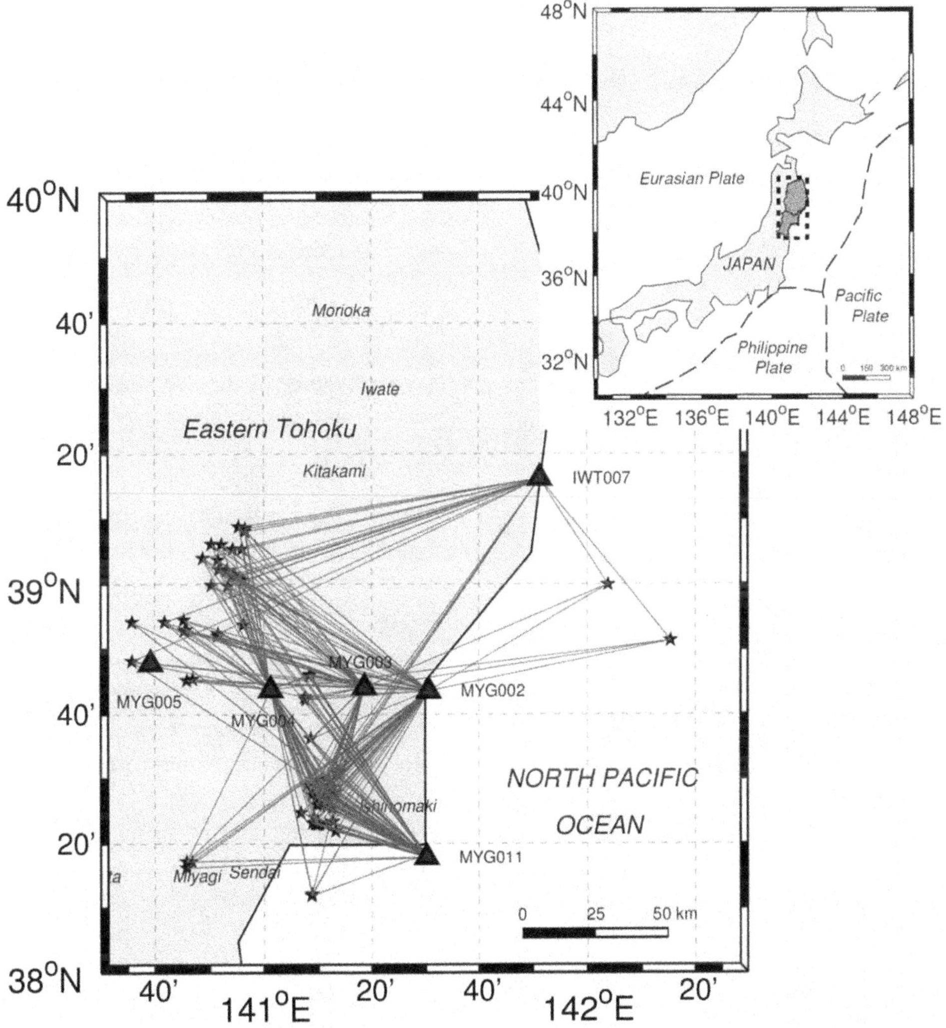

Fig. 3 Map of the Eastern Tohoku region that showing the distribution of the epicentres (*stars*) and the stations (*triangles*)

bandwidth. The signals to be analyzed have to satisfy stringent signal-to-noise ratio (SNR) requirements. Only data points with SNR greater than three were included in the final dataset. Figure 5 shows the raw and smoothed FAS of the chosen S-wave window from an accelerogram recorded at station IWT007. At lower frequencies, the database is slightly sparser due to SNR constraints as shown in Fig. 6.

4 Method

The generalized inversion technique (GIT) has been widely applied to crustal earthquake datasets (Andrews 1986; Castro et al. 1990). As a result of inversion, the frequency dependence of the attenuation of seismic waves as well as source characteristics and site response can easily be found (Parolai et al. 2000; Bindi et al. 2006).

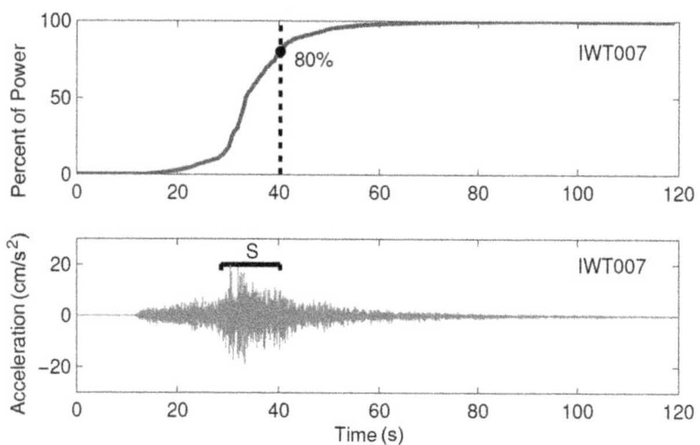

Fig. 4 Accelerogram of one of the earthquakes recorded at the station IWT007 (*lower panel*), highlighting the S-wave duration and the corresponding Husid plot (*upper panel*)

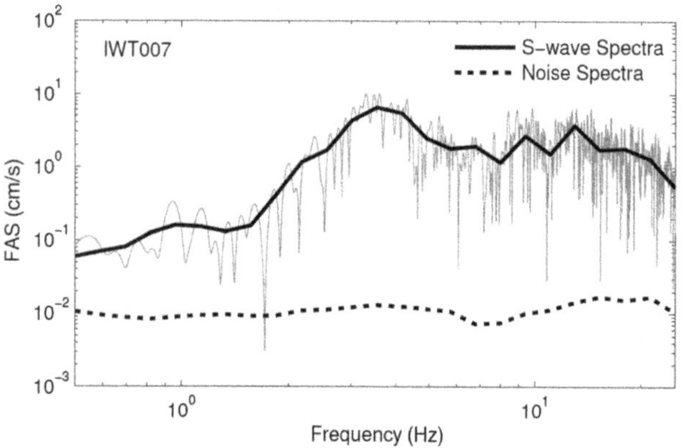

Fig. 5 Fourier amplitude spectrum (FAS) of the selected S-wave window from an accelerogram recorded at station IWT007. Raw and smoothed acceleration spectrums are shown by *solid grey* and *solid black lines*, respectively. Noise spectrum (*dashed black line*) is also shown for the same event to illustrate signal to noise levels

These dataset from Eastern Tohoku region allows us to make stable and reliable estimates of attenuation properties as it covers a reasonable range of magnitudes, distances and focal depths. Moreover, each event in the dataset was recorded by at least three stations, creating multiple crossing ray paths from sources to stations. Under these conditions, the effect of in-homogeneities in whole-path attenuation is thought to be effectively averaged out and GIT could successfully be applied to our dataset.

In this study, a non-parametric approach was employed (Castro et al. 1990), which considers attenuation phenomenon to be a smooth function, decreasing with distance. In non-parametric approach, the inversion is performed in two steps. Since, the attenuation characteristics $\hat{A}_{ij}(f, r_{ij})$ are determined from the first step, we restrict the procedure to the first step of inversion. However, a subsequent inversion can be performed to separate source and site effects (Andrews 1986; Oth et al. 2009).

4.1 Non-parametric approach

The dependence of the spectral amplitudes $U(f, r)$ at frequency f and on distance r may be written as

$$U_{ij}(f, r_{ij}) = \hat{A}_{ij}(f, r_{ij}) \cdot \hat{S}_i(f), \tag{1}$$

where $U_{ij}(f, r_{ij})$ is the spectral amplitude (acceleration) from the ith earthquake at the jth station resulting, r_{ij} is the source-site distance, the non-parametric attenuation function (NAF) $\hat{A}_{ij}(f, r_{ij})$ describes seismic attenuation with distance, and $\hat{S}_i(f)$ is a scaling factor depending on the size of the ith event. With this approach, the attenuation function remains independent of the size of an event. $\hat{A}_{ij}(f, r_{ij})$ is not supposed to have any specific shape and implicitly

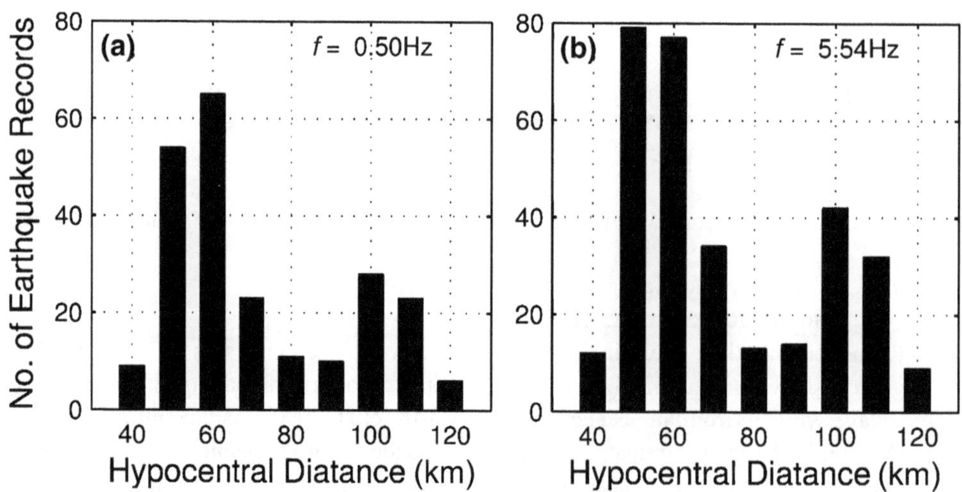

Fig. 6 Example of distribution of hypocentral distances at the two selected frequencies, 0.5 and 5.54 Hz, respectively. At 0.5 Hz, the database is slightly sparser than at 5.54 Hz due to signal-to-noise ratio constraints

contains all effects leading to attenuation along the travel path (geometrical spreading, anelasticity, scattering, etc.). Based on the idea that these properties vary slowly with distance, $\hat{A}_{ij}(f, r_{ij})$ is constrained to be a smooth function of distance i.e., with a small second derivative (Castro et al. 1990) and to take the value $\hat{A}_{ij}(f, r_0) = 1$ at some reference distance r_0.

The representation given by Eq. (1) does not include a factor related to the site, and hence the site effects are necessarily absorbed both in $\hat{A}_{ij}(f, r_{ij})$ and $\hat{S}_i(f)$. This is the reason that '^' is used to distinguish them from the 'uncontaminated' path and source terms $\hat{A}_{ij}(f, r_{ij})$ and S_i (f), respectively. $\hat{S}_i(f)$ contains, in fact, an average value of the site amplifications of all stations that recorded the ith earthquake. Thus, $\hat{S}_i(f)$ is not the true source spectrum of the ith event.

Equation (1) can be easily linearized by taking the logarithm:

$$\log_{10} U_{ij}(f, r_{ij}) = \log_{10} \hat{A}_{ij}(f, r_{ij}) + \log_{10} \hat{S}_i(f). \qquad (2)$$

Equation (2) represents an over-determined system of the form $\boldsymbol{Ax} = \boldsymbol{b}$, where $\boldsymbol{b}$ is the data vector containing the logarithmic spectral amplitudes, $\boldsymbol{x}$ is the vector containing the model parameters, and $\boldsymbol{A}$ is the system matrix relating the two of them. The distance range was subdivided into nine distance bins (N_D), each 10 km wide, and the value of $\hat{A}_{ij}(f, r_{ij})$ was computed in each bin. In matrix formulation, Eq. (2) takes the following form.

The left part of the system matrix in Eq. (3) contains the factors related to the attenuation parameters, whereas the right-hand side reflects those related to the source terms. The system matrix also includes rows relevant to constraints. The weighting factor w_1 is used to impose $\hat{A}_{ij}(f, r_0) = 1$ at the reference distance r_0. Since the dataset contains most of the events with hypocentral distance greater than 40 km, reference distance r_0 is set to 40 km. By setting $\hat{A}_{ij}(f, r_0) = 1$ at $r_0 = 40$ km is, in fact, equal to assuming that there is no attenuation over that distance from the source. Hence, there is a cumulative attenuation effect over these 40 km that is impossible to resolve, and therefore, the Q_s-model derived from the slopes of the attenuation functions, only reflects the attenuation characteristic over the remaining part of the travel path (i.e., more than 40 km away from the source). The weighing factor w_2 is implemented to achieve monotonically decaying attenuation curves with reasonable degree of smoothness to suppress the site-related effects and yet preserve variations of the attenuation characteristics with distance (Oth et al. 2009).

At each of the 24 selected frequencies, an inversion was performed in a least-square sense and a solution $x = (\boldsymbol{A}^{\mathrm{T}}\boldsymbol{A})^{-1}\ \boldsymbol{A}^{\mathrm{T}}\boldsymbol{b}$ for a numerically stable system was computed (Menke 1989). As a result of successful inversion, the modal matrix gives the NAFs $\hat{A}_{ij}(f, r_{ij})$ one for each bin and the values of $\hat{S}_i(f)$, one for each earthquake i. In this way, the unknown values which reduce the deviation between observations and model predictions can be

$$\underbrace{\begin{bmatrix} \log_{10} U_{11} \\ . \\ . \\ \log_{10} U_{ij} \\ \vdots \\ 0 \\ 0 \\ 0 \\ 0 \\ 0 \\ 0 \end{bmatrix}}_{\textit{Data Vector}} = \left[\underbrace{\begin{matrix} 1 & 0 & 0 & . & \cdots \\ 0 & 1 & 0 & . & \cdots \\ . & . & . & . & \cdots \\ 1 & 0 & 0 & . & \cdots \\ \vdots & \vdots & \vdots & \vdots & \vdots \\ w_1 & 0 & 0 & . & \cdots \\ -\frac{w_2}{2} & w_2 & -\frac{w_2}{2} & . & \cdots \\ 0 & -\frac{w_2}{2} & w_2 & -\frac{w_2}{2} & \cdots \\ . & . & . & . & \cdots \\ . & . & . & . & \cdots \end{matrix}}_{\textit{Attenuation Parameters}} \quad \underbrace{\begin{matrix} 1 & 0 & 0 & . & \cdots \\ 1 & 0 & 0 & . & \cdots \\ . & . & . & . & \cdots \\ 0 & 1 & 0 & . & \cdots \\ \vdots & \vdots & \vdots & \vdots & \cdots \\ . & . & . & . & \cdots \\ . & . & . & . & \cdots \\ . & . & . & . & \cdots \\ . & . & . & . & \cdots \\ . & . & . & . & \cdots \end{matrix}}_{\textit{Sources}} \right] \underbrace{\begin{bmatrix} \log_{10} A_1 \\ . \\ . \\ . \\ \log_{10} A_{ND} \\ \log_{10} \hat{S}_1 \\ . \\ . \\ . \\ \log_{10} \hat{S}_{NE} \end{bmatrix}}_{\textit{Model Parameters}} \qquad (3)$$

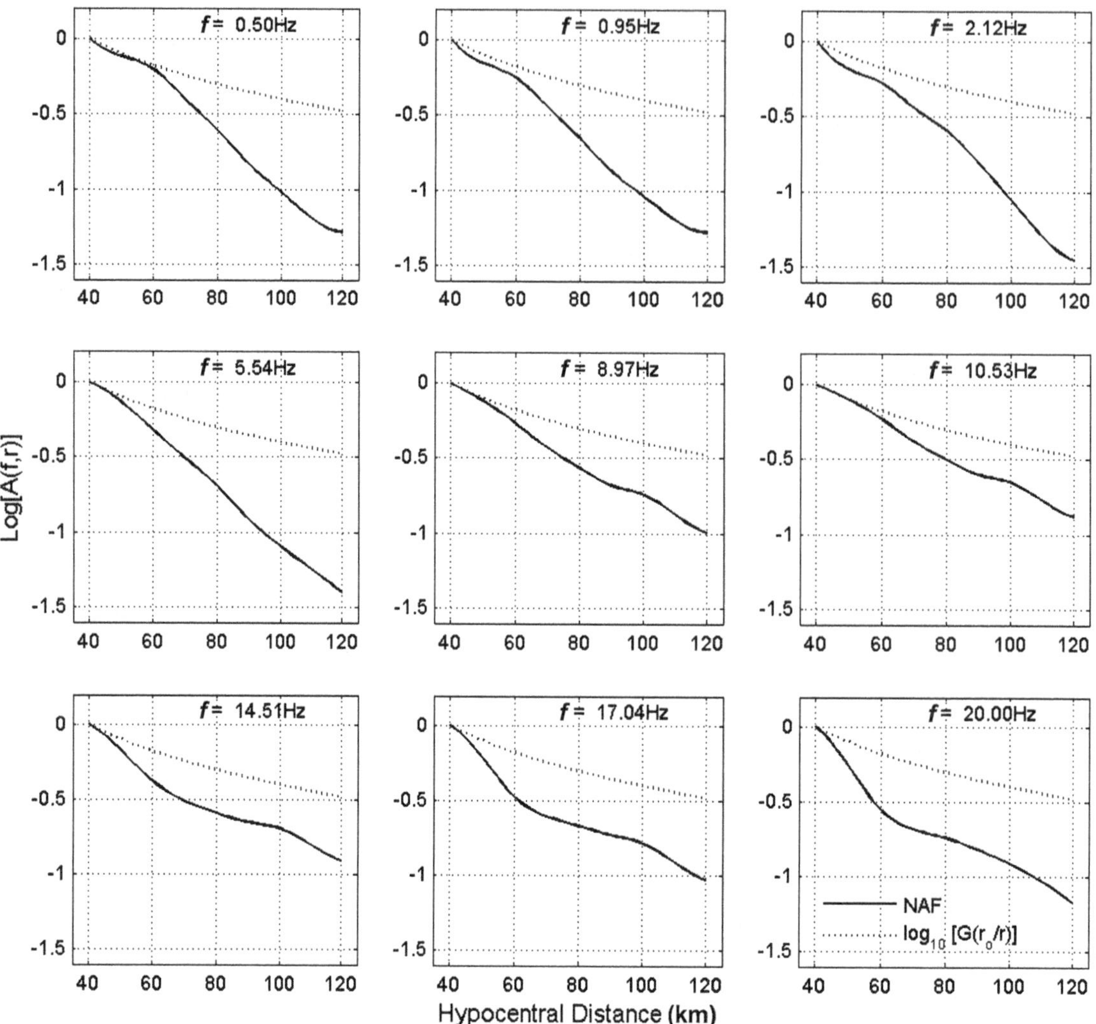

Fig. 7 Attenuation profiles observed at 9 of the 24 frequencies studied (*solid lines*) by fitting one function to the entire dataset (*dashed lines*). The functions are normalized to zero (in logarithm) at the reference distance r_0

evaluated in the form of sum of the squared residuals. The residual between the observed spectral amplitudes and those predicted by $U_{ij}(f, r_{ij}) = \hat{A}_{ij}(f, r_{ij}) \cdot \hat{S}_i(f)$, are interpreted as site effects. The attenuation curves evaluated at different frequencies are shown in Fig. 7.

4.2 Seismic quality factor (Q_s)

The attenuation function $\hat{A}_{ij}(f, r_{ij})$ combines the effect of the geometrical spreading and the anelastic attenuation in the same function. By using the non-parametric form of attenuation function, it becomes very easy to test it against any assumed geometrical spreading without repeating earlier computations.

The attenuation term $\hat{A}_{ij}(f, r_{ij})$ can be parameterized in terms of frequency-dependent quality factor Q_s (*f*) and geometric spreading *G* (*r*). Thus, we can express the attenuation function as

$$\hat{A}_{ij}(f, r_{ij}) = G(r_{ij}) \cdot \exp\left(\frac{-\pi f r_{ij}}{Q_s(f)\beta}\right). \tag{4}$$

The average S-wave velocity (β) estimated in the region is 3.2 km/s, measured from the S-P arrival times (Zhao et al. 1992; Kurahashi and Irikura 2011). The Q_s estimates are also sensitive to the choice of geometrical spreading function. Considering body wave propagation in an infinite homogeneous medium, the geometric spreading function *G* (*r*) was chosen to be $1/r$. The amplitudes are normalized to 40 km as most of the observed spectral amplitudes start at 40 km. Thus, for each frequency *f*, Eq. (4) is linearized correcting the empirical attenuation functions by the effect of geometrical spreading $G\ (r) = 40/r_{ij}$ and taking the logarithm. Thus, Eq. (4) is written as

$$a(r) = -mr, \tag{5}$$

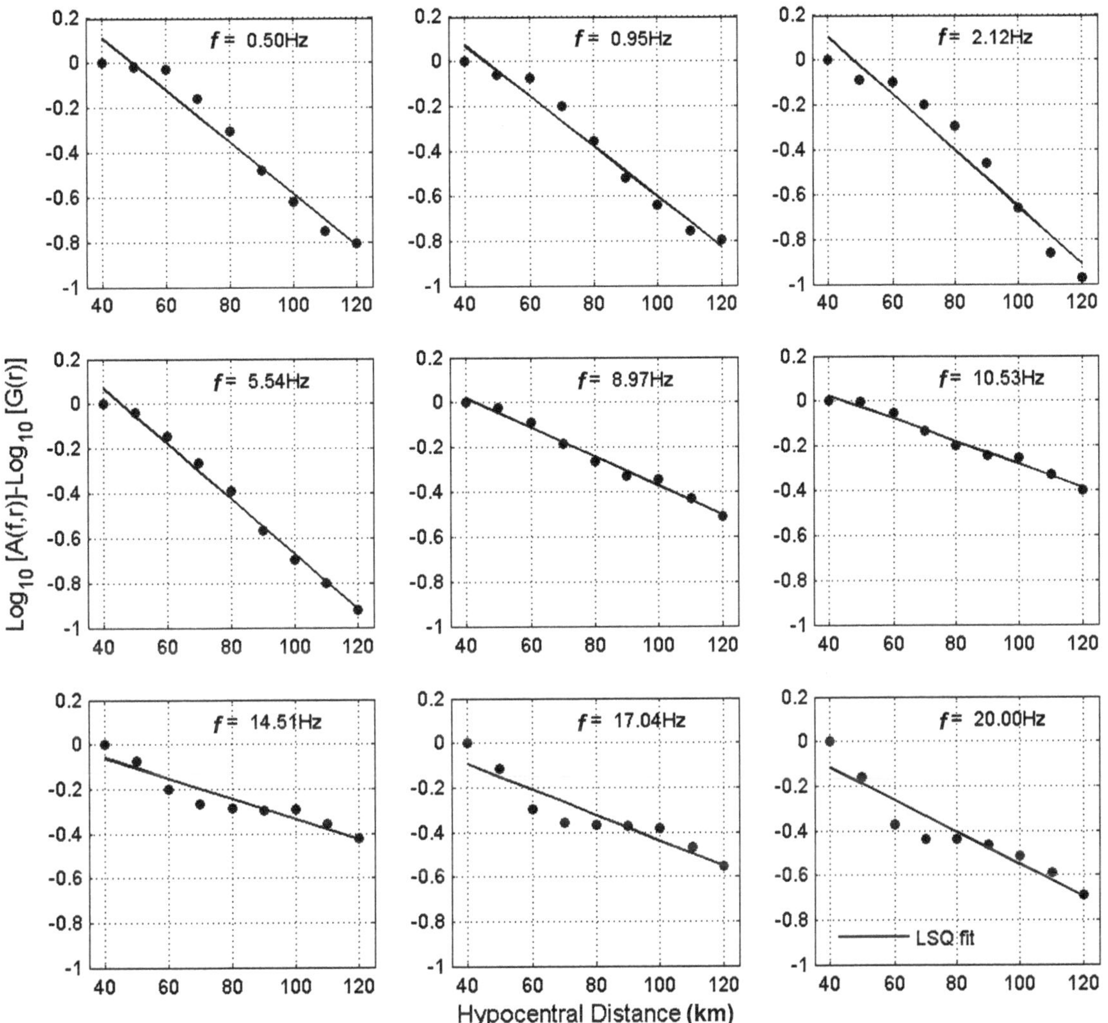

Fig. 8 Attenuation curves corrected for geometrical spreading [$\log_{10}A(f,r) - \log_{10}G(r)$ versus r] and fitted straight line (least squares fit) at 9 of the 24 frequencies studied

where $a(r) = \log_{10}\hat{A}_{ij}(f, r_{ij}) - \log_{10}G(r_{ij})$ and $m = \{[\pi fr/Q_s(f)\beta]\cdot\log_{10}(e)\}$. It is important to note that m is the resulting slope of the linear least-squares fit obtained between 40 and 120 km for each frequency analyzed. Therefore, Eq. (5) takes the following form:

$$\log_{10}\hat{A}_{ij}(f, r_{ij}) - \log_{10}G(r_{ij}) = -\frac{\pi f}{Q_s(f)\beta}r_{ij}\cdot\log_{10}e \qquad (6)$$

It becomes evident that by correcting the attenuation functions for geometrical spreading and plotting versus distance, $Q_s(f)$ can be evaluated from the slope of a linear least-squares fit. Figure 8 shows the attenuation functions corrected for the geometrical spreading and the regression used to estimate Q_s for different frequencies.

A homogeneous attenuation model for the studied region from a linear fit of the determined values of Q_s over the selected frequency band of 0.5 to 20 Hz takes the form $Q_s(f) = 33f^{1.22}$ as shown in Fig. 9. The regression error of the fit is shown by vertical error bars and can be significantly reduced by choosing multiple Q_s models over different frequency bands.

5 Results and discussion

The obtained NAFs show that in general the attenuation curves are well constrained. At higher frequencies ($f > 12$ Hz), initially the change in rate of amplitude decay is fast but gradually slows down beyond 70 km distance. A strong frequency dependence of Q_s is found between the frequency range of (0.5 and 20 Hz) and is given by the relation $Q_s(f) = 33f^{1.22}$. Below 6 Hz, Q_s can be best approximated by $Q_s(f) = 36f^{0.94}$ showing relatively weaker frequency dependence as compared to relation Q_s

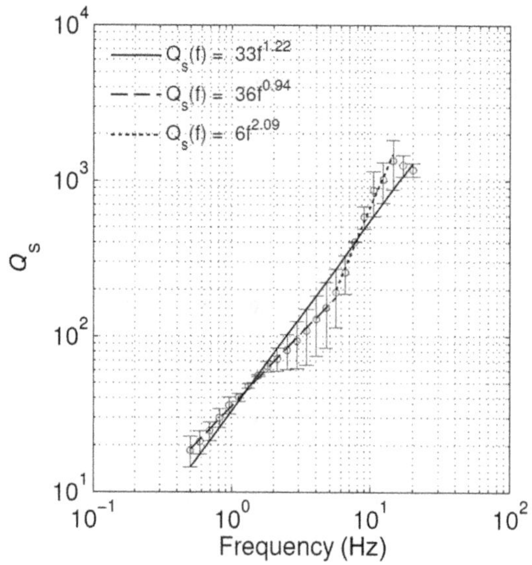

Fig. 9 Estimated Q_s-model for Eastern Tohoku obtained from the least square fit of Q_s between 0.5 and 20 Hz

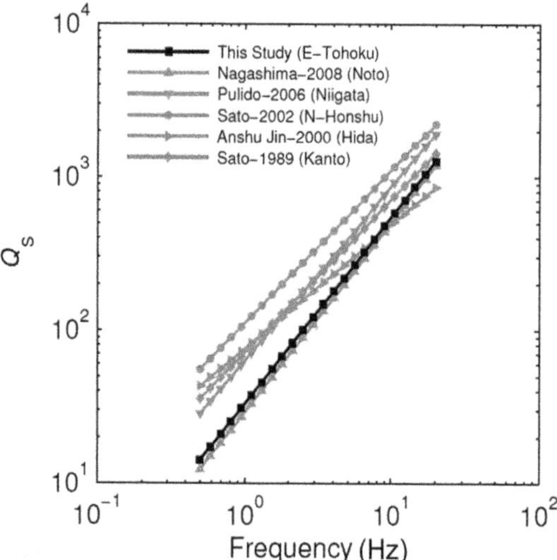

Fig. 10 Comparison of Q_s for crustal earthquakes in Japan

(f) = $6f^{2.09}$ for the frequency range between 6 and 15 Hz. This change in frequency behaviour of Q_s can be attributed to the localized discontinuities within the crust which are present in the form of faults or densely fractured zones in the region (Umino et al. 1990).

A strong attenuation in Tohoku region is quite obvious, as evident from several past studies. The Q_s estimates found in this study are comparable to those obtained for the Noto region in Japan (Itoi et al. 2008). Both functions are obtained from the crustal earthquake datasets. The slopes of the functions are quite similar, leading to the determination of almost identical Q_s (f)-model as shown in Fig. 10. The derived Q_s (f)-model is also comparable to those of crustal earthquakes at other regions in Japan.

6 Conclusions

Frequency-dependent seismic quality factor Q_s (f) has been evaluated by applying non-parametric inversion method to crustal earthquakes dataset. A strong attenuation in Eastern Tohoku region has been observed between (0.5 and 20 Hz) and is given by the relation Q_s (f) = $33\,f^{1.22}$ which is close to the background value in the region. This strong attenuation can be attributed to structural heterogeneities created by large earthquakes in the region. The estimates of Q_s could be further refined by adopting a bilinear Q_s (f)-model. The frequency dependence of Q_s in the frequency range between 0.5 and 6 Hz can be best approximated by Q_s (f) = $36\,f^{0.94}$ showing relatively weaker frequency dependence as compared to the relation Q_s (f) = $6\,f^{2.09}$ for the frequency range between 6 and 15 Hz. The observed estimates of Q_s could provide basic input for determining source and site parameters (Parolai et al. 2000). Therefore, more realistic estimates of strong ground motion parameters could be obtained for assessment of seismic hazards in the region.

Acknowledgments The study was carried out as a part of author's M.Sc Research under the project: "Strengthening of Earthquake Engineering Center", funded by Higher Education Commission, Government of Pakistan. The data used in this study were downloaded online from K-Net (Kyoshin Network) website: http://www.kik.bosai.go.jp/, managed under the framework of National Research Institute for Earth Science and Disaster Prevention (NIED). The author wishes to thank Professor Apostolos Papageorgiou for his valuable guidance and support to carry out this research.

Appendix

See Table 2.

Table 2 Earthquake coordinates and hypocentral distances

Event no.	Lat. (°)	Long. (°)	Depth (km)	M_w	IWT007 (km)	MYG002 (km)	MYG003 (km)	MYG004 (km)	MYG005 (km)	MYG011 (km)
1	39.025	140.910	13	5.0	89.38	–	62.70	63.14	105.21	105.21
2	38.456	141.160	13	4.8	111.18	101.37	52.96	62.33	–	–
3	38.458	141.150	14	5.0	111.92	–	53.93	62.83	56.70	56.70
4	38.520	141.230	11	4.4	102.00	96.09	46.31	58.82	–	53.65
5	38.466	141.190	12	4.4	108.90	99.50	51.04	61.42	–	53.59
6	38.855	142.260	18	5.1	64.57	117.45	94.98	122.56	–	–
7	39.067	140.810	9	4.6	94.85	113.26	67.73	64.33	–	106.12
8	38.998	142.060	19	3.9	46.19	113.15	85.49	–	–	–
9	39.090	140.910	13	4.6	87.74	–	66.88	67.24	–	–
10	39.142	140.940	11	5.2	82.93	–	67.69	68.95	–	–
11	39.147	140.920	11	4.3	84.47	113.47	68.89	69.51	–	–
12	38.995	140.890	10	4.8	91.22	108.19	60.06	59.19	–	96.22
13	39.010	140.940	12	4.0	87.38	108.34	–	61.09	–	–
14	39.010	140.940	12	4.0	95.33	111.27	63.49	61.12	–	99.11
15	38.997	140.840	11	5.3	83.15	112.20	–	68.44	–	–
16	39.135	140.940	11	4.6	90.04	109.62	62.47	61.58	–	–
17	39.035	140.890	10	4.0	98.15	–	54.01	50.68	–	–
18	38.873	140.860	6	4.1	110.07	–	67.97	60.54	–	–
19	38.903	140.700	10	5.4	108.61	–	62.93	52.58	–	–
20	38.903	140.700	0	4.8	122.33	119.47	73.00	–	104.82	104.82
21	38.803	140.600	10	5.0	88.90	109.63	61.52	61.94	–	104.17
22	39.020	140.910	12	4.3	–	93.79	39.99	50.22	72.33	72.33
23	38.765	141.140	9	4.5	–	93.63	41.44	51.14	59.83	59.83
24	38.603	141.150	8	4.1	–	93.69	39.89	50.32	72.41	72.41
25	38.768	141.140	9	4.1	–	92.24	38.55	47.92	67.56	67.56
26	38.711	141.130	7	4.2	–	103.48	–	–	50.12	50.12
27	38.365	141.220	13	3.5	–	101.26	53.75	62.78	53.59	53.59
28	38.433	141.160	12	5.5	–	102.76	–	–	53.18	53.18
29	38.400	141.150	12	3.5	–	102.23	–	–	51.86	51.86
30	38.388	141.160	11	3.5	–	100.73	–	63.17	52.60	52.60
31	38.433	141.190	12	3.9	–	102.55	–	64.67	52.28	52.28
32	38.386	141.150	11	3.9	–	101.47	–	–	54.66	54.66
33	38.458	141.190	14	3.6	–	102.61	56.76	67.17	50.96	50.96
34	38.390	141.210	13	4.0	–	101.15	–	61.99	56.45	56.45
35	38.483	141.180	14	4.0	–	100.26	–	61.62	53.99	53.99
36	38.456	141.170	12	4.0	–	100.59	52.52	61.87	54.01	54.01
37	38.450	141.170	12	4.2	–	97.91	49.84	61.30	–	–
38	38.465	141.220	11	4.1	–	100.68	–	–	56.28	56.28
39	38.481	141.160	13	3.6	–	101.16	–	62.17	55.15	55.15
40	38.518	141.230	10	3.5	–	95.21	–	58.05	52.79	52.79
41	38.465	141.190	13	3.9	–	100.55	–	62.24	54.51	54.51
42	38.483	141.130	14	4.1	–	102.24	52.76	61.32	58.35	58.35
43	38.513	141.220	11	3.5	–	96.33	–	59.08	53.38	53.38
44	38.485	141.190	12	3.8	–	98.96	–	60.40	54.45	54.45
45	38.383	141.170	11	3.8	–	102.18	–	65.17	51.26	51.26
46	38.381	141.190	12	3.5	–	102.63	–	–	51.01	51.01

Table 2 continued

Event no.	Lat. (°)	Long. (°)	Depth (km)	M_w	IWT007 (km)	MYG002 (km)	MYG003 (km)	MYG004 (km)	MYG005 (km)	MYG011 (km)
47	38.466	141.160	13	3.6	–	101.23	–	–	55.84	55.84
48	38.202	141.150	12	4.5	–	112.59	72.87	79.13	–	–
49	38.198	141.150	12	4.5	–	112.85	–	79.45	41.73	41.73
50	38.412	141.120	13	4.0	–	104.26	–	64.18	44.38	44.38
51	39.102	140.870	13	3.9	–	115.26	69.29	68.55	–	–
52	38.867	140.860	9	4.1	–	105.69	55.77	53.22	–	–
53	38.880	140.760	7	3.9	–	109.34	61.54	54.92	–	–
54	38.908	140.760	7	4.1	–	109.97	62.55	56.02	–	–
55	38.883	140.760	8	4.0	–	109.88	61.80	55.67	–	–
56	39.090	140.930	11	4.2	–	110.62	64.41	65.34	–	–
57	39.038	140.860	12	4.5	–	112.37	65.10	63.81	–	–
58	38.757	140.780	3	4.6	–	103.54	55.32	47.82	–	–
59	38.753	140.760	3	4.6	–	104.49	56.61	48.49	–	–
60	39.103	140.840	12	3.9	–	115.78	70.24	68.40	–	–
61	39.062	140.860	10	4.3	–	111.75	65.49	63.76	–	–
62	38.895	140.940	8	3.9	–	101.91	51.53	51.93	–	–
63	38.500	141.200	10	4.4	–	96.35	47.10	58.22	41.25	41.25
64	38.286	140.760	13	5.0	–	122.82	80.21	75.77	77.62	77.62
65	38.270	140.760	13	4.0	–	123.54	–	–	77.67	77.67
66	38.286	140.780	14	3.6	–	122.84	–	–	77.03	77.03
67	38.703	141.130	8	3.7	–	93.10	39.42	–	67.39	67.39
68	39.021	140.910	12	3.7	–	–	39.94	50.27	72.33	72.33
69	38.766	141.140	9	3.5	–	–	49.05	60.53	52.16	52.16
70	38.478	141.220	11	3.9	–	–	–	63.64	53.96	53.96
71	38.436	141.180	13	3.8	–	–	–	62.11	55.29	55.29
72	38.481	141.170	12	3.5	–	–	–	60.21	55.19	55.19
73	38.468	141.190	12	4.0	–	–	–	61.31	53.67	53.67
74	38.446	141.190	14	4.1	–	–	–	64.10	54.63	54.63
75	38.383	141.160	12	3.8	–	–	–	65.79	52.55	52.55
76	38.425	141.210	11	3.7	–	–	–	63.35	50.41	50.41
77	38.428	141.190	12	3.9	–	–	–	63.64	52.03	52.03
78	38.383	141.160	11	3.8	–	–	–	65.09	51.51	51.51
79	38.463	141.190	12	3.9	–	–	–	61.58	53.47	53.47
80	38.496	141.180	12	4.4	–	–	–	59.65	55.37	55.37
81	38.451	141.170	13	3.9	–	–	–	62.69	54.73	54.73
82	38.903	140.600	10	4.3	–	–	–	65.02	111.15	111.15
83	38.903	140.700	10	4.5	–	–	–	60.54	105.16	105.16
84	38.803	140.600	10	4.2	–	–	–	62.64	104.82	104.82
85	38.903	140.600	10	4.3	–	–	–	65.02	111.15	111.15

References

Andrews D (1986) Objective determination of source parameters and similarity of earthquakes of different size. Earthquake source mechanics, American Geophysical Monograph, Maurice Series 6, vol 37. American Geophysical Union, Washington, DC, pp 259–267

Bindi D, Parolai S, Grosser H, Milkereit C, Karakisa S (2006) Crustal attenuation characteristics in Northwestern Turkey in the range from 1 to 10 Hz. Bull Seismol Soc Am 96:200–214

Castro RR, Anderson JG, Singh SK (1990) Site response, attenuation and source spectra of S-waves along the Guerrero, Mexico, subduction zone. Bull Seismol Soc Am 80:1481–1503

Itoi T, Nagashima I, Uchiyama Y (2008) Spectral amplitude and phase characteristics of shallow crustal earthquake based on linear inversion of ground motion spectra and some engineering applications. In: The 14th world conference on earthquake engineering, October 12–17, 2008, Beijing, China

Kaminuma K, Aki K (1963) Crustal structure in Japan from the phase velocity of Rayleigh waves, part 2. Bull Earthq Res Inst 42:19–38

Kurahashi S, Irikura K (2011) Source model for generating strong ground motions during the 2011 off the Pacific coast of Tohoku Earthquake. Earth Planets Space 63:571–576

Menke W (1989) Geophysical data analysis: discrete inverse theory. Int. Geophys. Series, vol 45. Academic Press, New York, p 289

Okada Y, Kasahara K, Hori S, Obara K, Sekiguchi S, Fujiwara H, Yamamoto A (2004) Recent progress of seismic observation networks in Japan—Hi-net, F-net, K-NET and KiK-net—. Earth Planets Space 56:xv–xviii

Oth A, Parolai S, Bindi D, Wenzel F (2009) Source spectra and site response from S-waves of intermediate-depth Vrancea, Romania, earthquakes. Bull Seismol Soc Am 99:235–254

Oth A, Parolai S, Bindi D (2011) Spectral analysis of K-NET and KiK-net data in Japan. Part I: database compilation and peculiarities. Bull Seismol Soc Am 101(2):652–656

Parolai S, Bindi D, Augliera P (2000) Application of the generalized inversion technique (GIT) to a microzonation study: numerical simulations and comparison with different site-estimation techniques. Bull Seismol Soc Am 90:286–297

Pavlenko OV, Wen KL (2008) Estimation of nonlinear soil behavior during the 1999 Chi-Chi, Taiwan earthquake. Pure Appl Geophys 165:373–407

Smith SW (1999) The Scientist and Engineer's Guide to Digital Signal Processing, 2nd edn. California Technical Publishing, San Diego, pp 277–282

Umino N, Hasegawa A, Takagi A (1990) The relationship between seismicity patterns and fracture zones beneath northeastern Japan. Tohoku Geophys J 33(2):149–162

Zhao D, Horiuchi S, Hasegawa A (1992) Seismic velocity structure of the crust beneath the Japan islands. Tectonophysics 212:289–301

14

The strain seismograms of P- and S-waves of a local event recorded by four-gauge borehole strainmeter

Zehua Qiu · Shunliang Chi · Zhenming Wang · Seth Carpenter · Lei Tang · Yanping Guo · Guang Yang

Abstract At a sampling rate of 100 samples per second, the YRY-4 four-gauge borehole strainmeters (FGBS) are capable of recording transient strains caused by seismic waves such as P and S waves or strain seismograms. At such a high sampling rate, data from the YRY-4 strainmeters demonstrate fairly satisfactory self-consistency. The strain tensor seismograms demonstrate the senses of motion of P waves, that is, the type of seismic wave travels in the direction of the maximum normal strain change. The observed strain patterns of S waves significantly differ from those of P waves and should contain information about the source mechanism. Spectrum analysis shows that the strain seismograms are consistent with conventional broadband seismograms from the same site.

Keywords Four-gauge borehole strainmeter (FGBS) · Strain seismogram · P wave · S wave · Self-consistency · Spectrum analysis

Z. Qiu (✉) · L. Tang · Y. Guo
Institute of Crustal Dynamics, China Earthquake Administration, Beijing 100085, China
e-mail: qzhbh@163.com

S. Chi
Earthquake Administration of Hebi, Hebi 458000, He'nan Province, China

Z. Wang · S. Carpenter
Kentucky Geological Survey, University of Kentucky, Lexington, KY 40506, USA

G. Yang
Guza Seismic Station, Earthquake Administration of Sichuan Province, Kangding 626001, Sichuan Province, China

1 Introduction

Two-dimensional borehole tensor strainmeters have been well developed since the 1970s both in theory (Pan 1977; Su 1977; Gladwin and Hart 1985; Hart et al. 1996; Roeloffs 2010; Qiu et al. 2013) and in technique (Ouyang 1977; Gladwin 1984; Ishii 2001; Chi et al. 2009). Many such instruments have been deployed in countries such as China, the United States and Japan to monitor tectonic movements related to earthquakes, volcanoes, episodic tremors and slips, and other events (Linde et al. 1996; Qiu et al. 2007; Wang et al. 2008; Voight et al. 2010; Hodgkinson et al. 2010; Chardot et al. 2010; Qiu et al. 2011; Hawthorne and Rubin 2013).

Theoretically, seismic wave propagation in a continuous medium is governed by the stress–strain relationship (Timoshenko and Goodier 1951; Bullen 1963; Stein and Wysession 2003). Therefore, a tensor strainmeter may have some advantages for studying wave propagation over a conventional seismometer. To date, however, a systematic study on strain seismograph has not been carried out. A high-quality 2D strainmeter, the YRY-4 four-gauge borehole strainmeter (FGBS), has been successfully developed and deployed in China (Chi et al. 2009; Qiu et al. 2013). In the China Earthquake Administration's observatory, FGBS data are routinely sampled at a rate of one sample per minute. In recent years, however, experiments have shown that the FGBS is capable of working at a sampling rate as high as 100 samples per second, with good signal-to-noise ratio.

2 Strain seismograms

On March 17, 2008, a FGBS at the Guza site (Fig. 1) recorded a local earthquake of magnitude 3.5 in Sichuan,

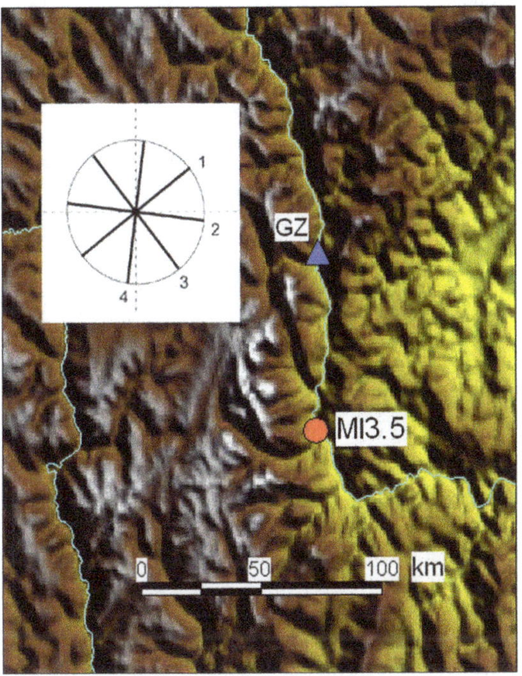

Fig. 1 Locations of the Guza site, and the *M*L3.5 earthquake. Directions of the four gauges in the FGBS are plotted

China, at 100 samples per second. The data from each component (Fig. 2) are high quality, with good signal-to-noise ratio.

In order to use the data of the four components to determine the strain tensor, we applied corrections based on relative in situ calibration, given in Qiu et al. (2013), to the raw time series, and the mean was removed from the data. Next, to satisfy a self-consistency criterion (Su 1977; Qiu et al. 2013), we verified equality of the sum of gauges S1 and S3 and the sum of gauges S2 and S4. As shown in Fig. 2, the seismograms of S1+S3 and S2+S4 are similar. We further examined self-consistency *via* plotting of S1+S3 *versus* S2+S4 by using the initial parts of the waveforms of the P wave and the S wave, separately (Fig. 3). The lines fitting separately from S1+S3 and S2+S4 are close (Fig. 3), which demonstrates a high degree of self-consistency of the FGBS data, even for such a high sampling rate. Some data are not very well self-consistent, especially for P wave. The reason is not yet known.

3 Strain pattern of P waves

Strain seismograms make it possible to plot deformation with time, while seismic waves pass through a site. When the data are self-consistent, the principal strain orientations are credible. Figure 4 shows the series of deformation patterns during the initial 0.24 s of the P wave (indicated in Fig. 2). The dotted ovals are drawn from a reference solid circle with exaggerated strains. Since the dominant period of the P wave here is nearly 0.2 s, the process plotted is a little longer than one cycle of the deformational vibration.

By observing this series of plots, instant by instant, we can see that the axes of the ovals are mainly along two constant directions. One is nearly NS when the oval is larger than the reference circle; the other is nearly EW

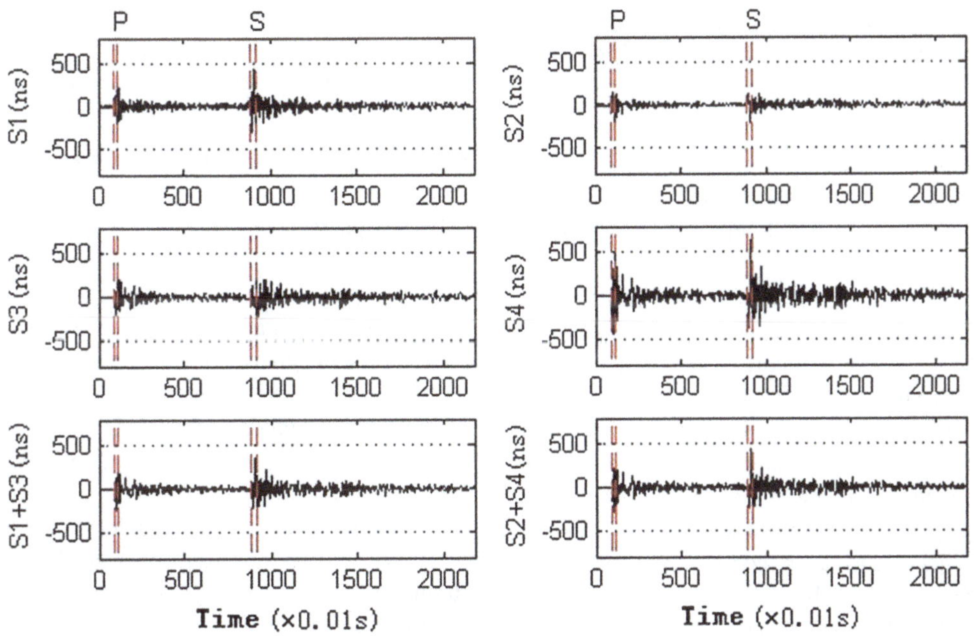

Fig. 2 Strain seismograms recorded at the site Guza from an *M*L3.5 earthquake of March 17, 2008. *Dashed lines* indicate the time intervals to be studied in detail for P waves and S waves

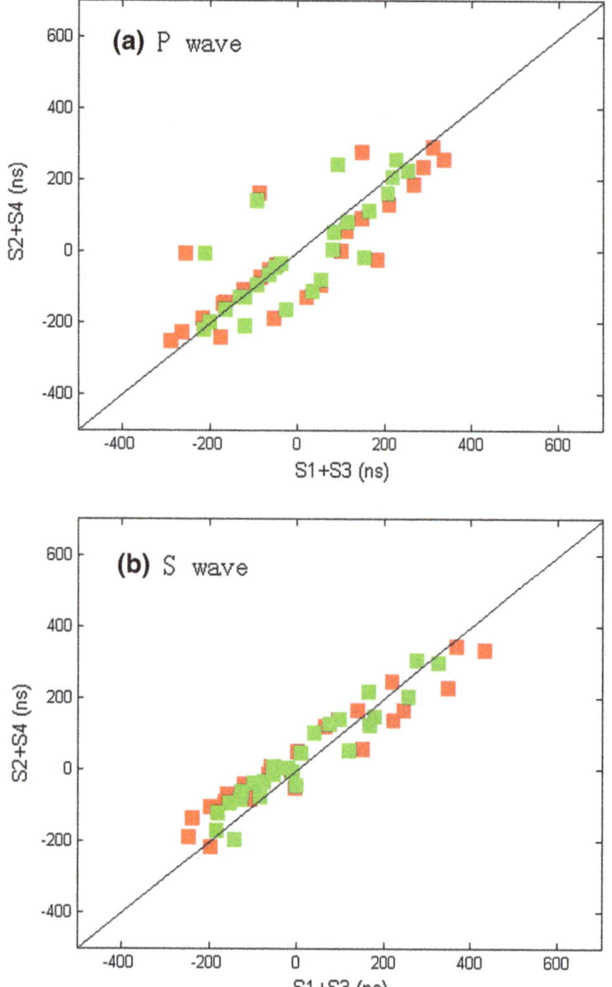

Fig. 3 Graphs of S1+S3 *versus* S2+S4. **a** and **b** are for the initial parts of the P wave and S wave, as indicated by *dashed lines* in Fig. 2. *Red* stand for before correction, *green* after correction. Corrections in detail are given in Qiu et al. (2013)

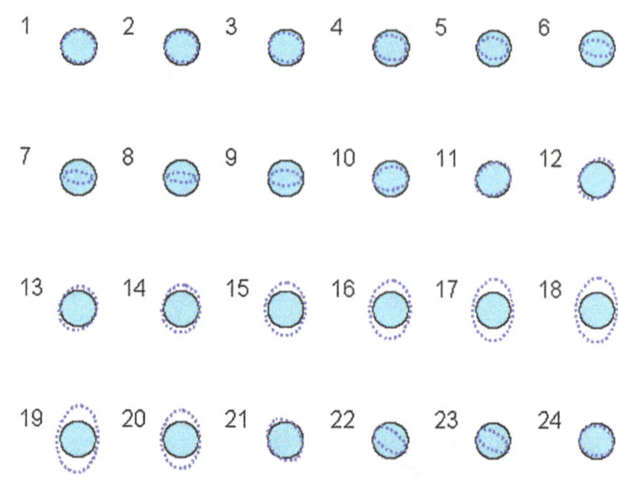

Fig. 4 Illustration of the recorded deformation for multiple instants at Guza while the initial part of the P wave passes through. The time interval is indicated in Fig. 2. The *dotted ovals* are drawn from a *reference circle* with exaggerated strains

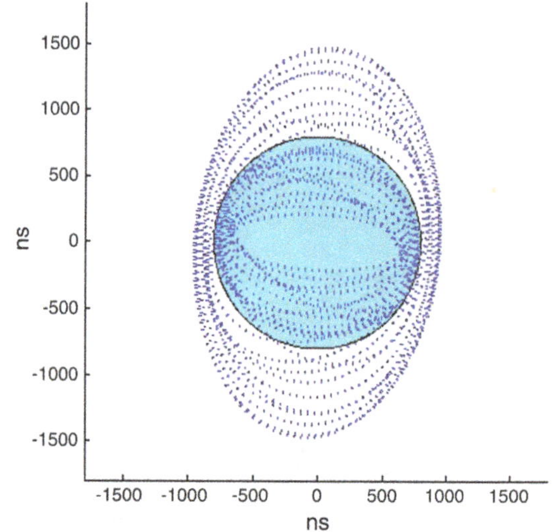

Fig. 5 *Stacked ovals* of the P wave deformations plotted in Fig. 4

when the oval is smaller than the circle. In fact, the NS axis changes much more in length than the EW axis. This can be clearly shown in Fig. 5 as all the ovals are stacked together.

Comparison of Figs. 1–5 shows that the most important feature of the P wave is that the diameter that changes most greatly coincides with the azimuth of the P wave's ray path, which here comes from the south.

The ovals actually indicate that there are shear strains in P wave. Qiu and Chi (2013) studied the graphs of S1–S3 and S2–S4 recorded by the YRY-4 of the identical event studied in this paper. The graphs show that P wave strain contains significant shear strain components, which are obviously bigger than possible error. According to the theory of seismology, P wave has been clarified to have no rotation but volumetric strain and shear strains.

4 Strain pattern of S wave

Figures 6 and 7 show the series of deformation patterns during the initial part of the S wave (indicated in Fig. 2, total 0.24 s). Dotted ovals are drawn in the same way as those in Fig. 5. The extension directions of the ovals are obviously different from those of the P wave. In this case, there are also two nearly constant directions: one is toward the northeast when the oval is larger; the other is toward the northwest when the oval is smaller.

The S wave is different from the P wave in terms of strain. Unlike the strain pattern of the P wave, the strain pattern of the S wave does not indicate the location of the epicenter.

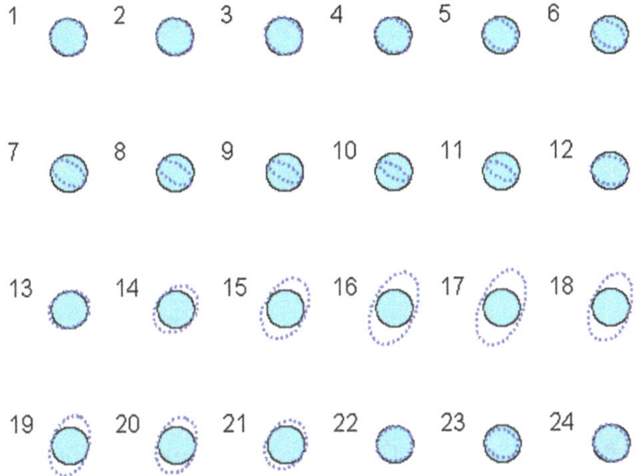

Fig. 6 Illustration of the recorded deformation for multiple instants at Guza while the initial part of the S wave passes through. The time interval is indicated in Fig. 2. The *dotted ovals* are drawn from a *reference circle* with exaggerated strains

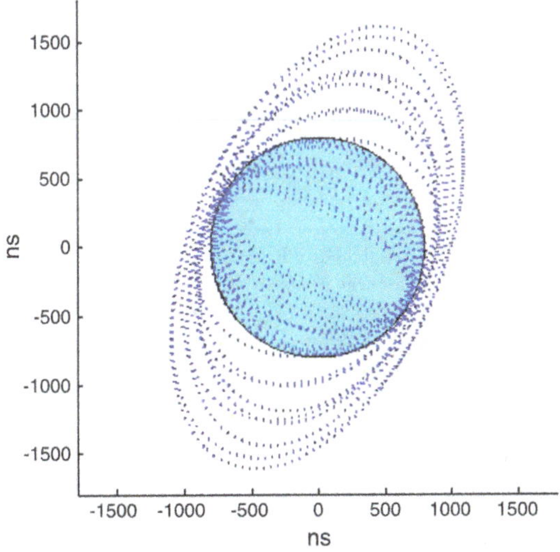

Fig. 7 *Stacked ovals* of the S wave deformations plotted in Fig. 6

5 Comparison to conventional seismograms

A tensor strainmeter is different from a conventional seismometer in that a conventional seismometer records the rigid, translational motion of a site, whereas a tensor strain seismograph observes the internal deformation at a site. The spectra of strain seismograms and conventional seismograms should be in agreement with each other, however, because they both are governed by wave propagation. Thus, strain seismograms and conventional seismograms should be comparable in terms of waveform and frequency content.

We compared the strain seismograms, rotated into cardinal directions according to Qiu et al. (2013), with seismograms recorded by a broadband CTS-1 seismograph (flat response from 120 s to 20 Hz), co-located at the Guza site for the March 17, 2008, *M*L3.5 event (Fig. 1). To the first order, the records strongly resemble each other (Fig. 8), though the instruments measure different physical quantities (strain *versus* velocity). Distinct P and S wave arrivals are visible in the strain and seismic waveforms: impulsive P and SH phases. In addition, we observed the same pattern of P and S phase polarities in the seismic data and the rotated strain data. To elaborate, P and S phase first motions observed on the north component of the seismic data (VN) are both positive, and the first motions on the north component of the strain data are both negative. Likewise, the P and S phase first motions on the east component of the seismic data (VE) and the strain data are all positive. These observations suggest that the sense of motion correlates with the strain being either compressive or extensional.

We also compared the frequency content within each signal by estimating the power spectral densities (PSD) using Welch's method (Fig. 9). The smoothed spectral density estimates show well-defined peaks in the seismic data and the north and east strain data, ε_N and ε_E, respectively, at 7 Hz, corresponding to the S wave. All PSDs start falling off at around 13 Hz, which corresponds to frequencies in the P wave, but show local maxima and minima at the same frequencies above 13 Hz. There is an elevated response in the strain data to frequencies below 5 Hz, compared to the seismic data. Power around 5 Hz is contained in both the P and S waves in the strain data and may reflect an increased sensitivity in the strainmeter at this frequency, perhaps a resonance.

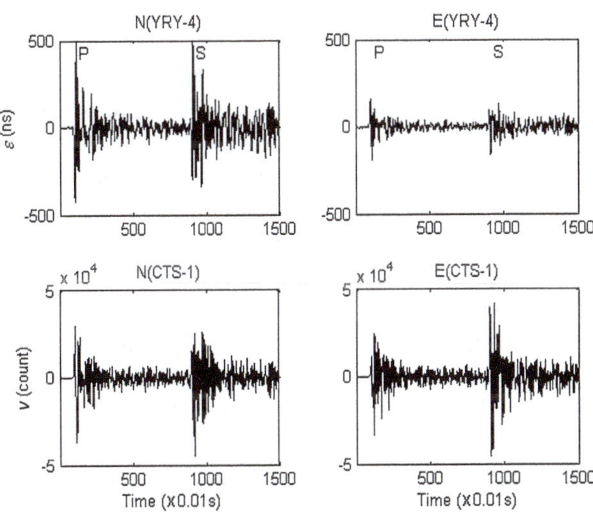

Fig. 8 Seismograms showing rotated nanostrain (*top row*) and velocity (*bottom row*; in counts) from the YRY-4 FGBS and the co-located CTS-1 seismograph, respectively, at station Guza. North–south nanostrain and seismic records are in the *left column*, and east–west records are in the *right column*

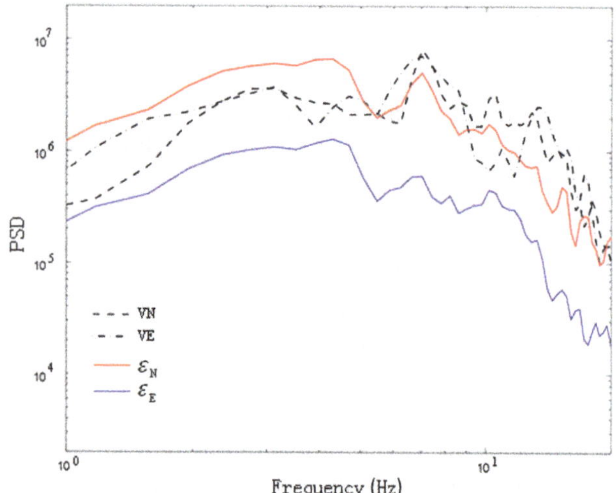

Fig. 9 Power spectral densities of the seismic (VN, VE) data and the rotated strain data (ε_N and ε_E). The units of the seismic PSDs are in counts^2/Hz, whereas the strain data PSDs are in nanostrain^2/Hz. PSDs are estimated using Welch's method with 2s windows, overlapping 50 % to calculate each periodogram

These comparisons show that, to the first order, the records strongly resemble each other and suggest that the FGBS is sensitive to seismic waves at high frequencies. Further study is needed in order to determine if the FGBS could serve as a replacement for a seismometer.

6 Conclusions

At high sampling rates (i.e., 100 samples per second), the FGBS could be used as a strain seismometer. The gauges record consistent spectra across a broadband of frequencies and are comparable to those of a conventional seismometer. The data are self-consistent, even at a high sampling rate. The FGBS strain recordings verify that a P wave can be defined as the type of seismic wave that always travels in the direction of maximum normal strain. The strain patterns of S waves are different from those of P waves. Strain seismograms might have some advantages in comparison with conventional seismograms for studying P and S waves, as well as the seismic source.

This paper presents the observation result of the YRY-4 instrument at Guza of the *M*L3.5 event. It provides new meaningful information about body seismic waves. Among the tens of sites in China equipped with this kind of instrument, Guza is the first where the test of 100 samples per second is carried out. More observations are needed to better understand this phenomenon.

The relation between strain records and translation records needs to be further explained. Some studies have been published concerning this issue. For instance, Bouchon and Aki (1982) simulated the time histories of strain, tilt, and rotation in the vicinity of earthquake faults in comparison to translational velocities. However, it is about near-field situation where the distance from the epicenter is less than the length of the seismic fault. Therefore, it does not fit the observation which is done 70 km away from the small event whose seismic fault should be no more than a few kilometers in size.

Acknowledgments This study was supported by the Special Fund for Earthquake Research in the Public Interest (No. 201108009).

References

Barbour AJ, Agnew DC (2012) Detection of seismic signals using seismometers and strainmeters. Bull Seismol Soc Am 102:2484–2490

Bouchon M, Aki K (1982) Strain, tilt, and rotation associated with strong ground motion in the vicinity of earthquake faults. Bull Seismol Soc Am 72(5):1717–1738

Bullen KE (1963) An Introduction to the Theory of Seismology. Cambridge University Press, Cambridge, p 381

Chardot L, Voight B, Foroozan R, Sacks S, Linde A, Stewart R, Hidayat D, Clarke A, Elsworth D, Fournier N, Komorowski JC, Mattioli G, Sparks RSJ, Widiwijayanti C (2010) Explosion dynamics from strainmeter and microbarometer observations, Soufrière Hills volcano, Montserrat, 2008–2009. Geophys Res Lett 37:6

Chi SL, Chi Y, Deng T, Liao CW, Tang XL, Chi L (2009) The necessity of building national strain-observation network from the strain abnormality before wenchuan earthquake. Recent Dev World Seismol 1:1–13 **(in Chinese with English abstract)**

Frank FC (1966) Deduction of earth strains from survey data. Bull Seismol Soc Am 56:35–42

Gladwin MT (1984) High precision multi-component borehole deformation monitoring. Rev Sci Instrum 55:2011–2016

Gladwin MT, Hart R (1985) Design parameters for borehole strain instrumentation. Pure appl Geophys 123:59–80. doi:10.1007/BF00877049

Hart R, Gladwin MT, Gwyther RL, Agnew DC, Wyatt FK (1996) Tidal calibration of borehole strain meters: removing the effects of small-scale heterogeneity. J Geophys Res 101:25553–25571. doi:10.1029/96JB02273

Hawthorne JC, Rubin AM (2013) Short-time scale correlation between slow slip and tremor in Cascadia. J Geophys Res 118(3):1316–1329

Hodgkinson K, Mencin D, Borsa A, Jackson M (2010) Plate boundary observatory strain recordings of the February 27, 2010, *M*8.8 Chile Tsunami. Seismol Res Lett 81:3

Ishii H (2001) Development of new multi-component borehole instrument. *Report of Tono Research Institute of Earthquake Science* 6:5–10 **(in Japanese)**

Linde AT, Gladwin MT, Johnston M, Gwyther RL, Bilham RG (1996) A slow earthquake sequence on the San Andreas fault. Nature 383:65–68

Ouyang ZX (1977) RDB-1 type electric capacity strainmeter. *Selected Papers of the National Conference on Stress Measurement*, Part 2: 337–348 **(in Chinese)**

Pan LZ (1977) On the formulae of ground stress measurement. *Selected Papers of the National Conference on Stress Measurement*, Part 1: 1–41 **(in Chinese)**

Qiu ZH, Chi SL (2013) Shear strains of P wave observed with an YRY-4 borehole strainmeter. Earthquake 33(4):64–70 **(in Chinese with English Abstract)**

Qiu ZH, Ma J, Chi SL, Liu HM (2007) Earth's free torsional oscillations of the great Sumatra earthquake observed with borehole shear strainmeter. Chin J Geophys 50(3):797–805 **(in Chinese with English Abstract)**

Qiu ZH, Zhang BH, Chi SL, Tang L, Song M (2011) Abnormal strain changes observed at Guza before the Wenchuan earthquake. Sci China Ser D-Earth Sci 54(2):157–314

Qiu ZH, Tang L, Zhang BH, Guo YP (2013) In situ calibration of and algorithm for strain monitoring using four-gauge borehole strainmeters (FGBS). J Geophys Res 118:1609–1618. doi:10.1002/jgrb50112

Roeloffs E (2010) Tidal calibration of plate boundary observatory borehole strainmeters: roles of vertical and shear coupling. J Geophys Res 115:B06405. doi:10.1029/2009JB006407

Stein S, Wysession M (2003) *An introduction to Seismology, Earthquake, and Earth Structure*. Blackwell Publishing, Oxford

Su KZ (1977) Methods of relative measurement of ground stress. In: *Selected Papers of the National Conference on Stress Measurement*, Part 1: 42–61 **(in Chinese)**

Timoshenko S, Goodier JN (1951) *Theory of Elasticity*, 2nd edn. McGraw-Hill, New York

Voight B, Hidayat D, Sacks S, Linde A, Chardot L, Clarke A, Elsworth D, Foroozan R, Malin P, Mattioli G, McWhorter N, Shalev E, Sparks RSJ, Widiwijayanti C, Young SR (2010) Unique strainmeter observations of Vulcanian explosions, Soufrière Hills Volcano, Montserrat, July 2003. Geophys Res Lett 37:L00E18. doi:10.1029/2010GL042551

Wang K, Dragert H, Kao H, Roeloffs E (2008) Characterizing an "uncharacteristic" ETS event in northern Cascadia. Geophys Res Lett 35:L15303. doi:10.1029/2008GL034415

15

Out-of-plane (SH) soil-structure interaction: a shear wall with rigid and flexible ring foundation

Thang Le · Vincent W. Lee · Hao Luo

Abstract Soil-structure interaction (SSI) of a building and shear wall above a foundation in an elastic half-space has long been an important research subject for earthquake engineers and strong-motion seismologists. Numerous papers have been published since the early 1970s; however, very few of these papers have analytic closed-form solutions available. The soil-structure interaction problem is one of the most classic problems connecting the two disciplines of earthquake engineering and civil engineering. The interaction effect represents the mechanism of energy transfer and dissipation among the elements of the dynamic system, namely the soil subgrade, foundation, and superstructure. This interaction effect is important across many structure, foundation, and subgrade types but is most pronounced when a rigid superstructure is founded on a relatively soft lower foundation and subgrade. This effect may only be ignored when the subgrade is much harder than a flexible superstructure: for instance a flexible moment frame superstructure founded on a thin compacted soil layer on top of very stiff bedrock below. This paper will study the interaction effect of the subgrade and the superstructure. The analytical solution of the interaction of a shear wall, flexible-rigid foundation, and an elastic half-space is derived for incident SH waves with various angles of incidence. It found that the flexible ring (soft layer) cannot be used as an isolation mechanism to decouple a superstructure from its substructure resting on a shaking half-space.

T. Le · V. W. Lee
Sonny Astani Civil & Environmental Engineering Department, University of Southern California, Los Angeles, CA 90089, USA

H. Luo (✉)
HNTB Corporation, 200 E Sandpointe Ave #200, Santa Ana, CA 92707, USA
e-mail: haoluo1979@gmail.com

Keywords Out-of-plane SH waves · Closed-form analytic solution · Rigid-flexible foundation · Fourier-bessel series · Soil-structure interaction

List of symbols

a Radius of the semi-circular rigid foundation
$\bar{a}$ Radius of the semi-circular flexible foundation
a_{n} Coefficients of the free-field waves
B Width of building
A_n Complex constants
$B_n^{(1)}, B_n^{(2)}$ Complex constants
C_β Shear wave velocity in the soil
$C_{\beta_{\mathrm{b}}}$ Shear wave velocity in the building
$C_{\beta_{\mathrm{f}}}$ Shear wave velocity in the flexible foundation
f_{f} Force per unit length acting on the rigid foundation from the flexible foundation
f_{b} Height of the building
$\mathrm{H}_n^{(1)}(x)$ Hankel function of the first kind with argument x and order n
$\mathrm{H}_n^{(2)}(x)$ Hankel function of the second kind with argument x and order n
i Imaginary unit
n Subscripts used for sequence number
$J_n(x)$ Bessel function of the first kind with argument x and order n
k Wave number in the soil $k = \omega^2/C_\beta$
k_{b} Wave number in the building $k_{\mathrm{b}} = \omega^2/C_{\beta_{\mathrm{b}}}$
M_{B} Mass of shear wall per unit length
M_{R} Mass of rigid foundation per unit length
M_{F} Mass of flexible ring per unit length
γ Angle of incidence for SH waves

Δ	Amplitude of the displacement of the foundation
w	Amplitude of the displacement of the total wave field in the soil
$w^{(ff)}$	Amplitude of the displacement of the free-field wave in the half-space soil
$w^{(B)}$	Amplitude of the displacement of the wave field in the building
$w^{(R)}$	Amplitude of the displacement of the wave field in the rigid foundation
$w^{(S)}$	Amplitude of the displacement of the scattered wave field in the soil
$w^{(i)}$	Amplitude of the displacement of the incident plane wave in the soil
$w^{(r)}$	Amplitude of the displacement of the reflected plane wave in the soil
μ	Shear modulus of the soil
μ_b	Shear modulus of the shear wall
ρ	Density of the soil
ρ_b	Density of the shear wall
ρ_r	Density of the rigid foundation
ρ_f	Density of the flexible ring
ω	Circular frequency of the incident SH waves
δ_n	Unit impulse function
ε	Dimensionless parameters

1 Introduction

The study of soil-structure interaction (SSI) began during the late 19th century. In the early part of the 20th century, it slowly evolved and developed due to research on the design of nuclear power plants and on improvements in the safety of building structures due to significant earthquake events around the world. More recently, the sophistication of SSI modeling has grown at a fast pace due to the increasing computational ability of computer technology and in response to the demands of research in the field of non-linearity of building materials and soil media.

Reissner (1936) observed the natural effect of soil inertia and established that it lay in the inertial properties of the soil media. He also discovered that the radiation damping into the soil contributed a great deal to the response of a structure, thus marking the beginning of SSI study using the analytical method. A number of years later, Housner (1957) demonstrated that the variation of density and elasticity in the soil media caused a change in seismic wave propagation velocity that led to the reflection and refraction of incoming seismic energy. This phenomenon is understood as wave passage or kinematic interaction. Additionally, the weight of a structure generates inertial forces that impinge on the soil media when responding to seismic waves, causing additional deformation in the soil known as inertial interaction. The integration of the dynamic response of a structure and the supporting soil media causes the inertia effects. The structure deforms to dissipate the energy caused by incoming seismic waves and, in turn, the waves scatter away from the structure, increasing the soil deformation.

The topic was taken up again in 1969 by Luco (1969), who focused on the diffraction of normal incidence plane SH waves by an elastic shear wall resting on a rigid semi-circular foundation embedded in soil media. Trifunac (1972) generalized Luco's analytical solution into cases for arbitrary oblique incidence angles of SH waves. Analytical solutions of the dynamic interaction of shear walls with circular rigid foundations embedded in the half-space were derived and evaluated. This analysis demonstrated that waves scattered from a rigid foundation contribute significantly to the surface ground motion near the shear wall and at distances at least one order of magnitude greater than the characteristic length of the foundation. Therefore, waves reflected and diffracted by the foundation must not be neglected when the Fourier amplitude ratios of accelerograms recorded in and around the building are used to study SSI. For this reason, when considering rigid shear walls founded on hard soil excited by low frequency waves (long period waves), it is possible to obtain base displacements and base shear forces which are higher when the values are computed neglecting SSI. However, when considering rigid shear walls founded on soft soil excited by higher frequency waves, the SSI has a significant effect on displacement and base shear force and should be carefully considered. In more flexible structures, the effect of interaction can vary widely for a single structure-soil couplet and must be considered as depending on the harmonic interaction of the native vibrational frequency of the superstructure with the transmitted frequency of specific seismic waves in the soil as affected by the soil's stiffness.

In the early 1970s, Lee and Westley (1973) investigated the influence of SSI effects on the seismic response of nuclear reactors using a three-dimensional model subjected to vertical propagation of SH waves. The normal incidence SH waves used the analytical method along the two orthogonal directions and a spring-mass model for structures attached to the foundations. Luco and Contesse (1973) presented closed-form analytic solutions for the two-dimensional out-of-plane problems related to the interaction between elastic shear walls fixed on rigid circular foundations that are subjected to vertical and oblique angles of incident harmonic SH waves. Wong and Trifunac (1975) extended the analytical solutions for the incident plane SH waves to shallow or deep elliptical rigid foundations, as well as to multiple buildings and foundations. The solutions to these problems showed that buildings that are closely spaced could affect the fundamental frequencies (or period) of the neighboring buildings due to SSI.

Lee (1979) studied three-dimensional analytical solutions for the interaction of a single degree-of-freedom oscillator supported by a semi-spherical foundation for the harmonic P-, SV-, and SH waves. Kobori and Kusakabe (1980) investigated rigid rectangular and circular foundations welded to the surface of elastic half-spaces, and subjected to harmonic seismic waves. Triantafyllidis and Prange (1988) studied the dynamic interaction of two rigid circular foundations embedded in elastic half-space, and subjected to Rayleigh waves impinging at an arbitrary angle. These studies demonstrated that forces react on the foundations perpendicular to the incidence of propagation, in addition to forces in the direction of the motion. Todorovska (1993) studied the in-plane foundation-soil interaction of circular foundations embedded in elastic soil media. The research mainly focused on the influences of wave passage and the depth of the foundation below the half-space for in-plane SSI. In the same year, Hryniewicz (1993) investigated two two-dimensional trip foundations based on a semi-infinite medium embedded in a homogeneous half-space excited by anti-plane SH waves.

Figure 1 is the realization of the two-dimensional mathematical model in this paper presenting the interaction

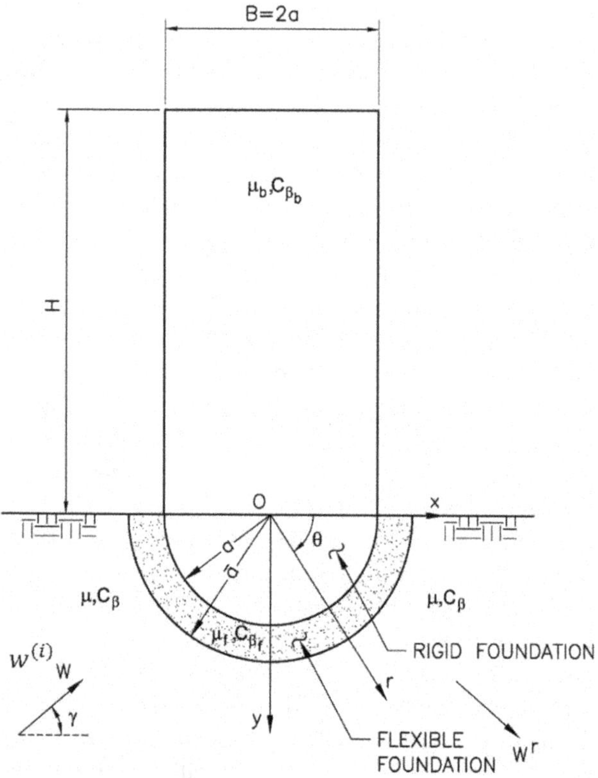

Fig. 1 The mathematical model SSI with semi-circular flexible and rigid foundation

of an elastic shear wall (structure), flexible-rigid foundation, and the elastic half-space for incident plane SH waves. The foundation consists of a semi-circular rigid foundation which is wrapped with an elastic semi-circular flexible foundation. The lower semi-circular section can be modeled as flexible for soft soil or as rigid for hard soil—such as bedrock—which provides a more accurate mathematical model for the shear wall or structure interacting with the soil media. This accuracy is important in modeling an SSI with a stiffer layer of soil overlaying a more flexible layer. An engineer goes through a decision making process when selecting the optimum type of foundation system for soil deposits that are soft and not suitable to support the superstructure. Selecting an optimum system is based on the principle that cost-effective alternatives such as soft ground improvements must be sought first before considering relatively costly foundation alternatives (mat or pile deep foundation). Soil treatment and stabilization are techniques to enhance some aspects of soil behavior and improve the strength and bearing capacity of soft ground conditions. Through the process of soil treatment, the property of soil supporting the superstructure foundation is modified and improved in comparison to the surrounding strata. This layer of modified soil will alter the seismic wave propagating from the half-space to the superstructure foundation. The model in this paper will accurately analyze this interaction and assist in selection of optimum foundation systems.

In this paper, the displacement of the shear wall structure and flexible/rigid foundation and the ground motion close to the subjected structure are investigated and compared the results with Luco (1969) and Trifunac (1972) which studied the interaction of a shear wall, rigid foundation, and the half-space for incident SH waves. Luo (2008) also studied this problem but the derivation and final equations here are much simplified with improved convergency and accuracy of numerical results. Moreover, the emphasis here is different, where the graphs computed here are confirmed with the results of Trifunac (1972) for only rigid foundation case, and the effect of shear wall structure response due to the flexible ring is studied. The base shear force of the wall structure is also studied with various thicknesses and stiffnesses of the flexible ring.

2 The mathematical model

2.1 The model

The model studied in this paper is a two-dimensional rectangular building resting on a semi-circular rigid foundation of radius a which is wrapped with an elastic semi-circular flexible foundation of radius $\bar{a}$ embedded in a half-space, as illustrated in Fig. 1. All materials here are assumed to be homogeneous, elastic, and isotropic. The material constants, namely shear modulus and wave speed of the half-space soil, building, and flexible foundation are denoted by μ, C_β and μ_b, C_{β_b}, and μ_f, C_{β_f}. The contacts between the soil, foundation, and building are assumed to be fixed with no slippage between them and with the assumption that the foundation is removable.

A train of parallel harmonic incident SH waves impinge on the foundation from half-space at an incidence angle γ with respect to the horizontal axis. The width and height of the structure are $2a$ and H, respectively. A Cartesian coordinate system (x, y) and a corresponding polar coordinate system (r, θ) have been defined with the origin at the center of the semi-circular foundation.

2.2 The free-field waves in the half-space

The incident wave field consists of a train of plane waves of unit amplitude with harmonic frequency ω, wave speed C_β, and wave number $k = k_\beta = \omega/C_\beta$. The incident waves can be expressed in both the rectangular and polar coordinates as follows:

$$w^{(i)}(x,y) = e^{i(k_x x - k_y y)} = e^{i(x\cos\gamma - y\sin\gamma)},$$
$$w^{(i)}(r,\theta) = e^{ikr(\cos\gamma\cos\theta - \sin\gamma\sin\theta)} = e^{ikr\cos(\gamma+\theta)}, \tag{1}$$

and the reflected plane waves can be written as

$$w^{(r)}(x,y) = e^{i(k_x x + k_y y)} = e^{i(x\cos\gamma + y\sin\gamma)},$$
$$w^{(r)}(r,\theta) = e^{ikr(\cos\gamma\cos\theta + \sin\gamma\sin\theta)} = e^{ikr\cos(\gamma-\theta)}. \tag{2}$$

The $e^{-i\omega t}$ harmonic time factor is present in all wave equations, and will be understood to be omitted from all equations. Here γ is the angle of incidence or reflection with respect to the horizontal axis; $k_x = k\cos\gamma$ and $k_y = k\sin\gamma$ represent the components of the SH wave number k along the x- and y-axes, respectively. Applying the Jacobi–Anger Expansion (Pao and Mow 1973),

$$e^{\pm ikr\cos\theta} = \sum_{n=0}^{\infty} \varepsilon_n (\pm i)^n J_n(kr)\cos n\theta, \tag{3}$$

where $i = \sqrt{-1}$ is the imaginary complex unit, and $J_n(.)$ is the Bessel function of the first kind with order n.

$$e^{ikr\cos(\gamma\pm\theta)} = \sum_{n=0}^{\infty} \varepsilon_n i^n J_n(kr)\cos[n(\gamma\pm\theta)],$$
$$e^{ikr\cos(\gamma\pm\theta)} = \sum_{n=0}^{\infty} \varepsilon_n i^n J_n(kr)[\cos n\gamma\cos n\theta \pm \sin n\gamma\sin n\gamma]. \tag{4}$$

The two formulas in Eqs. (1) and (2) can be expanded into infinite series by using polar coordinates (r, θ). The free-field wave field is then given by the sum of the represented waves that are finite everywhere in the half-space for $n = 0, 1, 2, 3\ldots$

$$w^{(ff)}(r,\theta) = w^{(i)} + w^{(r)} = e^{ikr\cos(\gamma+\theta)} + e^{ikr\cos(\gamma-\theta)},$$
$$w^{(ff)}(r,\theta) = \sum_{n=0}^{\infty} 2\varepsilon_n i^n J_n(kr)\cos n\gamma\cos n\theta = \sum_{n=0}^{\infty} a_n J_n(kr)\cos n\theta, \tag{5}$$

where $a_n = 2\varepsilon_n i^n \cos n\gamma$ are the coefficients of the free-field waves, and $\varepsilon_0 = 1$, $\varepsilon_n = 2$ for $n > 0$. The wave field in the half-space scattered from the flexible foundation is written as

$$w^{(S)}(r,\theta) = \sum_{n=0}^{\infty} A_n H_n^{(1)}(kr)\cos(n\theta) \text{ for } n \geq a, \tag{6}$$

where A_n are the unknown complex numbers to be determined by boundary conditions and the wave function, and $H_n^{(1)}e^{-i\omega t}$ represents outgoing waves toward infinity satisfying Sommerfeld's radiation condition. The expression of the wave inside the flexible foundation is

$$w^{(F)}(r,\theta) = \sum_{n=0}^{\infty} \left[B_n^{(1)} H_n^{(1)}(k_f r) + B_n^{(2)} H_n^{(2)}(k_f r)\right]\cos n\theta \tag{7}$$

for $a \leq r \leq \bar{a}$ and $0 \leq \theta \leq \pi$ where $H_n^{(1)}(k_f r)$ and $H_n^{(2)}(k_f r)$ are the Hankel functions of the first or second kind with argument $k_f r$ and order n; $B_n^{(1)}$ and $B_n^{(1)}$ are the unknown complex numbers to be determined by boundary conditions and the wave functions.

2.3 The wave field within the structure

As pointed out by Trifunac (1972), the displacement of the shear wall, in the z-direction (out-of-plane), has the same harmonic frequency ω as the rigid foundation, and $w^{(B)}$ must satisfy the Helmholtz wave equation with y, the axis pointing vertically down (Fig. 1):

$$\frac{\partial w^{(B)}}{\partial y^2} + k_b^2 w^{(B)} = 0 \text{ for } -H \leq y \leq 0 \tag{8}$$

with $k_b = \omega/C_{\beta_b}$ being the building shear wave number, and C_{β_b} the wave speed in the shear wall. The shear wall must satisfy the boundary conditions of

$$\sigma_{yz} = \mu_b \frac{\partial w^{(B)}}{\partial y} = 0 \text{ at } y = -H, \text{ at top of shear wall;}$$
$$w^{(B)} = \Delta e^{-i\omega t} \text{ at } y = 0, \text{ at bottom of shear wall.} \tag{9}$$

Dependence on x in the shear wall is eliminated in Eq. (8) by the assumption that the foundation is rigid. The solution of Eqs. (8) and (9) is then given by

$$w^{(\mathrm{B})}(y) = \Delta \mathrm{e}^{-\mathrm{i}\omega t}(\cos k_{\mathrm{b}}y - \tan k_{\mathrm{b}}H \sin k_{\mathrm{b}}y). \tag{10}$$

The shear stress along the interface of building and foundation could be derived as

$$\tau_{yz}\big|_{y=0} = \mu_{\mathrm{b}} \frac{\partial w^{(\mathrm{B})}}{\partial y}\bigg|_{y=0} = -(\mu_{\mathrm{b}}k_{\mathrm{b}} \tan k_{\mathrm{b}}H)\Delta. \tag{11}$$

The base shear force per unit length of the shear wall f_z^{b} can be expressed as

$$f_z^{\mathrm{b}} = -\Delta\mu_{\mathrm{b}}k_{\mathrm{b}}(2a)\tan k_{\mathrm{b}}H = -\omega^2 M_{\mathrm{B}}\left(\frac{\tan k_{\mathrm{b}}H}{k_{\mathrm{b}}H}\right)\Delta. \tag{12a}$$

From Eq. (12a), a dimensionless function proportional to the base shear force acting on the shear wall can be expressed as

$$\frac{f_z^{\mathrm{b}}}{\omega^2 M_{\mathrm{B}}} = -\left(\frac{\tan k_{\mathrm{b}}H}{k_{\mathrm{b}}H}\right)\Delta. \tag{12b}$$

2.4 The action of flexible foundation on the rigid foundation

The action of flexible foundation on the rigid foundation, f_z^{f}, can be expressed in term of stress as follows:

$$f_z^{\mathrm{f}} = -\int_0^{\pi} \tau_{rz}\big|_{r=a} a\mathrm{d}\theta$$

$$f_z^{\mathrm{f}} = -\mu_{\mathrm{f}}k_{\mathrm{f}}a \sum_{n=0}^{\infty}\left[B_n^{(1)}\mathrm{H}_n^{(1)\prime}(k_{\mathrm{f}}a) + B_n^{(2)}\mathrm{H}_n^{(2)\prime}(k_{\mathrm{f}}a)\right] \times \int_0^{\pi}\cos n\theta \mathrm{d}\theta, \tag{13}$$

where $\int_0^{\pi}\cos n\theta\mathrm{d}\theta = \begin{cases}\pi, & n=0\\ 0, & n\neq 0\end{cases}$. So f_z^{f} can be rewritten as

$$f_z^{\mathrm{f}} = \pi\mu_{\mathrm{f}}k_{\mathrm{f}}a\left[B_0^{(1)}\mathrm{H}_1^{(1)}(k_{\mathrm{f}}a) + B_0^{(2)}\mathrm{H}_1^{(2)}(k_{\mathrm{f}}a)\right]. \tag{14}$$

3 The boundary conditions

3.1 Displacement and stress continuity

The free-stress boundary conditions of the ground surface should be satisfied by the free-field waves $w^{(\mathrm{ff})}$ and the scattered waves $w^{(\mathrm{S})}$. The displacement and stress continuity equations along the semi-circular interface at $0 \leq \theta \leq \pi$ and $r = \bar{a}$, respectively are

$$w^{(\mathrm{ff})} + w^{(\mathrm{S})}\big|_{r=\bar{a}} = w^{(\mathrm{F})}\big|_{r=\bar{a}} \quad \text{for } 0 \leq \theta \leq \pi, \tag{15}$$

$$\mu\frac{\partial}{\partial r}\left(w^{(\mathrm{ff})} + w^{(\mathrm{S})}\right)\bigg|_{r=\bar{a}} = \mu_{\mathrm{f}}\frac{\partial}{\partial r}\left(w^{(\mathrm{F})}\right)\bigg|_{r=\bar{a}} \quad \text{for } 0 \leq \theta \leq \pi. \tag{16}$$

Substitution of Eqs. (5), (6), and (7) into Eqs. (15) and (16) leads to the following two boundary condition equations, for $n = 0, 1, 2, 3, \ldots$,

$$a_n J_n(k\bar{a}) + A_n\mathrm{H}_n^{(1)}(k\bar{a}) = B_n^{(1)}\mathrm{H}_n^{(1)}(k_{\mathrm{f}}\bar{a}) + B_n^{(2)}\mathrm{H}_n^{(2)}(k_{\mathrm{f}}\bar{a}), \tag{17}$$

$$\begin{aligned} &a_n J_n'(k\bar{a}) + A_n\mathrm{H}_n^{(1)\prime}(k\bar{a}) \\ &= \kappa\left[B_n^{(1)}\mathrm{H}_n^{(1)\prime}(k_{\mathrm{f}}\bar{a}) + B_n^{(2)}\mathrm{H}_n^{(2)\prime}(k_{\mathrm{f}}\bar{a})\right], \end{aligned} \tag{18}$$

where $\kappa = \mu_{\mathrm{f}}k_{\mathrm{f}}/\mu k$ is the material property ratio in the equation for the stress continuity boundary condition.The boundary condition at the interface of rigid and flexible foundations can be expressed as given below:

$$w^{(\mathrm{F})}\big|_{r=a} = w^{(\mathrm{R})}\big|_{r=a} = \Delta\mathrm{e}^{-\mathrm{i}\omega t} \tag{19}$$

Substitute Eq. (7) into Eq. (19) to solve for $B_n^{(2)}$,

$$\begin{aligned} &B_0^{(1)}\mathrm{H}_0^{(1)}(k_{\mathrm{f}}a) + B_0^{(2)}\mathrm{H}_0^{(2)}(k_{\mathrm{f}}a) = \Delta \quad \text{for } n = 0\\ &B_n^{(1)}\mathrm{H}_n^{(1)}(k_{\mathrm{f}}a) + B_n^{(2)}\mathrm{H}_n^{(2)}(k_{\mathrm{f}}a) = 0 \quad \text{for } n > 0\end{aligned} \tag{20}$$

$B_n^{(2)}$ can be written in terms of $B_n^{(1)}$ and Δ as follows:

$$B_n^{(2)} = \begin{cases}\dfrac{\Delta - B_0^{(1)}\mathrm{H}_0^{(1)}(k_{\mathrm{f}}a)}{\mathrm{H}_0^{(2)}(k_{\mathrm{f}}a)} & \text{for } n = 0\\ -\dfrac{B_n^{(1)}\mathrm{H}_n^{(1)}(k_{\mathrm{f}}a)}{\mathrm{H}_n^{(2)}(k_{\mathrm{f}}a)} & \text{for } n > 0.\end{cases} \tag{21}$$

3.2 The dynamic equation for the rigid foundation, $w^{\mathrm{f}} = \Delta\mathrm{e}^{-\mathrm{i}\omega t}$

As pointed out by Luco (1969) and Trifunac (1972), displacement of the foundation Δ can be determined by the kinetic equation for the rigid foundation,

$$M_{\mathrm{R}}\ddot{w}^{(\mathrm{R})} = -(f_z^{\mathrm{f}} + f_z^{\mathrm{b}})\mathrm{e}^{-\mathrm{i}\omega t}, \tag{22}$$

where $\ddot{w}^{(\mathrm{R})} = -\Delta\omega^2\mathrm{e}^{-\mathrm{i}\omega t}$, M_{R} is the mass of the rigid foundation per unit depth in the z-axis and $w^{(\mathrm{R})}$ represents the displacement function of the rigid foundation in terms of time factor t as described in Eq. (19).The foundation displacement Δ can be solved from Eqs. (12a), (14), and (22),

$$\Delta = \left[\frac{\mu_{\mathrm{f}}k_{\mathrm{f}}\pi a\left(\mathrm{H}_1^{(1)}(k_{\mathrm{f}}a) - \frac{\mathrm{H}_0^{(1)}(k_{\mathrm{f}}a)}{\mathrm{H}_0^{(2)}(k_{\mathrm{f}}a)}\mathrm{H}_1^{(2)}(k_{\mathrm{f}}a)\right)}{\omega^2 M_{\mathrm{R}} + \mu_{\mathrm{b}}k_{\mathrm{b}}(2a)\tan k_{\mathrm{b}}H - \mu_{\mathrm{f}}k_{\mathrm{f}}\pi a\left(\frac{\mathrm{H}_1^{(2)}(k_{\mathrm{f}}a)}{\mathrm{H}_0^{(2)}(k_{\mathrm{f}}a)}\right)}\right]B_0^{(1)}. \tag{23}$$

Equation (23) can be further simplified as

$$\Delta = \left[\frac{-\frac{4i}{\pi k_f a H_0^{(2)}(k_f a)}}{\frac{k_f \bar{a}^2}{2a}\left(\frac{M_R}{M_F}+\frac{M_B}{M_F}\frac{\tan k_b H}{k_b H}\right)-\frac{H_1^{(2)}(k_f a)}{H_0^{(2)}(k_f a)}}\right] B_0^{(1)} = \Delta_0 B_0^{(1)}. \quad (24)$$

where

$$\Delta_0 = \frac{-\frac{4i}{\pi k_f a H_0^{(2)}(k_f a)}}{\frac{k_f \bar{a}^2}{2a}\left(\frac{M_R}{M_F}+\frac{M_B}{M_F}\frac{\tan k_b H}{k_b H}\right)-\frac{H_1^{(2)}(k_f a)}{H_0^{(2)}(k_f a)}} \quad (25)$$

M_B, M_R, and M_F are the masses of the building, rigid foundation, and flexible ring, respectively; ρ_b, ρ_r, ρ_f stand for the density of those three media, sequentially; and the Wronskian $W\left[H_n^{(1)}(k_f a), H_n^{(2)}(k_f a)\right] = \left[H_n^{(1)\prime}(k_f a)H_n^{(2)}(k_f a) - H_n^{(1)}(k_f a)H_n^{(2)\prime}(k_f a)\right] = -4i/\pi k_f a$. Other terms can be found in the "Appendix" section.

As pointed out by Luco (1969), the natural frequencies of a shear wall on a fixed foundation correspond to $k_b H = (2n+1)\pi/2$. And Δ becomes zero at the values of $k_b H$ mentioned in the equation above.

Substitute Eq. (24) into Eq. (21), $B_n^{(1)}$ can be derived explicitly in terms of $B_n^{(2)}$,

$$B_n^{(2)} = \begin{cases} \frac{\Delta_0 B_0^{(1)} - B_0^{(1)} H_0^{(1)}(k_f a)}{H_0^{(2)}(k_f a)} & \text{for } n = 0 \\ -\frac{B_n^{(1)} H_n^{(1)}(k_f a)}{H_n^{(2)}(k_f a)} & \text{for } n > 0 \end{cases}$$

$$B_n^{(2)} = \left[\frac{\delta_n \Delta_0 - H_n^{(1)}(k_f a)}{H_n^{(2)}(k_f a)}\right] B_n^{(1)} \quad \text{for } n = 0,\ 1,\ 2,\ 3,\ 4\ldots \quad (26)$$

where $\delta_n = \begin{cases} 1 & \text{for } n = 0 \\ 0 & \text{for } n > 0 \end{cases}$.

By substituting Eq. (26) into Eqs. (17) and (18), we can solve for wave function coefficients A_n and $B_n^{(1)}$ explicitly.

$$a_n J_n(k\bar{a}) + A_n H_n^{(1)}(k\bar{a}) = \left[H_n^{(1)}(k_f \bar{a}) + \left(\frac{\delta_n \Delta_0 - H_n^{(1)}(k_f a)}{H_n^{(2)}(k_f a)} H_n^{(2)}(k_f \bar{a})\right)\right] B_n^{(1)}, \quad (27)$$

$$a_n J_n'(k\bar{a}) + A_n H_n^{(1)\prime}(k\bar{a}) = \kappa \left[H_n^{(1)\prime}(k_f \bar{a}) + \left(\frac{\delta_n \Delta_0 - H_n^{(1)}(k_f a)}{H_n^{(2)}(k_f a)} H_n^{(2)\prime}(k_f \bar{a})\right)\right] B_n^{(1)}. \quad (28)$$

From Eq. (27), A_n can be derived and expressed in terms of a_n and $B_n^{(1)}$.

$$A_n = \left[\frac{G_n^{(1)}}{H_n^{(1)}(k\bar{a}) H_n^{(2)}(k_f a)}\right] B_n^{(1)} - \left[\frac{J_n(k\bar{a})}{H_n^{(1)}(k\bar{a})}\right] a_n, \quad (29)$$

where $G_n^{(1)} = H_n^{(1)}(k_f \bar{a}) H_n^{(2)}(k_f a) - H_n^{(1)}(k_f a) H_n^{(2)}(k_f \bar{a}) + \delta_n \Delta_0 H_n^{(2)}(k_f \bar{a})$.

Substitute Eq. (29) into Eq. (28), $B_n^{(1)}$ can be solved explicitly.

$$B_n^{(1)} = \frac{\left[J_n'(k\bar{a}) H_n^{(1)}(k\bar{a}) - J_n(k\bar{a}) H_n^{(1)\prime}(k\bar{a})\right] H_n^{(2)}(k_f a) a_n}{\kappa G_n^{(2)} H_n^{(1)}(k\bar{a}) - G_n^{(1)} H_n^{(1)\prime}(k\bar{a})}, \quad (30)$$

where $G_n^{(2)} = H_n^{(1)\prime}(k_f \bar{a}) H_n^{(2)}(k_f a) - H_n^{(1)}(k_f a) H_n^{(2)\prime}(k_f \bar{a}) + \delta_n \Delta_0 H_n^{(2)\prime}(k_f \bar{a}) = -4i/\pi k_f a + \delta_n \Delta_0 H_n^{(2)\prime}(k_f \bar{a})$. Solve for $B_n^{(1)}$ from Eq. (30),

$$B_n^{(1)} = \left[\frac{\left(-\frac{2i}{\pi k\bar{a}}\right) H_n^{(2)}(k_f a)}{-G_n^{(1)} H_n^{(1)\prime}(k\bar{a}) + \kappa G_n^{(2)} H_n^{(1)}(k\bar{a})}\right] a_n, \quad (31)$$

where Wronskian $W\left(J_n(k\bar{a}), H_n^{(1)}(k\bar{a})\right) = -\left(J_n(k\bar{a}) H_n^{(1)\prime}(k\bar{a}) - J_n'(k\bar{a}) H_n^{(1)}(k\bar{a})\right) = -2i/\pi k\bar{a}$. Derive equation for $B_0^{(1)} (n = 0)$ from Eq. (31),

$$B_0^{(1)} = \frac{\left(-\frac{2i}{\pi k\bar{a}}\right) H_0^{(2)}(k_f a) a_0}{-G_0^{(1)} H_0^{(1)\prime}(k\bar{a}) + \kappa G_0^{(2)} H_0^{(1)}(k\bar{a})} = \frac{\left(-\frac{2i}{\pi k\bar{a}}\right) H_0^{(2)}(k_f a) a_0}{G_0^{(1)} H_1^{(1)}(k\bar{a}) + \kappa G_0^{(2)} H_0^{(1)}(k\bar{a})} \quad (32)$$

By combining Eqs. (28), (29), and (31), we can derive the expressions for wave function coefficients A_n and $B_n^{(2)}$ as the following:

$$A_n = \left[\frac{\left(-\frac{2i}{\pi k\bar{a}}\right)\left(\frac{G_n^{(1)}}{H_n^{(1)}(k\bar{a})}\right)}{-G_n^{(1)} H_n^{(1)\prime}(k\bar{a}) + \kappa G_n^{(2)} H_n^{(1)}(k\bar{a})} - \frac{J_n(k\bar{a})}{H_n^{(1)}(k\bar{a})}\right] a_n, \quad (33)$$

$$B_n^{(2)} = \left[\frac{\left(-\frac{2i}{\pi k\bar{a}}\right)\left[\delta_n \Delta_0 - H_n^{(1)}(k_f a)\right]}{-G_n^{(1)} H_n^{(1)\prime}(k\bar{a}) + \kappa G_n^{(2)} H_n^{(1)}(k\bar{a})}\right] a_n. \quad (34)$$

4 Numerical analysis of displacement

As pointed out by Trifunac (1972), the condition for the envelope of the rigid foundation displacement, Δ, corresponding to the case of which $\kappa = 1$, $\bar{a} \to a$, is given by

$$\Delta_e = \left[J_1(ka) - \frac{J_0(ka)}{H_0^{(1)}(ka)} H_1^{(1)}(ka)\right] \times \left[\frac{J_0^2(ka) + Y_0^2(ka)}{Y_0(ka) J_1(ka) - Y_1(ka) J_0(ka)}\right] a_0. \quad (35)$$

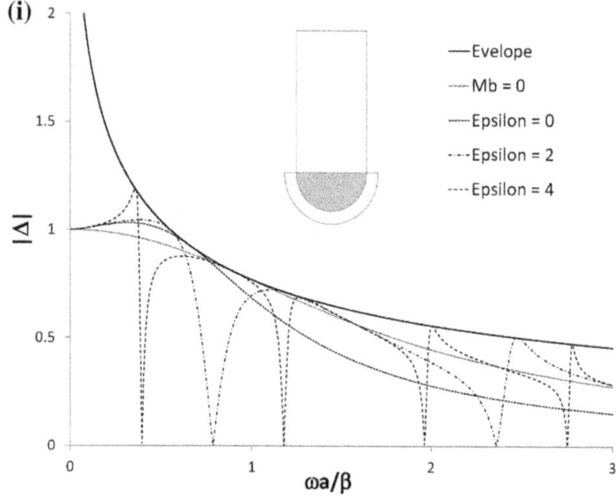

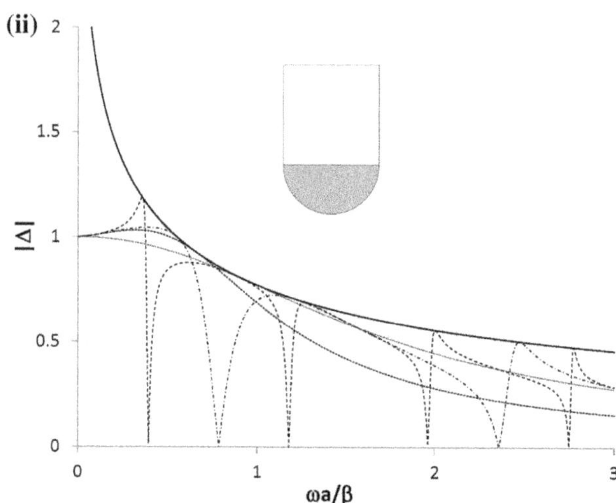

Fig. 2 Effect of interaction on Δ the amplitude of foundation vibration. **i** $\bar{a}/a = 1$, $\kappa = 1$, $M_B/M_S = 1$, $M_F/M_S = 1$, $\varepsilon = 0, 2, 4$. **ii** $M_B/M_S = 1$, $M_F/M_S = 1$, $\varepsilon = 0, 2, 4$ (Trifunac 1972)

The backbone curve of Δ could be understood as the displacement of the rigid foundation whose density is identical to that of the surrounding soil by setting $M_B/M_F = 0$ and $M_R/M_F = 1$.

$$\Delta_0 = \left[\frac{-\dfrac{4\mathrm{i}}{\pi ka \mathrm{H}_0^{(2)}(ka)}}{\dfrac{ka}{2}\left(\dfrac{M_R}{M_F}\right) - \dfrac{\mathrm{H}_1^{(2)}(ka)}{\mathrm{H}_0^{(2)}(ka)}} \right] a_0 \tag{36}$$

To characterize the problem, the dimensionless parameter is defined as $\varepsilon = k_b H / k_f a = \beta_f H / \beta_b a$. It is seen that the flexible, slender, and tall shear walls are described by large values of ε.

First, the correctness of the numerical results can be verified by comparing the results from the rigid semi-circular foundation case. This is done by setting $\kappa = \mu_f k_f / \mu k = 1$, $\bar{a} \to a$. Figure 2 represents the plots of the displacement on a Cartesian coordinate system with the x-axis being 'wave number' and the y-axis being 'displacement' and the initial conditions shown in the legends.

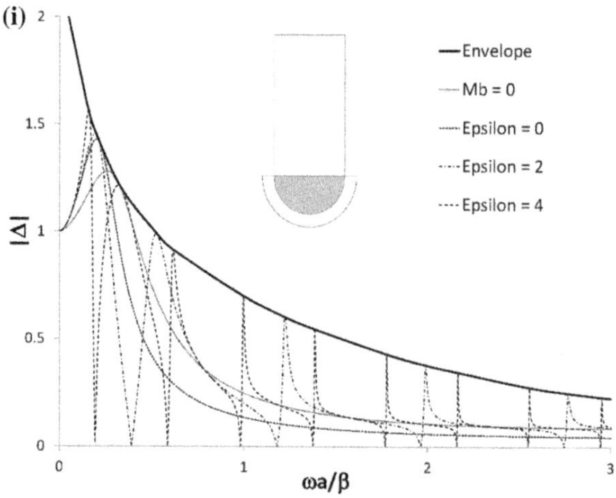

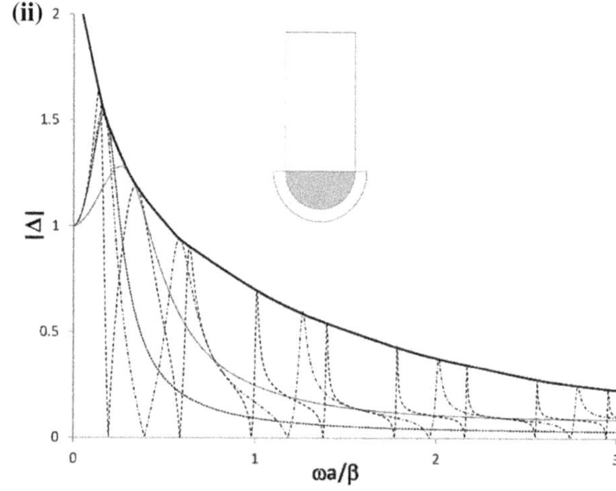

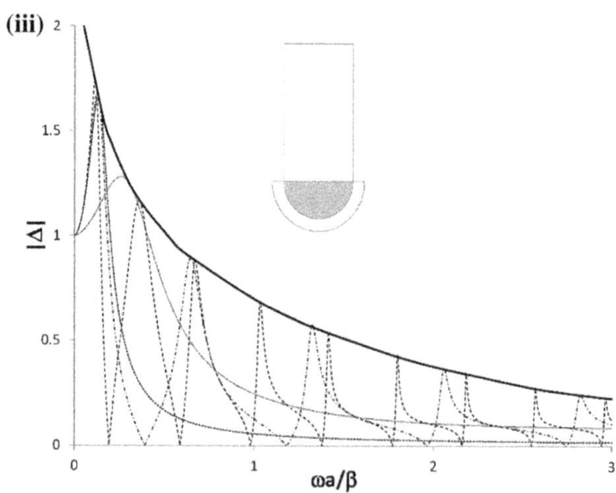

Fig. 3 Effect of interaction on Δ the amplitude of foundation vibration. **i** $\bar{a}/a = 1.25$, $\kappa = 2$, $M_B/M_S = 1$, $M_F/M_S = 1$, $\varepsilon = 0, 2, 4$. **ii** $\bar{a}/a = 1.25$, $\kappa = 2$, $M_B/M_S = 2$, $M_F/M_S = 1$, $\varepsilon = 0, 2, 4$ **iii** $\bar{a}/a = 1.25$, $\kappa = 2$, $M_B/M_S = 4$, $M_F/M_S = 1$, $\varepsilon = 0, 2, 4$

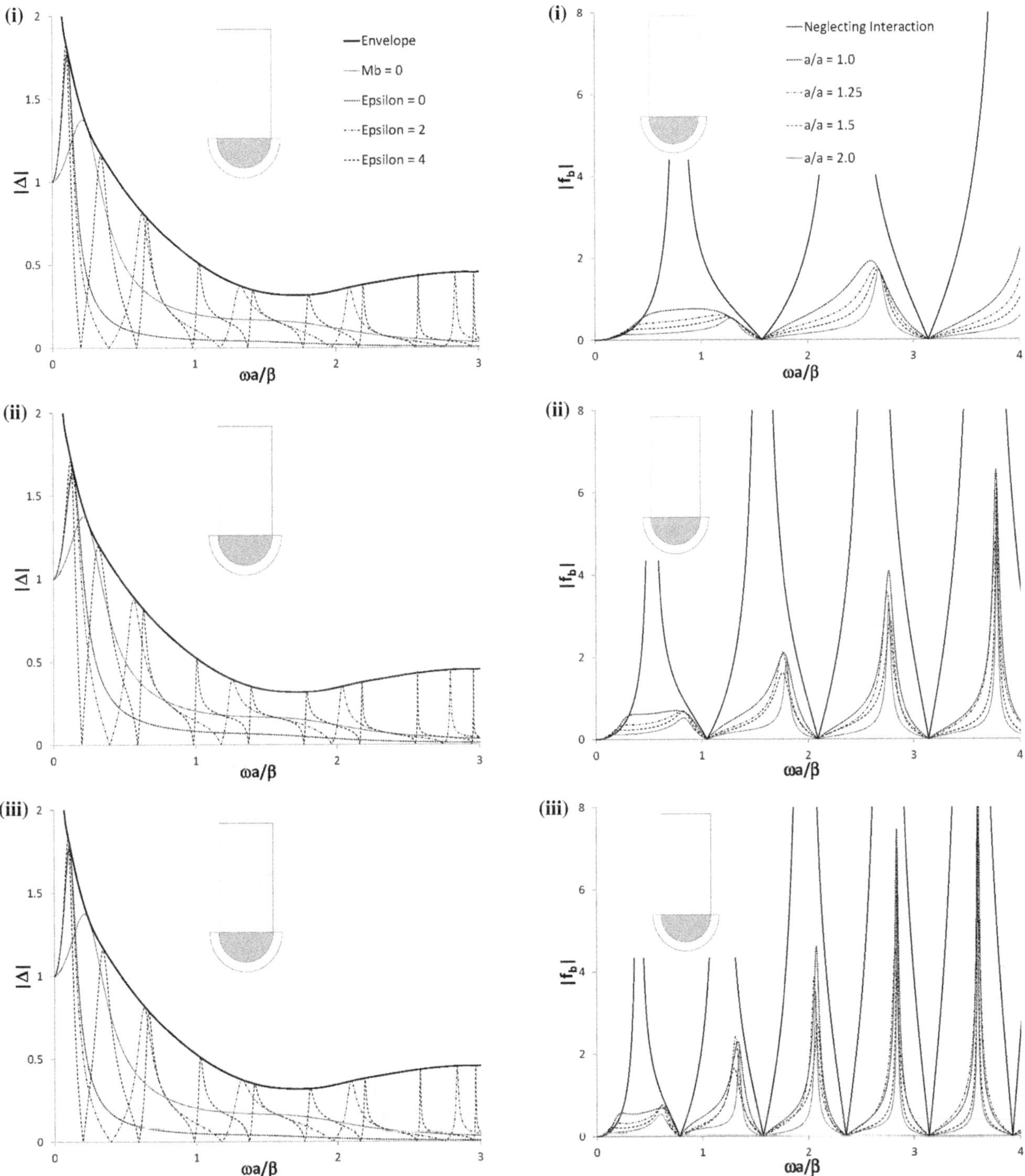

Fig. 4 Effect of interaction on Δ the amplitude of foundation vibration. **i** $\bar{a}/a = 1.50$, $\kappa = 2$, $M_B/M_S = 1$, $M_F/M_S = 1$, $\varepsilon = 0, 2, 4$. **ii** $\bar{a}/a = 1.50$, $\kappa = 2$, $M_B/M_S = 2$, $M_F/M_S = 1$, $\varepsilon = 0, 2, 4$. **iii** $\bar{a}/a = 1.50$, $\kappa = 2$, $M_B/M_S = 4$, $M_F/M_S = 1$, $\varepsilon = 0, 2, 4$

Fig. 5 The base shear force. **i** $\kappa = 1$, $M_B/M_S = 4$, $M_F/M_S = 1$, $\varepsilon = 2$. **ii** $\kappa = 1.5$, $M_B/M_S = 4$, $M_F/M_S = 1$, $\varepsilon = 2$. **iii** $\kappa = 2$, $M_B/M_S = 4$, $M_F/M_S = 1$, $\varepsilon = 2$

The abscissa in these figures is the dimensionless frequency $\omega a/\beta$ and the ordinate is the foundation displacement Δ. The displacement Δ would be equal to one if the movement of the rigid foundation does not depend on the shear wall. This serves as a reference value which shows the influence of the interaction of structure and soil on the movement of the foundation and consequently on the base shear force. The results and plots are in line with Luco (1969) and Trifunac (1972).

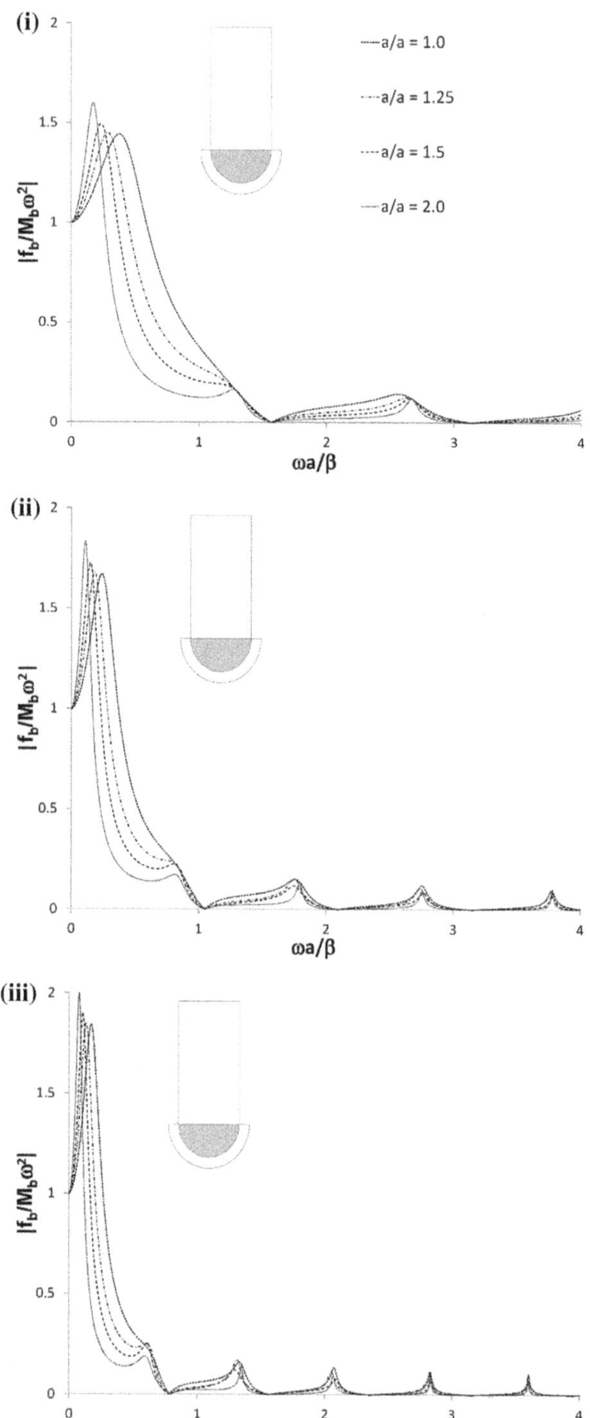

Fig. 6 The dimensionless base shear. **i** $\kappa = 1$, $M_B/M_S = 4$, $M_F/M_S = 1$, $\varepsilon = 2$. **ii** $\kappa = 1.5$, $M_B/M_S = 4$, $M_F/M_S = 1$, $\varepsilon = 2$. **iii** $\kappa = 2$, $M_B/M_S = 4$, $M_F/M_S = 1$, $\varepsilon = 2$

Figures 3 and 4 represent the displacement curves for $\bar{a}/a = 1.25$ and $\bar{a}/a = 1.50$, respectively. The zeros correspond to the fixed-base natural frequencies of the shear wall. It can be seen that the thickness of the flexible ring has strong effect on the displacement amplitude of the rigid foundation. As the ratio of $\bar{a}/a$ becomes large, peaks of the displacement amplitude increase significantly. The displacement amplitude increases greatly for low frequencies and is dependent on the amplitude of the wave in the flexible ring coefficient $B_0^{(1)}$. It is also noticed that the structure response is independent of the angle of incidence and only depending on the foundation displacement Δ.

Figure 5 represents the base shear force acting on the flexible shear wall for $\varepsilon = 2$. When there is no interaction, the base shear force is infinite for the fixed-base natural frequencies of the shear wall. When the interaction is considered, the base shear force is bounded for all frequencies. It can be seen that the thickness of the flexible ring has little effect on the peaks of the base shear force.

Figure 6 describes the dimensionless base shear coefficient of the flexible wall structure. For the lower frequencies range which is of special importance for earthquake engineering, the peaks of the base shear forces are increased in proportion to the ratio of $\kappa = \mu_f k_f/\mu k$ and decreased for larger $\bar{a}/a$. It is also indicated that the base shear force is bounded faster for higher value of material property ratio.

5 Conclusions

The analytical solution of the interaction of a shear wall, flexible-rigid foundation, and an elastic half-space is derived for incident SH waves with various angles of incidence. Comparison of these displacement amplitude plots and the ones in which the superstructure sits only on the rigid foundation shows that the flexible ring has the effect of diminishing the ground displacement amplitude as the building absorbs wave energy and scatters it back into the half-space that results in increased displacement amplitude of the foundation and decreased amplitude of ground displacement. This is an important phenomenon that differs from the typical SSI models in the absence of the flexible ring. It also found that the flexible ring (soft layer) cannot be used as an isolation mechanism to decouple a superstructure from its substructure resting on a shaking half-space due to the fact that waves are transmitted into and scattered from the flexible foundation.

Interesting results are shown in the graphs of displacement amplitude. The foundation displacement is zero for a rigid foundation with a fixed-base excited at the natural frequency of the shear wall. This concurs with the Luco (1969) and Trifunac (1972) results. However, for a heavy and flexible structure the peaks of displacement increase for low frequency waves as the thickness of the flexible foundation layer increases.

Base shear forces increase as the rigidity of the flexible foundation layer increases in comparison to the surrounding soil medium. Interestingly, for high frequency waves, peaks of the base shear force are the same for all ratios of $\bar{a}/a$. For shear walls founded on soft soil and for low frequencies, the base shear forces are higher than values computed for shear walls supported on hard soil. The structure response is independent of the angle of incidence and only depending on the foundation displacement.

All of the above analytical solutions are for cases where the foundations are rigid, non-elastic movable foundations. Analytic solutions are possible because the foundation has rigid displacement characterized by only one parameter. A more realistic assumption would be to allow the foundation to be elastic. As such, analytic solutions for soil-structure interaction of a building and shear wall on an elastic foundation deserve investigation, but this, of course, is a very challenging research problem.

This paper serves as an intermediate step for such a goal. It considers the soil-structure interaction of a shear wall on a semi-elastic foundation, with the rigid semi-circular foundation being supported by an elastic ring wrapped around it as the outside layer of the foundation. However, the solution presented in this paper, though analytic, cannot be directly adapted to solutions representing a fully flexible foundation. A sequel to this paper will be to present a new model with the same foundation geometry, namely an elastic ring around a rigid foundation, with a new shape of the superstructure where the analytic solution of the SSI problems can later be more easily adapted to the SSI of a flexible foundation.

Appendix

The mass of the soil in place of foundation, M_S per unit length is

$$M_S = \frac{\rho\pi a^2}{2} = \left(\frac{\mu\pi a^2}{2C^2}\right)\left(\frac{k}{k}\right)^2 = \frac{\mu\pi a^2 k^2}{2\omega^2}. \tag{A.1}$$

The mass of the building, M_B per unit length is

$$M_B = 2\rho_b aH = \left(\frac{2\mu_b aH}{C_b^2}\right)\left(\frac{k_b}{k_b}\right)^2 = \frac{2\mu_b k_b^2 aH}{\omega^2} \tag{A.2}$$

$$\frac{\mu_b k_b R_2 \nu\pi}{\mu k\pi a} = \left(\frac{\mu_b k_b R_2 \nu\pi}{\mu k\pi a}\right)\left(\frac{k_b H}{k_b H}\right) = \frac{\mu_b k_b^2 (R_2 \nu\pi H)}{\mu k\pi a(k_b H)} = \frac{\mu_b k_b^2 (2aH)}{\mu k\pi a(k_b H)} \tag{A.3}$$

$$\frac{\mu_b k_b R_2 \nu\pi}{\mu k\pi a} = \frac{\omega^2 M_b}{\left(\frac{2\omega^2 M_s}{\pi k^2 a^2}\right)(k_b H)} = \left(\frac{ka}{2}\right)\left(\frac{M_b}{M_s}\right)\left(\frac{1}{k_b H}\right). \tag{A.4}$$

The mass of the rigid foundation, M_R per unit length is

$$M_R = \frac{\rho_r \pi a^2}{2}. \tag{A.5}$$

The mass of the flexible ring foundation, M_F per unit length is

$$M_F = \frac{\rho_f \pi \bar{a}^2}{2} = \left(\frac{\mu_f \pi \bar{a}^2}{2C_f^2}\right)\left(\frac{k_f^2}{k_f^2}\right) = \frac{\mu_f k_f^2 \pi \bar{a}^2}{2\omega^2}, \tag{A.6}$$

where $H_0^{(1)\prime}(k_f a) = -H_1^{(1)}(k_f a)$, $H_0^{(2)\prime}(k_f a) = -H_1^{(2)}(k_f a)$, $\mu_b = \omega^2 M_B / 2k_b^2 aH$, $\omega^2 = \mu_f k_f^2 \pi \bar{a}^2 / 2M_F$, and $2a\mu_b k_b \tan k_b H = 2ak_b(\omega^2 M_B / 2k_b^2 aH) \tan k_b H = (\omega^2 M_B / k_b H) \tan k_b H$.

References

Housner GW (1957) Interaction of buildings and ground during an earthquake. Bull Seismol Soc Am 47(3):179–186

Hryniewicz Z (1993) Dynamic response of coupled foundations on layered random medium for out-of-plane motion. Int J Eng Sci 31(2):221–228

Kobori T, Kusakabe K (1980) Cross-interaction between two embedded structures in earthquakes. In: Proceedings of the Seventh World Conference on Earthquake engineering, Istanbul, Turkey, pp. 65–72

Lee VW (1979) Investigation of three dimensional soil-structure interaction. Department of Civil Engineering Report No. CE 79-11, University of Southern California, Los Angeles

Lee TH, Westley DA (1973) Soil-structure interaction of nuclear reactor structures considering through-soil coupling between adjacent structures. Nucl Eng Des 24(3):374–387

Luco JE (1969) Dynamic interaction of a shear wall with the soil. J Eng Mech Div Am Soc Civil Eng 95:333–346

Luco JE, Contesse L (1973) Dynamic soil-structure interaction. Bull Seismol Soc Am 63(4):1289–1303

Luo H (2008) Diffraction of SH-waves by surface or sub-surface topographies with application to soil-structure interaction on shallow foundations, Advisor: Vincent W. Lee, Ph.D., USC, Los Angeles, CA. ISBN: 978-0-54-997304-1

Pao YH, Mow CC (1973) The diffraction of elastic waves and dynamic stress concentrations. Report R-482-PR. RAND Corporation, Santa Monica, USA

Reissner E (1936) Stationare axialsymmetrische durch eine schuttelnde masse erregte Schwingungen eines homogenen elastischen, Halbraumes. Ingenieur-Arch 7(6):381–396

Todorovska MI (1993) In-plane foundation-soil interaction for embedded circular foundation. Soil Dyn Earthq Eng 12(5): 283–297

Triantafyllidis T, Prange B (1988) Rigid circular foundation: dynamics effects of coupling to the half-space. Soil Dyn Earthq Eng 7(1):40–52

Trifunac MD (1972) Interaction of a shear wall with the soil for Incident Plane SH-waves. Bull Seismol Soc Am 62(1):63–68

16

Global SH-wave propagation in a 2D whole Moon model using the parallel hybrid PSM/FDM method

Xianghua Jiang · Yanbin Wang · Yanfang Qin · Hiroshi Takenaka

Abstract We present numerical modeling of SH-wave propagation for the recently proposed whole Moon model and try to improve our understanding of lunar seismic wave propagation. We use a hybrid PSM/FDM method on staggered grids to solve the wave equations and implement the calculation on a parallel PC cluster to improve the computing efficiency. Features of global SH-wave propagation are firstly discussed for a 100-km shallow and 900-km deep moonquakes, respectively. Effects of frequency range and lateral variation of crust thickness are then investigated with various models. Our synthetic waveforms are finally compared with observed Apollo data to show the features of wave propagation that were produced by our model and those not reproduced by our models. Our numerical modeling show that the low-velocity upper crust plays significant role in the development of reverberating wave trains. Increasing frequency enhances the strength and duration of the reverberations. Surface multiples dominate wavefields for shallow event. Core–mantle reflections can be clearly identified for deep event at low frequency. The layered whole Moon model and the low-velocity upper crust produce the reverberating wave trains following each phases consistent with observation. However, more realistic Moon model should be considered in order to explain the strong and slow decay scattering between various phases shown on observation data.

X. Jiang · Y. Wang (✉)
Department of Geophysics, School of Earth and Space Sciences, Peking University, Beijing 100871, China
e-mail: ybwang@pku.edu.cn

Y. Qin
Equipe de Géosciences Marines, Institut de Physique du Globe de Paris, 4 Place Jussieu, 75252 Paris Cedex 05, France

H. Takenaka
Department of Earth Sciences, Faculty of Science, Okayama University, 3-1-1 Tsushima-Naka, Kita-ku, Okayama 700-8530, Japan

Keywords Whole Moon model · Seismic wavefield · SH-wave propagation · Hybrid method · Parallel computing

1 Introduction

The Apollo Passive Seismic Experiment consisted of a network of four seismometers deployed on the near side of the Moon during the Apollo missions between 1969 and 1972, which operated continuously until 1977 and recorded more than 12,000 moonquakes (Lognonné 2005). These data provided the basis for seismic studies of the lunar interior. Analyses of the lunar seismic data revealed the one-dimensional (e.g., Nakamura 1983; Khan and Mosegaard 2002; Lognonné et al. 2003; Gagnepain-Beyneix et al. 2006), lateral heterogeneous (e.g., Chenet et al. 2006; Zhao et al. 2008) lunar seismic velocity structure in the crust and mantle, the possible mechanism of moonquakes (e.g., Nakamura 1978; Nakamura et al. 1979; Koyama and Nakamura 1980) and their implications on the composition and physical properties of the lunar interior (e.g., Khan et al. 2006, 2007). Recent reanalysis of lunar seismograms (Weber et al. 2011; Garcia et al. 2011) determined the velocity structure of the lunar core and core–mantle transition zone. Based on both seismological and geodesic data, Garcia et al. (2011) constructed a very preliminary reference Moon model (VPREMOON).

Lunar seismograms differ greatly from the typical terrestrial seismic signals. They are characterized with reverberating wave trains of very long duration and a slow decay of amplitude. Except for the direct P wave and S wave for some events, various secondary seismic phases and surface waves are difficult to identify from the intense scattering and reverberating waveforms. Therefore, the process of seismic wave propagation inside the whole Moon is still not well understood from the observed lunar seismograms. However, numerical modeling of seismic wave propagation in the whole Moon model had been conducted to improve our understanding of lunar seismic wavefield. Dainty et al. (1974) and Nakamura (1977) applied the diffusion theory to a Moon model with randomly distributed scatters near the surface and closely reproduced the envelope of typical lunar seismograms from impacts and moonquakes. Their results explained some features of the scattering and reverberating lunar seismic signals of long duration and suggested the existence of a shallow scattering zone near the surface. Lawrence and Johnson (2010) and Blanchette-Guertin et al. (2012) applied a seismic phonon method to calculate global synthetic lunar seismograms and investigated the effects of seismic scattering in a highly heterogeneous regolith layer on high-frequency lunar signals. Wang et al. (2013) recently performed P-SV seismic wave propagation modeling for a 2D lateral heterogeneous whole Moon model with a pseudospectral (PSM) and finite difference (FDM) hybrid method. They calculated P-SV body wave propagation in the whole Moon radiated from both shallow and deep moonquakes, compared synthetics with observed Apollo seismic data, and found that the near-surface low-velocity layer contributes significantly to the development of waveform reverberation.

In an isotropic 2D whole Moon model, the horizontally polarized SH-waves are recoded in the transverse component. SH-waves are decoupled from P and SV waves; therefore, the SH-wavefield contains only the direct SH-waves and their interactions with interfaces are much simpler than the P and SV wavefields recorded in the radial and vertical components. For a similar 2D whole-Earth model, several numerical methods had been applied to simulate global SH-wave propagation. Cummins et al. (1994) proposed a direct solution method to calculate global SH-wavefield for a spherically symmetric model. Igel and Weber (1995) calculated 2D global SH-wave propagation with a high-order FDM for a 2D axisymmetric whole-Earth model. Igel and Gudmundsson (1997) investigated the frequency-dependent effects of S and SS waves with the similar FDM method. Thorne et al. (2007) studied lateral heterogeneity in the lower mantle beneath the Cocos Plate with a 2D global SH-wavefield simulation based on FDM. Jahnke et al. (2008) performed 2D global SH-wave modeling on a parallel computer. Toyokuni et al. (2005) presented a new quasi-spherical FDM approach for 2D global wavefield simulation. Wang and Takenaka (2011) applied a PSM method to calculate 2D global SH-wave propagation with arbitrary lateral heterogeneities. Wang et al. (2014) developed a parallel PSM/FDM hybrid method for calculating global SH-wave propagation on PC cluster. The SH-wave component observation data had often been used to study the localized shear-velocity heterogeneities in the Earth's interior for its simple waveforms (e.g., Wen 2002; Wang and Chen 2009).

In this paper, we attempted to simulate global SH-wave propagation in a 2D whole Moon model proposed by Weber et al. (2011) and Garcia et al. (2011). We apply a hybrid PSM/FDM method on staggered grids to solve the 2D SH-wave equations. We implemented parallel computing on a PC cluster for the hybrid method to improve computing efficiency. The 2D whole Moon model is decomposed into sub-domains in the radial direction and assigned to different processors of a PC cluster. We applied the parallel hybrid method to a lateral homogeneous whole Moon model for both deep and shallow moonquakes to show the process of SH-wave propagation in the whole Moon. Comparisons between the synthetics and observed Apollo seismograms from typical moonquake events were performed to illustrate wavefield features that our simulations reproduced and those that were not achieved. We also investigated the effects of lateral varying crustal thickness and frequency content on synthetic waveforms with various numerical models.

2 SH-wave equations and the parallel PSM/FDM scheme

2.1 Elastodynamic wave equations

Considering a cylindrical coordinate system (r, θ, z) and assuming invariance for all field variables in z, we get a 2D cylindrical coordinate system (r, θ). For an isotropic and elastic medium, SH-wave propagation in such 2D coordinates can be formulated by the following equations of momentum conservation in velocity-stress form:

$$\rho \frac{\partial v_z}{\partial t} = \frac{\partial \sigma_{zr}}{\partial r} + \frac{1}{r}\left(\frac{\partial \sigma_{z\theta}}{\partial \theta} + \sigma_{zr}\right) + f_z, \tag{1}$$

where v_z is the velocity component in anti-plane direction, ρ is the mass density, $\sigma_{zr}, \sigma_{z\theta}$ are components of stress tensor, and f_z is the body force component in z direction. The constitutive relations between stress and velocity are given by:

$$\begin{aligned} \frac{\partial \sigma_{zr}}{\partial t} &= \mu \frac{\partial v_z}{\partial r} \\ \frac{\partial \sigma_{z\theta}}{\partial t} &= \frac{\mu}{r}\frac{\partial v_z}{\partial \theta}, \end{aligned} \tag{2}$$

where μ is the shear modulus of the medium.

2.2 The parallel hybrid PSM/FDM scheme

We calculate SH-wave propagation in a 2D whole Moon model, which corresponds to a cross section of the 3D Moon cutting through a great circle. The 2D model is defined from the Moon's surface down to a depth of 1485.0 km in the liquid outer core in the radial direction and 0°–360° in the lateral direction. In order to solve Eqs. (1) and (2) for the 2D model, we discretize the whole Moon long r and θ directions with staggered grids as shown in Fig. 1. The two stress components are discretized into half grid spacing between the anti-plane velocities in both directions. In radial direction, the grids are distributed uniformly with constant grid spacing Δr. In lateral direction, the number of grids is the same for all radii and hence the grid spacing reduces gradually with increasing depth.

We apply a PSM/FDM hybrid scheme to solve Eqs. (1) and (2) for the discretized model on the staggered grids (Wang et al. 2014). In the lateral direction, derivatives with respect to θ are calculated with the PSM method. Let $f(j\Delta\theta)(j=0,1,2,\ldots,N-1)$ represent discretized field variable in θ direction, we need to evaluate its derivative half-way between grids at $(j\pm 1/2)\Delta\theta$. The derivative is calculated in the wavenumber domain by multiplication and inverse Fourier transformation:

$$\frac{d}{d\theta}f\left[\left(j\pm\frac{1}{2}\right)\Delta\theta\right]=\frac{\Delta k}{2\pi}\sum_{l=0}^{N-1}i(l\Delta k)e^{\pm il\Delta k\Delta\theta/2}F(l\Delta k)e^{i2\pi jl/N},\quad j=0,1,2,\ldots,N-1, \tag{3}$$

where $F(l\Delta k)(l=0,1,2,\ldots,N-1)$ is the Fourier transformation of $f(j\Delta\theta)$. In the radial direction, derivatives with respect to r are calculated with a fourth-order accuracy FDM on staggered grids. For field variable $f(i\Delta r)(i=0,1,2,\ldots,M-1)$, its derivative half-way between grids at $(i\pm 1/2)\Delta r$ is calculated with:

$$\frac{d}{dr}f(i\Delta r)=\frac{1}{\Delta r}\left\{c_1\left[f\left(\left(i+\frac{1}{2}\right)\Delta r\right)-f\left(\left(i-\frac{1}{2}\right)\Delta r\right)\right]+c_2\left[f\left(\left(i-\frac{3}{2}\right)\Delta r\right)-f\left(\left(i+\frac{3}{2}\right)\Delta r\right)\right]\right\}, \tag{4}$$

where $c_1=9/8$, $c_2=1/24$. The time domain calculation of v_z is performed with a second-order FDM scheme (Furumura et al. 1998). Time interval Δt is determined from the minimum grid spacing and the maximum seismic velocity in the model by satisfying stability condition for both PSM and FDM schemes (Wang et al. 2014).

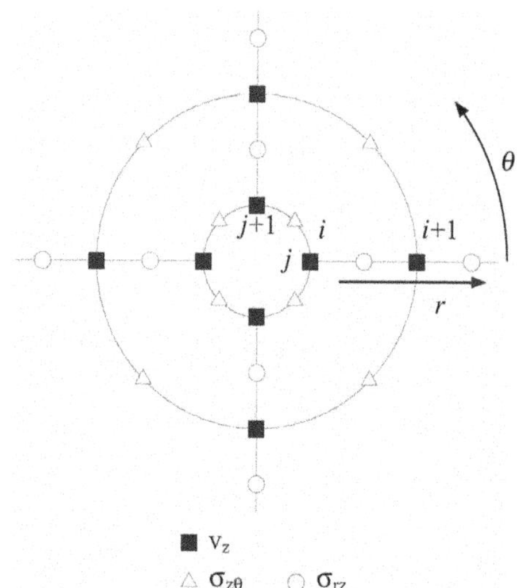

Fig. 1 Discretization of stress and velocity components for the 2D global model on staggered grids

The artificial boundary condition at the bottom of the model located in the outer core is treated by introducing a damping zone of 20 gird points (Cerjan et al. 1985). The free surface boundary condition at the Moon's surface is introduced into the calculation by satisfying a traction-free condition. Body force component f_z in Eq. (1) is calculated by a combination of corresponding moment tensor components as given by Wang et al. (2014). Distribution of body force component in space and source-time function is approximated with a bell-shaped Herrmann's function of unit area (Herrmann 1979).

Lunar seismic observation and numerical modeling for P-SV wavefield (Wang et al. 2013) showed that the lunar seismograms are dominated with high-frequency waves. Hence, it is important to model high-frequency wavefield in order to enhance our understanding of lunar seismic wave propagation. Comparing with the common hybrid method used by Wang et al. (2013), the parallel PSM/FDM algorithm used in this study enables us to improve calculating efficiency and hence calculate for higher frequency wavefield. FDM used in the radial direction is a localized derivative operator and hence allows us to divide the whole Moon model into sub-domains along the radial direction. In each sub-domain, the number of grids in both radial and lateral directions is the same. Each sub-domain is assigned to a processor of a PC cluster. In each time step, spatial derivatives are calculated in each processor simultaneously. Since the fourth-order staggered grid FDM is used in the radial direction, the values of the top and bottom two layers in each sub-domain are exchanged between neighboring processors. Overlap regions of two layers of grids are added to both the top and bottom of each sub-domain to store the exchanged data. The data exchange and inter-process communication are implemented by the Message Passing Interface. The good accuracy and efficiency enhancement of the parallel PSM/FDM scheme were discussed for 3D regional model by Qin et al. (2012) and SH-wave propagation modeling for 2D whole-Earth model by Wang et al. (2014). The following is the procedure to solve Eqs. (1) and (2) with the parallel PSM/FDM scheme at each time step: Firstly, the whole Moon model is divided into sub-domains along the radial direction and assigned to processors; In each sub-domain, the spatial derivatives of velocity in Eq. (2) are then evaluated with PSM or FDM and data in the overlap regions are exchanged; Stress components are calculated in Eq. (2) with time differencing and then their spatial derivatives in Eq. (1) are evaluated with PSM or FDM and data in the overlap regions are exchanged; Finally the velocity component in Eq. (1) are calculated with time domain differencing.

3 SH-wave propagation in the whole moon

We first show characteristics of global propagation of SH-wave in the whole Moon model. Recent reanalysis of Apollo seismic waveforms suggested the existence of a liquid outer core and a solid inner core. Our 1D whole Moon model is derived from the recently proposed velocity and density model by Weber et al. (2011) and the Q_s value given in VPREMOON model by Garcia et al. (2011) as shown in Fig. 2. The Moon's crust in this model is 39 km thick that includes the upper and lower layers. The upper crust is thinner but has a very low S-wave velocity (1.8 km/s) as compared to that in the lower crust (3.2 km/s). The Moon's mantle extends to a depth of 1257.1 km with a discontinuity at 738.0 km deep. Variation of S-wave velocity in the mantle is smoother than that for the P-wave velocity. A transition zone of 150.0 km thick was proposed by Weber et al. (2011) between the mantle and outer core. It has S-wave velocity as low as that in the lower crust. The 1-km layer just below the free surface is considered as a fractured and fragmented zone that has very low S-wave velocity.

The whole Moon is discretized into 8192 grids in the lateral direction that leads to a maximum lateral grid spacing of 1.33 km at the surface and a minimum grid spacing of 0.19 km at the base of the model. In the radial direction, the model is discretized into 1650 grids with uniform grid spacing of 0.9 km. Width of the bell-shaped source-time function is 3.0 s. The minimum wavelength in the calculation is 5.4 km considering the minimum S-wave velocity in the upper crust. The number of grid points per minimum wavelength equals to or great than 4.0 in the lateral direction and equals to 6.0 in the radial direction that is sufficient for PSM and fourth-order staggered grid FDM, respectively. The time interval determined from the minimum grid spacing, the maximum wave velocity, and the stability condition is 0.015 s in the calculation. We calculated 200,000 time steps for wave propagation of 3000 s in the whole Moon. The whole model is divided into 30 sub-domains along the radial direction and assigned to 30 processors on a PC cluster. Comparing with the modeling for global P-SV wave propagation in the whole Moon (Wang et al. 2013), the parallel modeling for SH-wave in this paper requires less parameters and less computer memory and computation time for the same number of grid points. Therefore, it is possible to discretize the whole Moon with finer grid spacing and extend the higher

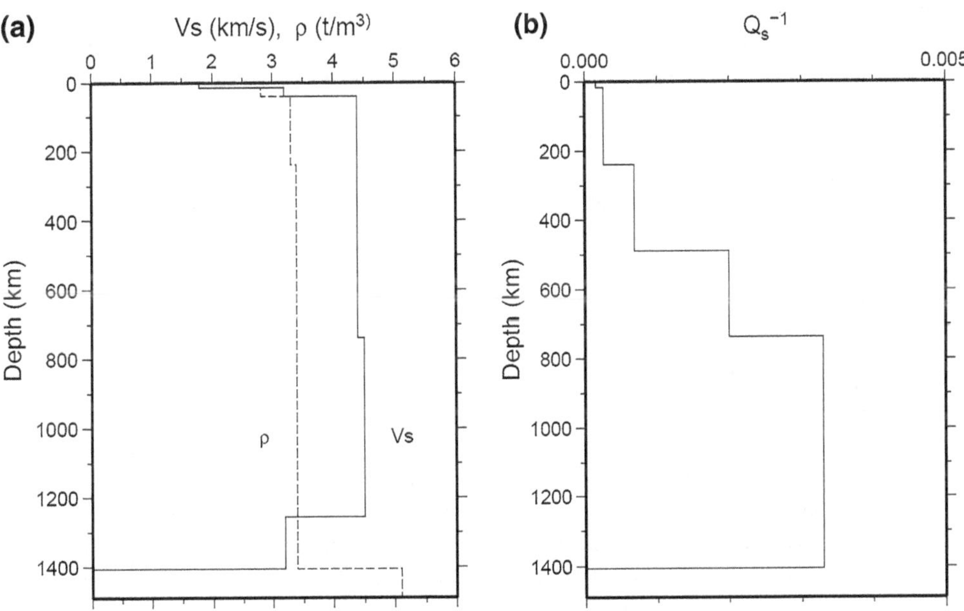

Fig. 2 The whole Moon model used in the modeling. **a** S-wave velocity and density distribution along depth (Weber et al. 2011), **b** anelastic attenuation factor for S wave (Garcia et al. 2011)

frequency range in the modeling. Characteristics of global SH-wave propagation can be investigated at higher frequency range than that for global P-SV wave propagation.

3.1 The shallow moonquake

Except for the thermal moonquakes and meteoroid impacts, most identified moonquakes occurred inside the Moon are classified into shallow moonquakes and deep moonquakes. Focal depth of the 28 detected shallow moonquakes is between 50 and 220 km in the upper mantle. Number of shallow moonquakes is much less than the deep moonquakes, but their magnitudes are larger. Although the focal mechanisms of shallow events are not well understood from the Apollo observation, it is inferred that they are comparable to intraplate earthquakes and are quite likely caused by tectonic reason (Nakamura et al. 1979). Wang et al. (2013) applied a double-couple line source corresponding to a 45° dipping fault for P-SV wave modeling. For SH-wave modeling in this study, a seismic source with moment tensor component $M_{rz} = 1.0$ is used that is commonly adopted in whole-Earth seismic wave propagation modeling (Wang et al. 2014). The focal depth of the simulated moonquake is 100 km as shown in Fig. 3.

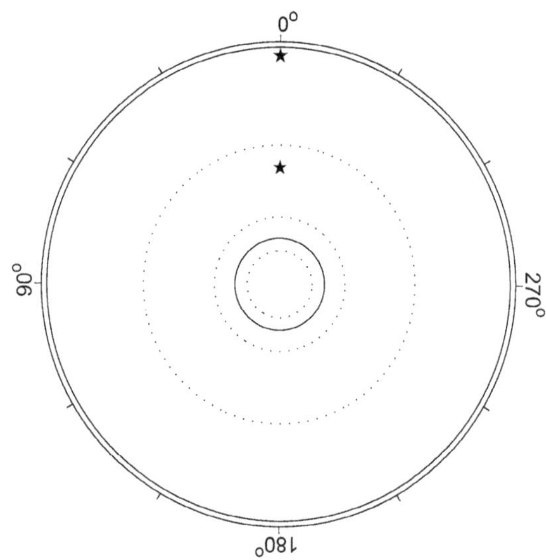

Fig. 3 Configuration of the whole Moon model used in the modeling. *Shallow* and *deep circles* show crust–mantle and core–mantle boundaries. *Dashed circles* are mantle discontinuity, *top* boundary of core–mantle transition zone, and inner–outer core boundary. *Stars* are locations of the shallow and deep moonquakes used in the modeling

Figure 4a shows synthetic seismograms at the Moon's surface for the shallow moonquake together with theoretical travel time curves for major phases. Direct S wave with large amplitude can be seen up to epicentral distance of 145°. Surface reflections and multiple reflections such as sS, SS, sSS, and SSS appear at epicentral distance greater than 45°. Interactions between the direct S and these reflections cause large amplitude between 45° and 150°. Core–mantle reflection ScS is strong and clear at all distances followed by sScS. Reflection from the top of the core–mantle transition zone is also visible before ScS. The core–mantle reflections $(ScS)_2$ and $(sScS)_2$ are still clearly identified at about 1300 and 1400 s. The most important features of wave propagation at later time are the development of wave trains of surface multiples such as SS, SSS, and SSSS. Comparing with SH-wave propagation in the whole Earth (Wang et al. 2014), waveforms of all phases for the Moon show strong reverberations following the first arrivals. The duration of the reverberating wave train increases with the propagation distance that is very obvious for SS, SSS, and SSSS. Such phenomenon is observed for global P-SV wave propagation in the Moon as well (Wang et al. 2013). This can be explained as the trapping wave effect in the low-velocity upper crust. The S-wave velocity in the upper crust is much lower than that in the lower crust and mantle. Transmission of energy from lower crust into this layer is rather efficient than that propagates back into the lower crust. Multiple reflections occurred in this layer causes reverberations that are observed on the synthetic waveforms. With increasing epicentral distance, constructive interference between the multiple reflections and later incident waves enhance the reverberating wave trains. This suggests that the trapping effect and the reverberations occurred within the low-velocity upper crust which contributes partly to the long wave trains observed for the moonquake.

3.2 The deep moonquake

Most of the identified moonquakes are deep events occurred repeatedly at depths of 700–1200 km within about 300 distinct foci. Magnitudes of the deep events are smaller than that for the shallow events. Their occurrence is supposed to be related to the tidal force raised on the Moon by the Earth and could be slipping along a nearly horizontal plane (Nakamura 1978; Koyama and Nakamura 1980). Zhao et al. (2012) presented a new tomographic model for the Moon. They found a correlation between the distribution of deep moonquakes and seismic wave velocity variations in the deep lunar mantle that is similar to earthquakes. They suggested that the occurrence of deep moonquakes is affected by the lunar structural heterogeneity in addition to the tidal stresses that is quite similar to the shallow and deep earthquakes occurred in the crust and upper mantle of the Earth. In global P-SV wave propagation modeling for a deep moonquake, Wang et al. (2013) applied a double-couple source corresponding to a horizontal plane. For SH-wave propagation in this paper, we adopt a seismic source with moment tensor component

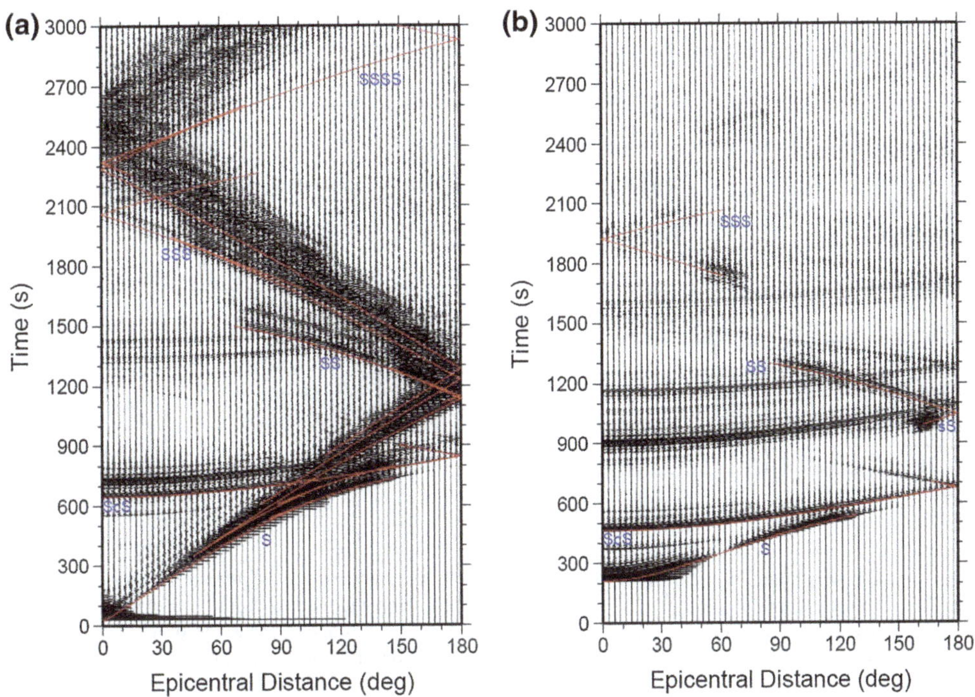

Fig. 4 Synthetic seismograms at the surface for the whole Moon model. **a** is for a 100-km and **b** is for a 900-km deep moonquake. The ray-theoretical arrival times for major phases are shown in red curves. The amplitude of all seismograms are multiplied by a factor of 1.0×10^6

$M_{\theta z} = 1.0$ that describes a horizontal slipping. The focal depth of the simulated deep moonquake is 900 km (Fig. 3).

Figure 4b shows synthetic seismograms for the deep moonquake with theoretical arrival times for major phases. We see very clear direct S waves with strong energy. Comparing with shallow moonquake, surface reflections and multiples do not interfere with direct S wave that allows S waveforms to be identified clearly at all distances. The core–mantle reflection ScS is rather obvious for deep event as compared to shallow one. Reflection from the top of core–mantle transition zone can be clearly discerned before ScS. Later reflections from core–mantle boundaries such as sScS and $(\mathrm{ScS})_2$ are also clearly identified at later time. Comparing with the waveforms for shallow events, reverberations following direct S and core–mantle reflections are weaker. The most significant differences between two events are the very weakly developed surface reflections and multiples for the deep one. This can be explained by the very small incidence angle for all epicentral distance for deep event for which the waves are efficiently transmitted into and out of the low-velocity upper crust layer. Hence the wave trapping effect and multiple reflections within the low-velocity layer is much less developed for deep event than that for shallow one. This is observed in case of P-SV wave propagation as well. The most important feature for the deep event is the very clearly shown reflections from the core–mantle boundary. Comparing with the shallow event (Fig. 4a) and the P-SV wave cases (Wang et al. 2013), the ScS for SH-wave is the most clear and identifiable for a large range of epicentral distance. Garcia et al. (2011) showed that the amplitude of the core–mantle reflection for vertically polarized S wave (ScSV) is smaller than that for horizontally polarized S wave (ScSH). They suggested that ScSH can be detected from the Apollo data and can be used to support the existence of the liquid core and constrain the radius of the core rather than ScSV. Our modeling for the deep moonquake shows that the ScSH is the most clearly identified core–mantle reflection and thus supports their suggestions. Hence ScSH observed from deep moonquake should be the most efficient seismic phase for the study of the Moon's core.

4 Frequency effects on synthetic waveforms

The observed moonquake waveforms are dominated with high-frequency reverberation and coda. For global P-SV wave propagation in the Moon, numerical modeling showed that amplitude and duration of reverberations increase with dominated frequency (Wang et al. 2013). In this section, we study frequency effects on synthetic waveforms for SH-wave propagation. In order to model higher frequency wavefield, we reduced the range of the model used in previous modeling from 360° to 180° in the lateral direction. The number of grids in the lateral direction is still 8192 and lateral grid spacing ranges from 0.67 km at the surface to 0.09 at the bottom of the model. The number of radial grids is 3000 with uniform grid

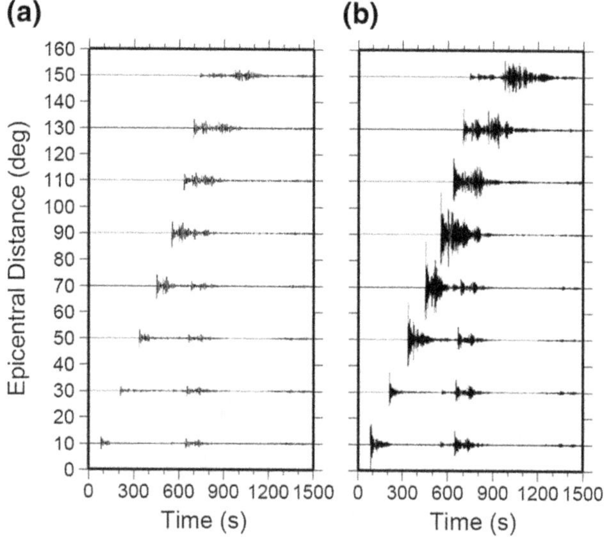

Fig. 5 Comparing of synthetic seismograms for different frequency for a shallow moonquake with focal depth of 100 km. The duration of source-time function for **a** and **b** is 3.0 and 1.67 s, respectively. The amplitude of all seismograms are multiplied by a factor of 5.0×10^5

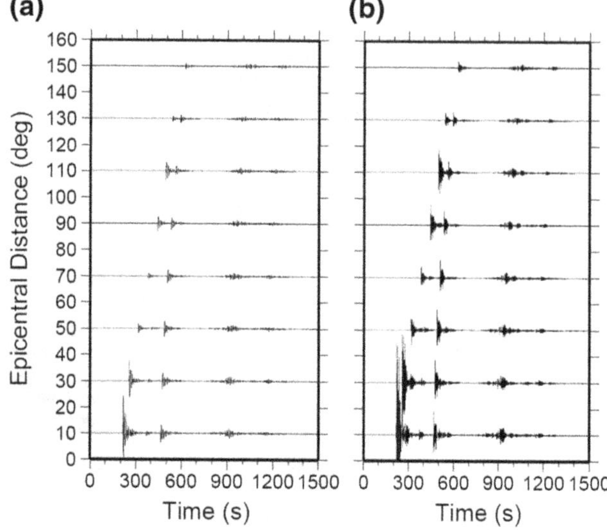

Fig. 6 Comparing of synthetic seismograms for different frequency for a deep moonquake with focal depth of 900 km. The duration of source-time function for **a** and **b** is 3.0 and 1.67 s, respectively. The amplitude of all seismograms are multiplied by a factor of 5.0×10^5

spacing of 0.5 km. With the reduced grid spacing, the width of the bell-shaped source-time function reduces to 1.67 s. The minimum wavelength in the calculation is 3.0 km. Time interval determined from stability condition is 0.01 s. The calculation is carried out for 150,000 time steps to show wave propagation of a duration of 1500 s. The whole model is divided into 30 sub-domains along radial direction and assigned to 30 processors. In order to compare with waveforms of the previous models, the same focal mechanisms are used in the modeling.

Figure 5 shows comparing of synthetic waveforms up to epicentral distance of 150°. Figure 5a shows results calculated for the 360° model with source-time function of 3.0 s duration in previous section. Figure 5b is the result for the 180° model with source-time function of 1.67 s duration calculated in this section. We see that the amplitude of the direct wave and the following reverberations increases with increasing frequency. For higher frequency model, the reverberations decay slower and their duration becomes longer. Hence the reverberations enhance with increasing frequency. Similar phenomenon is observed for P-SV wave propagation in the Moon as well (Wang et al. 2013). The reason is that for shorter wavelength component, the trapping effect and multiple reflections occurred within the low-velocity layer is stronger. The comparison indicates that the seismic phases could be identified more clearly on lower frequency waveforms.

Comparing of synthetic waveforms of different frequency for deep moonquake is shown in Fig. 6. The strength and duration of reverberations following each phase increase with frequency. However, identification of direct wave and reflections from core–mantle boundary become less distinct with increasing frequency. Comparing with shallow moonquake, the enhancement of reverberation with frequency is weaker. This is because waves from deep moonquake propagate into the low-velocity layer with very small incidence angle at all epicentral distances. The trapping effect and multiple reflections occurred within the low-velocity layer is weaker for deep event than that for shallow one. Garcia et al. (2011) compared observed waveforms at two different frequency ranges and found that the energy of ScSH is lower at high frequencies than at low frequencies and suggested that the search for this seismic phase should be performed at frequencies as low as possible. Results of our comparison show similar phenomenon and support their suggestion.

5 Effects of crustal thickness on synthetic waveforms

For seismic wave propagation on the Earth, numerical modeling showed that the low-velocity near-surface layer has significant effects on seismic body wave and generates seismic coda (e.g., Robertsson and Holliger 1997). Analysis of observations and numerical modeling suggested that the lateral variation of crust thickness and velocity has pronounced effects on regional seismic wave propagation and the thick zone of lower seismic velocities in the continental crust acts to trap S-wave energy and generates crustally guided phases (Furumura and Kennett 1997; Kennett and Furumura 2001). The most recent research shows that the lateral variation of crust thickness on the

Moon is much larger than that on the Earth, especially across some young impact basins where crust varies from near zero to about 60 km drastically (Hikida and Wieczorek 2007; Wieczorek et al. 2013). Hence, it is very interesting to investigate the effects of lateral variation of crust thickness on wave propagation on the Moon.

Similar to the modeling performed in Sect. 3, we calculated SH-wave propagation for a 100-km shallow moonquake and a 900-km deep moonquake. The whole Moon model, the number of grids, grid spacing, time interval, width of the source-time function, and focal mechanism are the same as used in Sect. 3 except for that the crust thickness varies laterally. Considering the average crust thickness of 40 km on the Moon, the crust incorporated in the modeling varies from 20 to 60 km laterally. The top 8 km and 20 km layers are set to be the upper crust for the 20 and 60 km crust, respectively. For the shallow moonquake modeling, the crust is 20 km thick between 0° and 180° and 60 km thick from 180° to 360°. For the deep moonquake modeling, thickness of the crust is 60 km between 0° and 180° and 20 km between 180° and 360°. The seismic source is located at 180° for both models. With such models, it is possible to show difference of wave propagation caused by varying crust thickness.

Synthetic seismograms for both the shallow and deep moonquakes are shown in Fig. 7. For shallow moonquake, the basic features of synthetic waveforms for both the thin and thick crust parts are similar to that for the whole model shown in Fig. 3. We see clear direct wave and core–mantle reflections. Surface reflections and multiples dominate the latter of the wavefield with increasing strength. Comparing between the thin and thick crust regions shows that the reverberations following each phases become stronger in the thick crust region. This suggests that wave trapping effects and multiple reflections are more efficient in the thick crust region. Such phenomenon is similar to the S-wave trapping effects in thick zone of lower velocity continental crust on the Earth (Kennett and Furumura 2001). One interesting phenomenon that is not observed in Fig. 3 is the very efficient transmission of waves from mantle into the low-velocity thick crust region occurred at the boundaries between two regions at 180° and 360°. The transmitted waves are very strong and propagate as trapped waves and dominate the whole wavefield above the 60-km thick crust region. Hence the identification of each phase is not so easy at this region as compared to the 20-km crust region. For deep moonquake, the basic features of wave propagation are similar to the model shown in Fig. 4 as well. We still observe the enhancement of reverberations following each phase at the thick crust region. Transmission of waves at the boundaries at 180° and 360° from mantle into the thick crust region are efficient and cause multiple reflections that contaminate the wavefield.

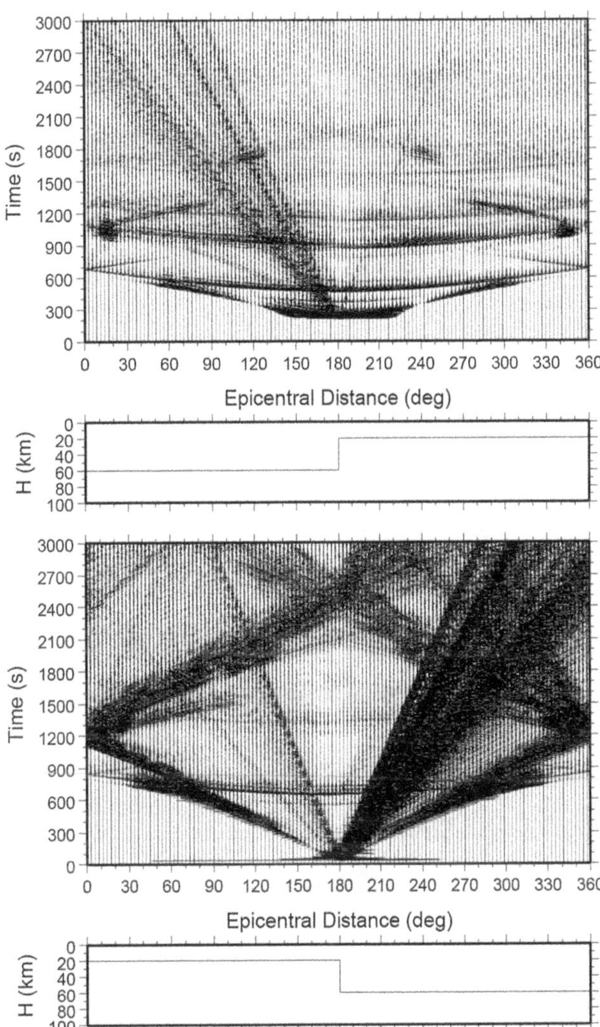

Fig. 7 Synthetic seismograms at the surface for the whole Moon model with lateral varying crust thickness. The *upper* and *lower panels* show results for 900 km deep and 100 km shallow moonquakes with crust thickness distribution, respectively. "H" means the thickness of the crust thickness. The seismic source is located at 180°. The amplitude of all seismograms are multiplied by a factor of 1.0×10^6

However, the effects of crust thickness on wave propagation are not so obvious for deep moonquake than that for shallow one and suggest that each phase could be more easily identified on deep moonquake waveforms even considering the lateral variation of crust thickness.

6 Comparison with the observed waveforms

In this section, we try to make comparison between our synthetics with Apollo observation data tentatively to show features of wave propagation that exhibited in our results and those not reproduced by our modeling. We compared with two shallow (SH1, SH2) and one deep (A1) events typically recorded on three Apollo stations (S12, S14, S16).

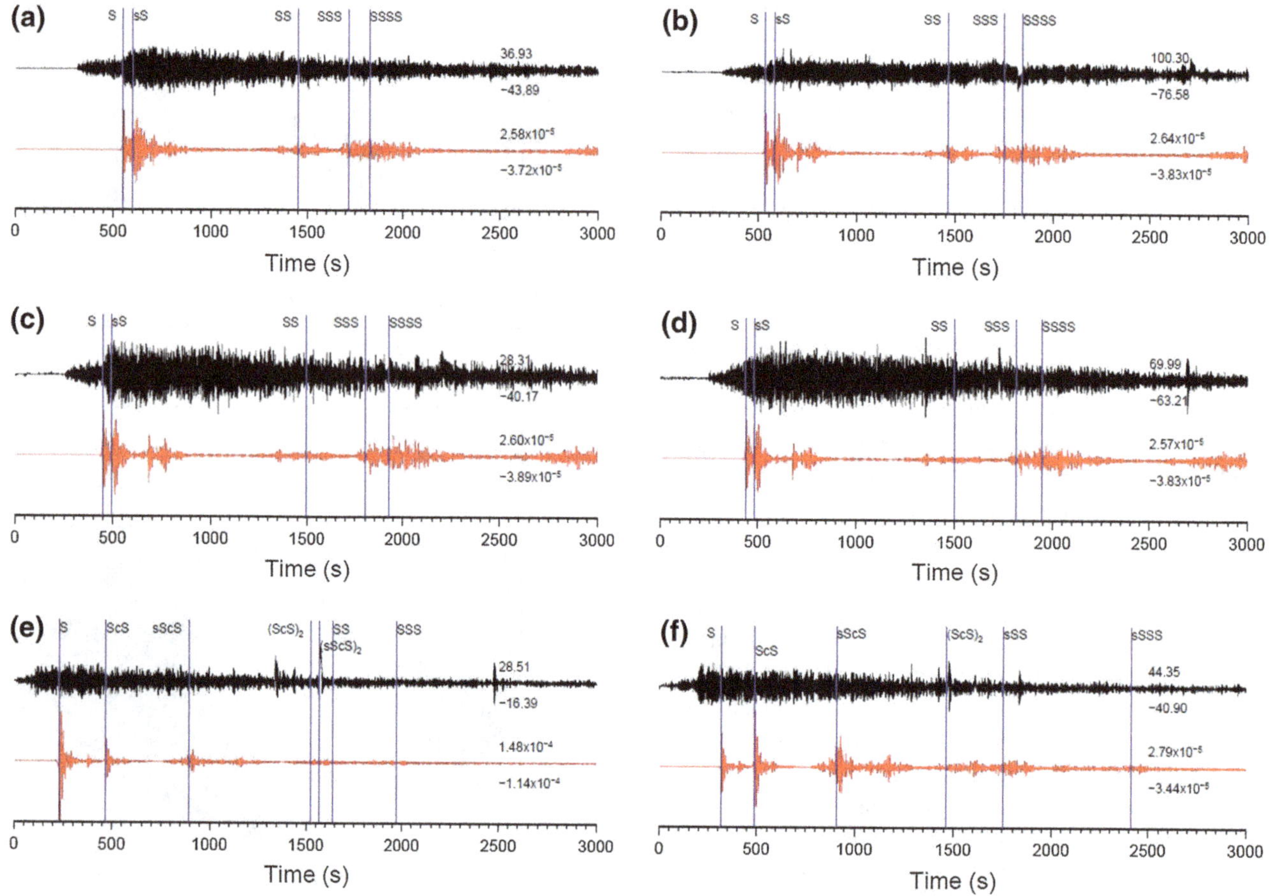

Fig. 8 Comparison between observed transverse component Apollo seismograms (*black*) with synthetic SH-waveforms (*red*). *Blue lines* show theoretical arrival times for major annotated phases. The maximum amplitude of observation (unit in DU) and synthetics (unit in cm/s) are annotated for each waveform. The event–station pair for each figure is: **a** SH1-S12, **b** SH1-S16, **c** SH2-S12, **d** SH2-S14, **e** A1-S12, **f** A1-S16. Epicentral distance for each pair: SH1-S12: 89.13°, SH1-S16: 85.73°, SH2-S12: 69.73°, SH2-S14: 67.43°, A1-S12: 18.15°, A1-S16: 51.22°

These events were used for comparisons with global P-SV wave propagation modeling as well (Wang et al. 2013). The original waveform data are processed with the same procedure as used by Wang et al. (2013). We first band-pass filter the data between 0.1 Hz and the Nyquist frequency (3.3127 Hz), remove large amplitude noise by a despiking algorithm, and then correct the gain difference between the two horizontal components. The two corrected horizontal components are finally rotated into the radial and transverse components for comparison. Wang et al. (2013) compared the radial and vertical components for P-SV waves. In this paper, we compare the obtained transverse component with our synthetic SH-waveform. Since our calculation is performed for a 2D model, we firstly corrected our synthetic waveforms from 2D to 3D for geometric spreading and source radiation and then band-pass filtered them with the same frequency band as used for observations before comparing.

Figure 8 shows comparison of waveforms at each event–station pair. For shallow events, the calculated direct S wave corresponds to a sharp increase of amplitude in observation that was identified as S wave. Because of the trapping effects and multiple reflections within the low-velocity surface layer, the direct S wave, the following sS and core–mantle reflections form reverberating wave trains lasting for about 400 s and they are not clearly separated. The wave train following the direct S wave is consistent with the observations and suggests that it contributes mainly to this portion of the observations. The second wave train lasting for about 1000 s is formed by the reverberating surface multiples SS, SSS, and SSSS. They contribute mainly to the middle portion of the observations and are consistent with data. At later time of the synthetics, reverberations caused by trapped waves in the surface layer agree well with observations and suggest their contribution at later portion. For deep event, the most interesting feature is the clearly identified and separated core–mantle reflections on synthetics. However, the corresponding arrivals cannot be identified on observations that are similar to the case for shallow events. Each phase shows reverberating waveform caused by surface low-velocity layer that is consistent with observation. With increasing epicentral distance, strength and duration of the reverberating wave train and their contribution to the waveform increase.

Although our modeling produced the reverberating features of each wave train that are consistent with the observations, the decay of our synthetic wave train is not as slow as that shown in the data. The strong scattering effects between seismic phases presented on observation are not reproduced in our synthetics. A similar phenomenon is also shown for global P-SV wave propagation modeling (Wang et al. 2013). This suggests that the low-velocity surface layer presented in our model contributes significantly to the development of reverberating waveforms, but it is not enough to produce the strong scattering and slow decay features of the observations with the present layered model. Similar phenomenon can be observed for earthquake as well. The spindle-shaped waveforms frequently appeared in the crustal earthquakes due to wave trapping effect in low-velocity layer (e.g., Furumura and Kennett 1997; Kennett and Furumura 2001). These features also appeared due to strong seismic wave scattering from small-scale heterogeneity with characteristic scale of several kilometers (e.g., Nielsen et al. 2003; Furumura and Kennett 2005; Takahashi et al. 2007; Kennett and Furumara 2008; Sato et al. 2012; Takemura and Yoshimoto 2014). This implies that small-scale heterogeneities should be required for modeling of realistic Moon structure. For earthquake observation, there is no wave before the direct SH-wave on transverse component for isotropic medium. However, transverse components for moonquakes in Fig. 7 show very strong energy before direct SH-wave and after direct P wave that do not appear on our synthetics. This discrepancy again suggests that the present isotropic 1D layered model is not enough to produce the characteristics of observations and more realistic models should be considered in order to well understand seismic wave propagation in the Moon.

7 Conclusions

We applied a hybrid PSM/FDM method to model global SH-wave propagation in the whole Moon and try to improve our understanding of lunar seismic wave propagation. In order to model high-frequency wavefield as presented on lunar seismic observations, we implemented parallel modeling on a PC cluster to improve the computing efficiency. The modeling is performed for shallow and deep moonquakes, respectively, to show SH-wave propagation in the whole Moon. Effects of frequency range and lateral variation of crust thickness on wave propagation are investigated with several models. Finally, we compared our synthetics with Apollo observation data and showed major features of wave propagation that are produced by our modeling and the limitations of our present models.

For shallow moonquake, the direct wave and core–mantle reflections can be identified at a limited range of epicentral distance because of interactions between them and other phases. Waveform of each phase shows strong reverberations caused by wave trapping effects and multiple reflections occurred in the surface low-velocity upper crust. Strong reverberating surface multiples dominate later in the wave propagation. For deep moonquake, the direct wave and core–mantle reflections can be clearly identified over wide range of epicentral distance. Reverberations in waveform and surface multiples are weaker than those for shallow event.

Increasing frequency enhances the strength and duration of waveform reverberation and reduces the possibility of phase identification. This suggests that body wave phases such as core–mantle reflections can be more clearly identified for deep moonquake at low-frequency range. Development and deployment of very broad band seismometers in the future lunar mission could provide further data to constrain the inner structure of the Moon. Lateral variation of crust thickness significantly influences lunar seismic wave propagation. Wavefield above the thick crust region tends to be more complicated because of transmitted wave from mantle into the crust.

Comparing with Apollo observation data that show that, our modeling produces the reverberating features of each phase but cannot reproduce the strong and slow decay scattering between each phase. For earthquake observation, similar phenomenon can be explained with wave trapping effect in low-velocity layer and scattering from small-scale heterogeneity. This suggests that more realistic Moon model should be considered in the future numerical modeling in order to improve our understanding of lunar seismic wave propagation.

Acknowledgments This research was supported by the National Natural Science Foundation of China (Grants 41374046 and 41174034).

References

Blanchette-Guertin J-F, Johnson CL, Lawrence JF (2012) Modeling seismic waveforms in a highly scattering Moon 43rd lunar and planetary science conference Abstract 1473 Lunar & Planetary Institute Huston

Cerjan C, Kosloff D, Kosloff R, Reshef M (1985) A nonreflecting boundary condition for discrete acoustic and elastic wave equation. Geophysics 50:705–708

Chenet H, Lognonné P, Wieczorek M, Mizutani H (2006) Lateral variations of lunar crustal thickness from the Apollo seismic data set. Earth Planet Sci Lett 243:1–14

Cummins PR, Geller RJ, Hatori T, Takeuchi N (1994) DSM complete synthetic seismograms: SH, spherically symmetric, case. Geophys Res Lett 21:533–536

Dainty AM, Toksöz MN, Anderson KR, Pines PJ, Nakamura Y, Latham G (1974) Seismic scattering and shallow structure of the moon in Oceanus Procellarum. Moon 91:11–29

Furumura T, Kennett BLN, Furumura M (1998) Seismic wavefield calculation for laterally heterogeneouswhole Earth models using the pseudospectral method. Geophys J Int 135:845–860

Furumura T, Kennett BLN (1997) On the nature of regional seismic phasesd—II. On the influence of structural barriers. Geophys J Int 129:221–234

Furumura T, Kennett BLN (2005) Subduction zone guided waves and the heterogeneity structure of the subducted plate: intensity anomalies in northern Japan. J Geophys Res 110:B10302. doi:10.1029/2004JB003486

Garcia RF, Gagnepain-Beyneix J, Chevrot S, Lognonné P (2011) Very preliminary reference Moon model. Phys Earth Planet Inter 188:96–113. doi:10.1016/j.pepi.2011.06.015

Gagnepain-Beyneix J, Lognonné P, Chenet H, Lombardi D, Spohn T (2006) A seismic model of the lunar mantle and constraints on temperature and mineralogy. Phys Earth Planet Inter 159:140–166

Herrmann RB (1979) SH-wave generation by dislocation source-a numerical study. Bull. Seismol Soc Am 69:1–15

Hikida H, Wieczorek MA (2007) Crustal thickness of the moon: new constraints from gravity inversions using polyhedral shape models. Icarus 192:150–166

Igel H, Weber M (1995) SH-wave propagation in the whole mantle using high-order finite differences. Geophys Res Lett 22:731–734

Igel H, Gudmundsson O (1997) Frequency-dependent effects on travel times and waveforms of long-period S and SS waves. Phys Earth Planet Inter 104:229–246

Jahnke G, Thorne MS, Cochard A, Igel H (2008) Global SH-wave propagation using a parallel axisymmetric spherical finite-difference scheme: application to whole mantle scattering. Geophys J Int 173:815–826. doi:10.1111/j.1365-246X.2008.03744.x

Kennett BLN, Furumura T (2001) Regional phases in continental and oceanic environments. Geophys J Int 146:562–568. doi:10.1046/j.1365-246x.2001.01467.x

Kennett BLN, Furumara T (2008) Stochastic waveguide in the lithosphere: Indonesian subduction zone to Australian craton. Geophys J Int 172:363–382. doi:10.1111/j.1365-246X.2007.03647.x

Khan A, Mosegaard K (2002) An inquiry into the lunar interior: a nonlinear inversion of the apollo seismic data. J Geophy Res 107. doi:10.1029/2001JE001658

Khan A, Maclennan J, Taylor SR, Connolly JAD (2006) Are the Earth and the Moon compositionally alike? Inferences on lunar composition and implications for lunar origin and evolution from geophysical modeling. J Geophys Res 111:E05005. doi:10.1029/2005JE002608

Khan A, Connolly JAD, Maclennan J, Mosegaard K (2007) Joint inversion of seismic and gravity data for lunar composition and thermal state. Geophys J Int 168:243–258. doi:10.1111/j.1365-246X.2006.03200.x

Koyama J, Nakamura Y (1980) Focal mechanism of deep moonquakes. In: Bedini (ed) Lunar and Planetary Science Conference Proceedings, vol. 11 of Lunar and Planetary Science Conference Proceedings, pp 1855–1865

Lawrence JF, Johnson CL (2010) Synthetic seismograms with high-frequency scattering for the Moon 41st Lunar and Planetary Science Conference Abstract 2701 Lunar & Planetary Institute Huston

Lognonné P, Gagnepain-Beyneix J, Chenet H (2003) A new seismic model of the moon: implications for structure, thermal evolution and formation of the moon. Earth Planet Sci Lett 211:27–44

Lognonné P (2005) Planetary seismology. Ann Rev Earth Planet Sci 33:191–1934

Nakamura Y (1977) Seismic energy transmission in an intensively scattering environment. J Geophys Res 43:389–399

Nakamura Y (1978) A1 moonquakes—source distribution and mechanism In: Lunar and Planetary Science Conference Proceedings, vol. 9 of Lunar and Planetary Science Conference Proceedings, pp 3589–3607

Nakamura Y (1983) Seismic velocity structure of the lunar mantle. J Geophys Res 88:677–686

Nakamura Y, Latham GV, Dorman HJ, Ibrahim A, Koyama J, Horvath P (1979) Shallow moonquakes—depth, distribution and implications as to the present state of the lunar interior. In: Hinners (ed) Lunar and Planetary Science Conference Proceedings, vol. 10. Pergamon Press, New York, pp 2299–2309

Nielsen L, Thybo H, Levander A, Solodilov N (2003) Origin of upper-mantle seismic scattering—evidence from Russian peaceful nuclear explosion data. Geophys J Int 154:196–204

Qin Y, Wang Y, Takenaka H, Zhang X (2012) Seismic ground motion amplification in a 3D sedimentary basin: effect of the vertical velocity gradient. J Geophys Eng 9:761–772. doi:10.1088/1742-2132/9/6/761

Robertsson JOA, Holliger K (1997) Modeling of seismic wave propagation near the Earth's surface. Phys Earth Planet Inter 104:193–211

Sato H, Fehler M, Maeda T (2012) seismic wave propagation and scattering in the heterogeneous Earth structure, 2nd edn. Springer-Verlag, New York

Takahashi T, Sato H, Nishimura T, Obara K (2007) Strong inhomogeneity beneath quaternary volcanoes revealed from the peak delay analysis of S-wave seismograms of microearthquakes in northeastern Japan. Geophys J Int 168:90–99. doi:10.1111/j.1365-246X.2006.03197.x

Takemura S, Yoshimoto K (2014) Strong seismic wave scattering in the low-velocity anomaly associated with subduction of oceanic plate. Geophys J Int 197:1016–1032. doi:10.1093/gji/ggu031

Thorne MS, Lay T, Garnero EJ, Jahnke G, Igel H (2007) Seismic imaging of the laterally varying D″ region beneath the cocos plate. Geophys J Int 170:635–648. doi:10.1111/j.1365-246X.2006.03279.x

Toyokuni G, Takenaka H, Wang Y, Kennett BLN (2005) Quasi-spherical approach for seismic wave modeling in a 2D slice of a global Earth model with lateral heterogeneity. Geophys Res. Lett 32:L09305. doi:10.1029/2004GL022180

Wang T, Chen L (2009) Distinct velocity variations around the base of the upper mantle beneath northeast Asia. Phys Earth Planet Inter. doi:10.1016/j.pepi.2008.09.021

Wang Y, Takenaka H (2011) SH-wavefield simulation for a laterally heterogeneous whole-Earth model using the pseudospectral method. Sci China Earth Sci 54:1940–1947. doi:10.1007/s11430-011-4244-8

Wang Y, Takenaka H, Jiang X, Lei J (2013) Modelling two-dimensional global seismic wave propagation in a laterally heterogeneous whole-Moon model. Geophys J Int. doi:10.1093/gji/ggs094

Wang Y, Luo L, Qin Y, Zhang X (2014) Global SH-wavefield calculation for a two dimensional whole-Earth model with the parallel hybrid PSM/FDM algorithm. Earthq Sci. doi:10.1007/s11589-014-0085-9

Weber RC, Lin P, Garnero EJ, Williams Q, Lognonné P (2011) Seismic detection of the lunar core. Science 331:309–312

Wen L (2002) An SH hybrid method and shear velocity structures in the lowermost mantle beneath the central Pacific and South

Atlantic Oceans. J Geophys Res 107:2055. doi:10.1029/2001JB000499

Wieczorek MA, Neumann GA, Nimmo F, Kiefer WS, Taylor GJ, Melosh HJ, Phillips RJ, Solomon SC, Andrews-Hanna JC, Asmar SW et al (2013) The crust of the Moon as seen by GRAIL. Science 339:671–675

Zhao D, Lei J, Liu L (2008) Seismic tomography of the Moon. Chin Sci Bull 53:3897–3970

Zhao D, Arai T, Liu L, Ohtani E (2012) Seismic tomography and geochemical evidence for lunar mantle heterogeneity: comparing with Earth. Global Planet Change 90:29–36

17

Ambient noise surface wave tomography of the Makran subduction zone, south-east Iran: Implications for crustal and uppermost mantle structures

Mahsa Abdetedal · Zaher Hossein Shomali · Mohammad Reza Gheitanchi

Abstract Seismic ambient noise of surface wave tomography was applied to estimate Rayleigh wave empirical Green's functions (EGFs) and then to study crust and uppermost mantle structure beneath the Makran region in south-east Iran. 12 months of continuous data from January 2009 through January 2010, recorded at broadband seismic stations, were analyzed. Group velocities of the fundamental mode Rayleigh wave dispersion curves were obtained from the empirical Green's functions. Multiple-filter analysis was used to plot group velocity variations at periods from 10 to 50 s. Using group velocity dispersion curves, 1-D v_S velocity models were calculated between several station pairs. The final results demonstrate significant agreement to known geological and tectonic features. Our tomography maps display low-velocity anomaly with SW-NE trend, comparable with volcanic arc settings of the Makran region which may be attributable to the geometry of Arabian Plate subducting beneath the overriding the Lut block. The northward subducting Arabian Plate is determined by high-velocity anomaly along the Straits of Hormuz. At short periods (<20 s), there is a sharp transition boundary between low- and high-velocity transition zone with the NW trending at the western edge of Makran which is attributable to the Minab fault system.

Keywords Ambient seismic noise · Cross-correlation · Empirical Green's functions (EGFs) · Surface wave tomography

M. Abdetedal (✉) · Z. H. Shomali · M. R. Gheitanchi
Institute of Geophysics, Tehran 14155-6466, Iran
e-mail: mahsa.etedal@gmail.com

1 Introduction

The Iranian plateau is subject to several tectonic episodes, including active stages of intense folding (e.g., in the Zagros region), faulting and different types of tectonic domains. The Makran subduction zones is located in the SE of Iran, extending from the Main Zagros Thrust (MZT) to the western end of the Makran wedge and to the Ornach-Nal and Chaman fault zones in southwestern Pakistan (see Fig. 1). The transition between the Zagros continent-continent collision zone and the western Makran subduction zone is marked as the Minab fault (see Fig. 1b). This complex was formed by subduction of the oceanic crust of the Arabian Plate under the Eurasian Plate (Fig. 1a). The Arabian Plate convergence rate, which occurs in a north direction towards Eurasia, is ∼23–25 mm/a according to GPS measurements (Bayer et al. 2003; McClusky et al. 2003; Vernant et al. 2003; Masson et al. 2007). The Makran subduction zone subducts at a higher convergence rate compared with the continental collision portion of the Arabian Plate boundary in the Zagros Suture Zone. In the Makran region, the convergence rate increases from west to east (Vernant et al. 2003). Geodetic data provide selected evidence that nearly all of the convergence of the Arabian Plate is accommodated in the eastern boundary of Makran, whereas only half of the convergence is accommodated in the west. At the eastern boundary of Makran, the convergence rate is ∼42.0 mm/a (Fig. 1a), as estimated by DeMets et al. (1990).

Notably little seismic tomography has been performed in the Makran subduction zone, especially on the structure of the upper mantle. Compared with the remainder of Iran, relatively little is known about the structure of the Makran subduction zone. Most of the tomographic studies performed in the Makran region are limited to global tomography surveys with low resolution. Few shallow seismic

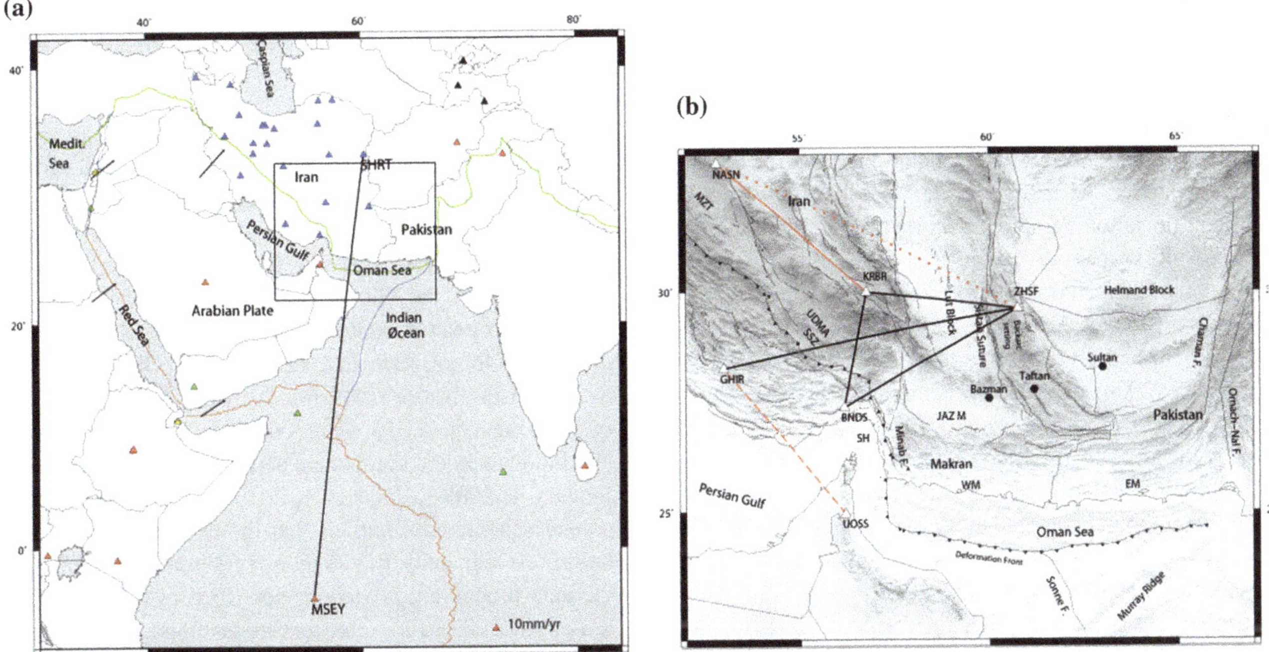

Fig. 1 **a** Locations of broadband seismic stations used in this study, marked by *triangles*. There are 21 stations from the International Institute of Earthquake Engineering and Seismology (IIEES) denoted by the *blue triangles* and the *red triangles* show the locations of Global Seismic Network (GSN) stations. The *green triangles* denote the GEOFON program of GFZ Potsdam (GEOFON). *Black* and *yellow triangles* show Tajikistan National Seismic Network (TJ) and the Virtual European Broadband Seismic Network (VEBSN). Ridge, trench, and transform boundaries are indicated by *red*, *green*, and *blue lines*, respectively. Plate motions are calculated in www.unavco.org based on APKIM2005 plate motion model (Drewes, 2009). **b** Topography map of the study area. *WM* Western Makran, *EM* Eastern Makran, *SSZ* Sanandaj-Sirjan Zone, *UDMA* Urumieh–Dokhtar Magmatic Arc, *MZT* Main Zagros Thrust, *JAZ M*. Jaz Murian., *SH* Straits of Hormuz

investigations have examined the sedimentary structure of the Makran belt to reveal the heterogeneity of the crust and upper mantle, and few seismic tomography studies have been conducted on the crustal and upper mantle structure of the Makran subduction zone. Giese et al. (1984) studied the Moho depth using a refraction profile consisting of sparse recordings along a line from central Iran to the Straits of Hormuz and indicated a crustal thickness of 40 km beneath central Iran. Using gravity measurements in the seismic results of Giese et al. (1984), Dehghani and Makris (1984) prepared a Moho map of the Iranian plateau and found that the crust beneath the Lut depression is less than 40 km thick. Snyder and Barazangi (1986) used the same data and found that the Moho depth was nearly 40 km beneath the Persian Gulf (Maggi and Priestley 2005). The crustal thickness of the Makran region is less well known. Few studies of the deep structure of the upper mantle exist for this area. Recent surface waveform tomography (Shad Manaman et al. 2011) indicated that the crustal thickness beneath the Oman seafloor and Makran fore-arc setting is ~25–30 km and is increasing to the volcanic arc. The Moho depth increases up to ~48–50 km under the Taftan-Bazman volcanic arc (see Fig. 1b), where the subducting plate bends. Similarly, from the fore-arc setting to the volcanic arc in eastern Makran, the Moho depth increases to ~40 km. Regional tomographic studies of the Iranian plateau do not provide detailed information on the structures in the crust and upper mantle due to the lack of well-documented earthquakes. The lateral resolution is usually limited in these studies, e.g., on the order of 200-km horizontal grid spacing (Maggi and Priestley 2005) and 4° (Shad Manaman et al. 2011) in the Iranian plateau. In the Makran region, the resolution is even more limited because the path coverage is not dense due to the inconvenient distribution of stations and limited seismicity. Compared with our understanding of active tectonics, much less is known about the Makran subduction zone, and the seismic behavior of the Makran subduction zone has remained largely unknown. Additionally, basic limitations exist in the earthquake-based surface wave tomography because of limited seismicity and poor station coverage.

Recent studies use ambient noise to extract the surface wave empirical Green's functions (EGFs). Rayleigh and Love waves obtained from these EGFs can provide important information on the 3D shear wave velocity structure in the crust and uppermost mantle on both a global and regional scales. Shapiro and Campillo (2004) inferred Rayleigh wave dispersion curves from ambient noise data recorded in station pairs separated by distances from ~100 to >2000 km. Shapiro et al. (2005) and Sabra et al. (2005) extracted time-domain Green's functions from ambient seismic noise and estimated the group velocity of Rayleigh waves at a local scale. Certain studies applied ambient noise tomography at short- and intermediate-band (e.g., 5–40 s) periods (Zheng et al. 2010; Yao 2012). However, other studies performed ambient noise tomography over longer periods, e.g., up to 60 s in China (Sun et al. 2010) and up to 70 s in the US (Bensen et al. 2008). Lin et al. (2008) presented seismic ambient noise tomography (ANT) for Love waves. Nishida et al. (2008) also analyzed Love waves to study seismic ambient noise tomography, including the crustal overtones.

Recent theoretical studies demonstrate that under the assumption, that the seismic noise wavefield is diffused, the empirical Green's functions between two stations can be estimated by correlating the noise recordings from these two sites (Weaver and Lobkis 2001; Derode et al. 2003; Snieder 2004). If sufficient stations exist, the study of ambient noise seismic waves yields results with a resolution higher than that of traditional surface wave tomography methods (e.g., Yang et al. 2007). The earthquake-based surface wave tomography has certain limitations, such as difficulty in obtaining high-quality short-period (<20 s) dispersion measurements from teleseismic events and inaccuracy in earthquake hypocentral locations and moment tensors, especially for small events. These limitations affect the inversions of seismic surface waves. Additionally, seismic surface waves only sample certain azimuths due to the uneven distribution of earthquakes. The purpose of this study is to produce Rayleigh wave group velocity maps with higher resolution than previous surface wave maps produced in the Makran region to study the crust and the upper mantle structures in the Makran subduction zone.

In this study, the ANT is applied using data from periods of ~10–50 s collected from the recordings of 41 stations between January 1, 2009 and January 1, 2010. Dispersion curves were measured using the fundamental mode of Rayleigh waves extracted from the ambient noise and inverting them to obtain 2D group velocity maps for the crustal and upper mantle structures of the region. We also investigated the directionality and seasonal variations of the noise sources. The difference between the causal and acausal components of the cross-correlation results of station pairs was examined to measure the main direction of the energy flux across the region. Using the group velocity dispersion curves, 1-D v_S velocity models were calculated between several station pairs. Finally, the resulting group velocity maps were used to infer crust and upper mantle structures in the Makran region.

2 Method

2.1 Data processing

This study is based on a variety of seismic sensor data collected from digital broadband instrument (BH) recordings from the International Institute of Earthquake Engineering and Seismology (IIEES), which is equipped with a CMG-3T broadband sensor (0.01–100 s), as well as seismic data from the Global Seismic Network (GSN), the Virtual European Broadband Seismic Network (VEBSN), the GEOFON program of GFZ Potsdam (GEOFON), and the Tajikistan National Seismic Network (TJ), as depicted in Fig. 1a. The EGFs were calculated for one year of continuous vertical component seismograms recorded from January 1, 2009 to January 1, 2010. Use of vertical component seismic data implies that the resulting cross-correlations contain only Rayleigh wave signals. We followed the data processing procedure described in detail by Bensen et al. (2007). First, the continuous noise data were cut into one-day data files, and those data with gaps of less than 10 s were chosen. The instrumental responses were subsequently removed from all data, followed by decimation of the data to one sample per second. Decimation reduced the amount of storage space and computational time required. The next step involved removing the trend and mean value, zero-phase Butter-worth high-pass filtering with a corner frequency of 0.01 Hz, and whitening and bandpass filtering around the target frequency (period from 10–50 s) as a function of the inter-station distance (Cho et al. 2007; Pedersen and Krüger 2007). The next processing step included temporal or time-domain normalization to remove further contaminating effects of earthquakes on the noise correlations (Bensen et al. 2007). The results were stacked over the total time period available for each pair to produce the resulting time series.

To quantitatively evaluate the quality of the stability of the stacking process, we calculated the signal-to-noise ratio (SNR) for each cross-correlation. The SNR is defined as the ratio of the peak amplitude within a time window surrounding the expected arrival time of the fundamental mode Rayleigh waves at a given period to the root-mean-square of noise trailing the signal arrival window (Bensen et al. 2007). The signal window is determined using the arrival times of Rayleigh waves at the minimum and maximum periods of the chosen pass-band frequency. All empirical Green's functions in the 10- to 50-s period bands are plotted in Fig. 2 to evaluate the quality of the cross-correlation functions. Based on cross-correlations of broadband seismic records obtained at stations within or adjacent to the Pacific Basin, Lin et al. (2006) indicated that broadband ambient noise propagates coherently between island stations and between island and continental stations. The Green's functions observed for fundamental mode Rayleigh

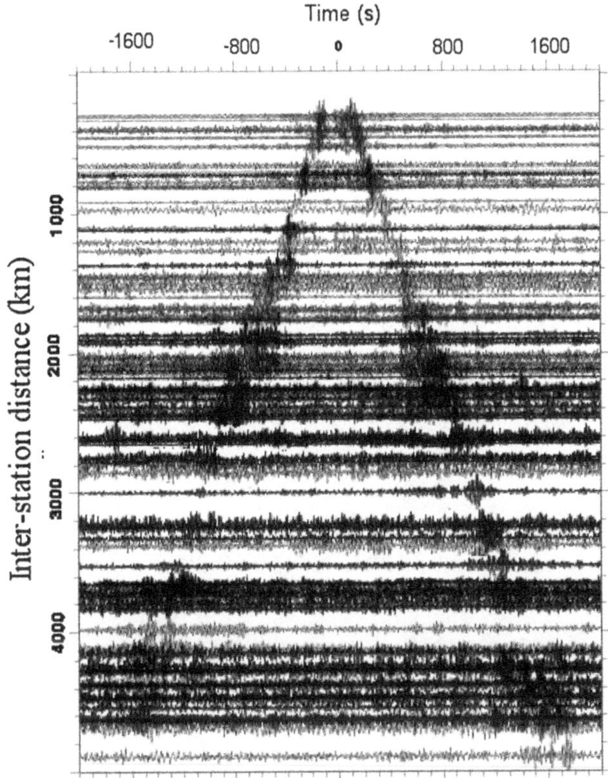

Fig. 2 Empirical Green's functions in the 10- to 50-s period band plotted as a function of distance

waves with high SNR establishes the physical basis for ANT across the Pacific, and any non-physical arrivals caused by long ocean-crossing paths were successfully rejected. Both positive and negative correlation lags show clear surface waves with an average apparent velocity of ~3.0 km/s. The group velocities used to predict arrival times were calculated from the AK135 velocity model, (Kennett 1995), and the RMS noise level was measured in a 500-s noise window at the end of the signal.

2.2 Directionality

To obtain a better understanding of the distribution of the noise source in space and time, the seasonal variability of the relatively continuous noise was studied, a process necessary for optimization of the noise-based seismic tomography. Seasonal variations of the cross-correlations for periods of 10–20 and 25–50 s were computed for two station pairs located perpendicular to each other (BNDS-KRBR with the NS direction and KRBR-ZHSF with the EW direction) for different seasons (Fig. 3). A positive time delay indicates waves propagating from the coastlines to the continent (BNDS to KRBR). For the other pair of stations, the positive lags indicate signals propagating from KRBR to ZHSF. Considering a period band of 10–20 s, the cross-correlations exhibit a clear seasonal variation (Figs. 3). For the NS station pair, the apparent asymmetry of the data indicates that the energy originating from the coastlines is much larger than that from the continent. This result shows that an important contribution of the noise observed in the Makran region originates from the south and the coastlines, likely from the Persian Gulf and Oman Sea. For another pair of stations (KRBR-ZHSF), the directionality is significant but with different characteristics. In spring and summer, a preferred directionality does not appear; however, a clear pattern in autumn indicates that the energy flux flows from the east to the west, and this trend is completely the opposite in winter.

At longer period bands (25–50 s), the behavior of the noise varies in different seasons. The cross-correlation in this band is symmetric in spring and summer for BNDS-KRBR (Fig. 3b), whereas in the period band of 10–20 s, the amplitude of the positive lag of the cross-correlations is much larger (Fig. 3a). Again, for KRBR-ZHSF, contrary to the observations in the period band of 10–20 s, the resulting using cross-correlation is not symmetric in spring (Fig. 3c, d). Moreover, in autumn, Fig. 3d indicates a wave propagation direction opposite to the results calculated in the period band of 10–20 s (Fig. 3c). This result suggests that the primary microseism is not recorded by the same process that generates the longer period noise (Stehly et al. 2006). The primary microseism has periods similar to those of the main swell (10–20 s). Therefore, it is believed that the primary microseism is related to the interaction of the sea waves with the coast (Gutenberg 1951). The long-period noise or "the hum" has been attributed to the so-called infragravity waves, i.e., the ocean wave mode that exists at long periods (Webb et al. 1991). Thus, the results of our analysis show that the sources of the primary microseism exhibit seasonal variability that differs from that of the long period noise (hum).

In practice, for different frequency bands, the observed distribution of ambient noise can be far from homogeneous (Stehly et al. 2006). Therefore, it is necessary to determine whether the ambient noise is sufficiently distributed in azimuth to return unbiased dispersion measurements for use in tomography. To quantify the effect of the strongly anisotropic background noise source distribution, Yang et al. (2008) performed synthetic experiments and found that in the presence of low level and homogeneously distributed ambient noise, <0.5 % of the measured phase velocities are affected by much stronger ambient noise in an off-axis direction. Therefore, we must show that in all period ranges studied, the useful amount of ambient noise signals in all azimuths is greater than 50 %.

To investigate the directions of the incoming ambient noise, we plotted the azimuthal distribution of the SNR for the positive and negative components of each cross-correlation for the four period bands of 10–20 s, 20–30 s, 30–40 s, and 40–50 s in the northern winter (Oct. to Mar.) and northern summer (May to Sept.) of 2009 (Fig. 4). The length

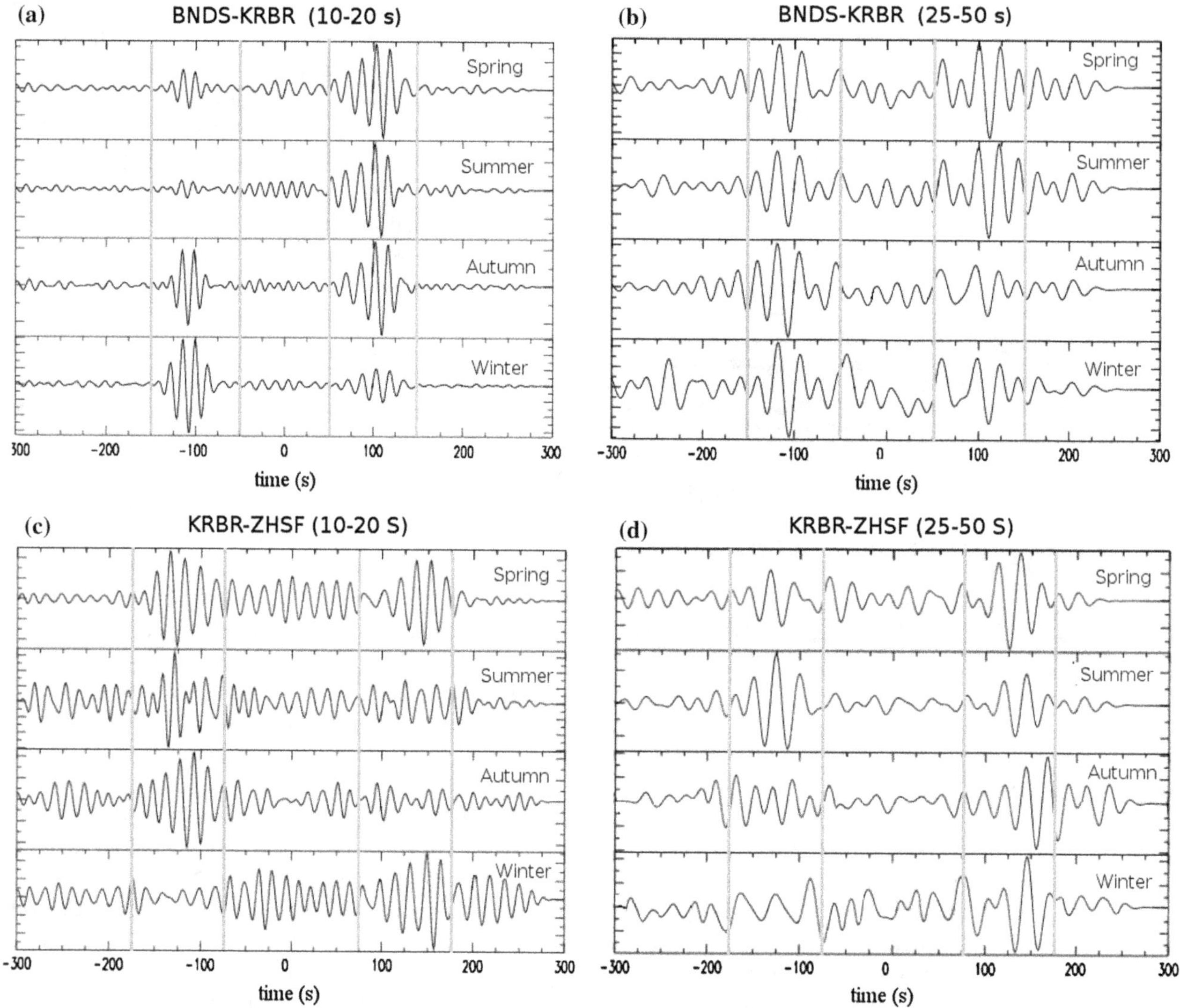

Fig. 3 Cross-correlation of the period in the range of 10–20 s (**a**) and 25–50 s (**b**) of one year, 2009, of noise recorded on BNDS-KRBR. The inter-station distance is 297 km. Cross-correlation of the period in the range of 10 and 20 s of one year, 2009, of noise recorded on KRBR-ZHSF with the inter-station distance of 389 km. Cross-correlation of the period in the range of 25–50 of one year, 2009, of noise recorded on KRBR-ZHSF. Locations of station pairs BNDS-KRBR, KRBR-ZHSF are shown in Fig. 1b

of each line is the amplitude of signal and the angle points in the direction from which the energy arrives. Each 20° azimuth bin shows the number of paths for both the inter-station azimuth (causal) and back-azimuth (acausal) components of the cross-correlation functions. Following the work of Bensen et al. (2008), the average Rayleigh wave EGFs with SNR >10 was computed at all four periods, and to compute the average fraction of yearly EGFs, the number of paths with SNR >10 in a given 20° azimuth bin was divided by the total number of paths in that bin. The average results over all azimuths at the four period bands of 10–20 s, 20–30 s, 30–40 s, and 40–50 s were 0.53, 0.64, 0.69, and 0.51, respectively. In other words, these values reveal that the fractions of relatively high SNR paths in all azimuths are greater 50 % in all period ranges studied, and hence, the useful amount of ambient noise signals is sufficiently distributed in different azimuths.

3 Group velocity measurements

In the next step, multiple-filter analysis (Herrmann and Ammon 2013) was used to measure group velocity dispersion curves. Each of the frequency components of the surface wave is sensitive to a different depth interval. In general, wave components with longer wavelength that propagate deeper will travel faster than the shallower ones because the seismic velocity of the Earth increases radially downwards.

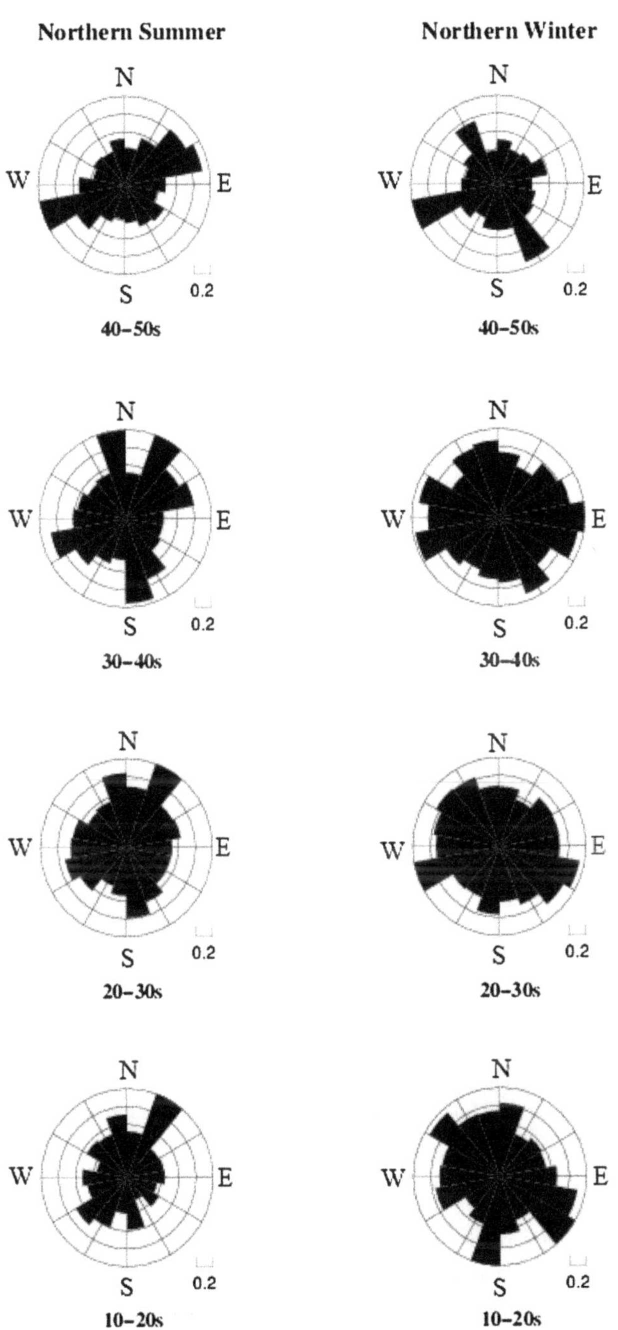

Fig. 4 Azimuthal distribution of SNR during the (*left*) northern summer and (*right*) northern winter at four periods 10–20 s, 20–30 s, 30–40 s, and 40–50 s

A technique known as phase-matched filtering was applied to determine the correct dispersion curve. The waveforms were narrow-bandpass-filtered with the Gaussian filter $G(f) = \exp[-\alpha(f - f_c)^2/f_c^2]$ where f_c is the center frequency. A trade-off exists between resolution in the time and frequency domains that is caused by filtering, i.e., larger values of α enhance the resolution in the frequency domain, whereas it decreases resolution in the time-domain (Herrmann 1973; Levshin et al. 1989). For distances

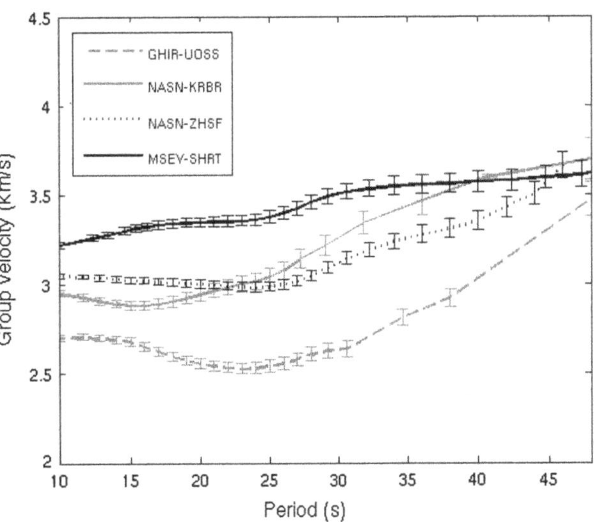

Fig. 5 Group velocity dispersion curves measured from the paths shown in Fig. 1a, b

between 200 and 3000 km and 3000–5000 km, $\alpha = 25$ and 50 s were found suitable for our measurements, respectively. Selected dispersion curves are plotted in Fig. 5, and the corresponding paths are depicted in Fig. 1a, b.

The estimated uncertainties for the group velocities are based on seasonal variability because dispersion measurements from cross-correlations of ambient noise are naturally repetitive. To analyze the uncertainty, we selected 12 overlapping three-month time series for each station pair. The three-month time windows are reliable for obtaining dispersion measurements and also contain the seasonal variation. Fig 6 shows group velocity measurements for four station pairs (BNDS-ZHSF, GHIR-UOSS, MSEY-SHRT, and NASN-KRBR) obtained in twelve three-month cross-correlations that were bandpass filtered from 10 to 50 s periods. The paths for the station pairs are depicted in Fig. 1a, b. The one-year measurement is plotted as a black line with error bars, indicating the computed standard deviation. The standard deviation is computed on all sequential three-month stacks, and following the work of Yang et al. (2007), it was computed for a station pair if four SNR values of the three-month stacks exceed the criterion (in our study, SNR >10). The measurement was rejected if the standard deviation was either so large or so small that it could not be obtained in the three-month stacks. The uncertainty tends to increase with the period, possibly because of the decreasing amplitude of ambient noise at periods greater than 20 s (Yang et al. 2007). If the standard deviation was more than three times the average of the standard deviations taken over all measurements, it was rejected because this indicates instability in the measurement (Bensen et al. 2008).

The 1-D crustal and upper mantle v_S structures were subsequently calculated between different station pairs using the surf96 package (Herrmann and Ammon 2013). The initial v_S model was parameterized into 4-km-thick

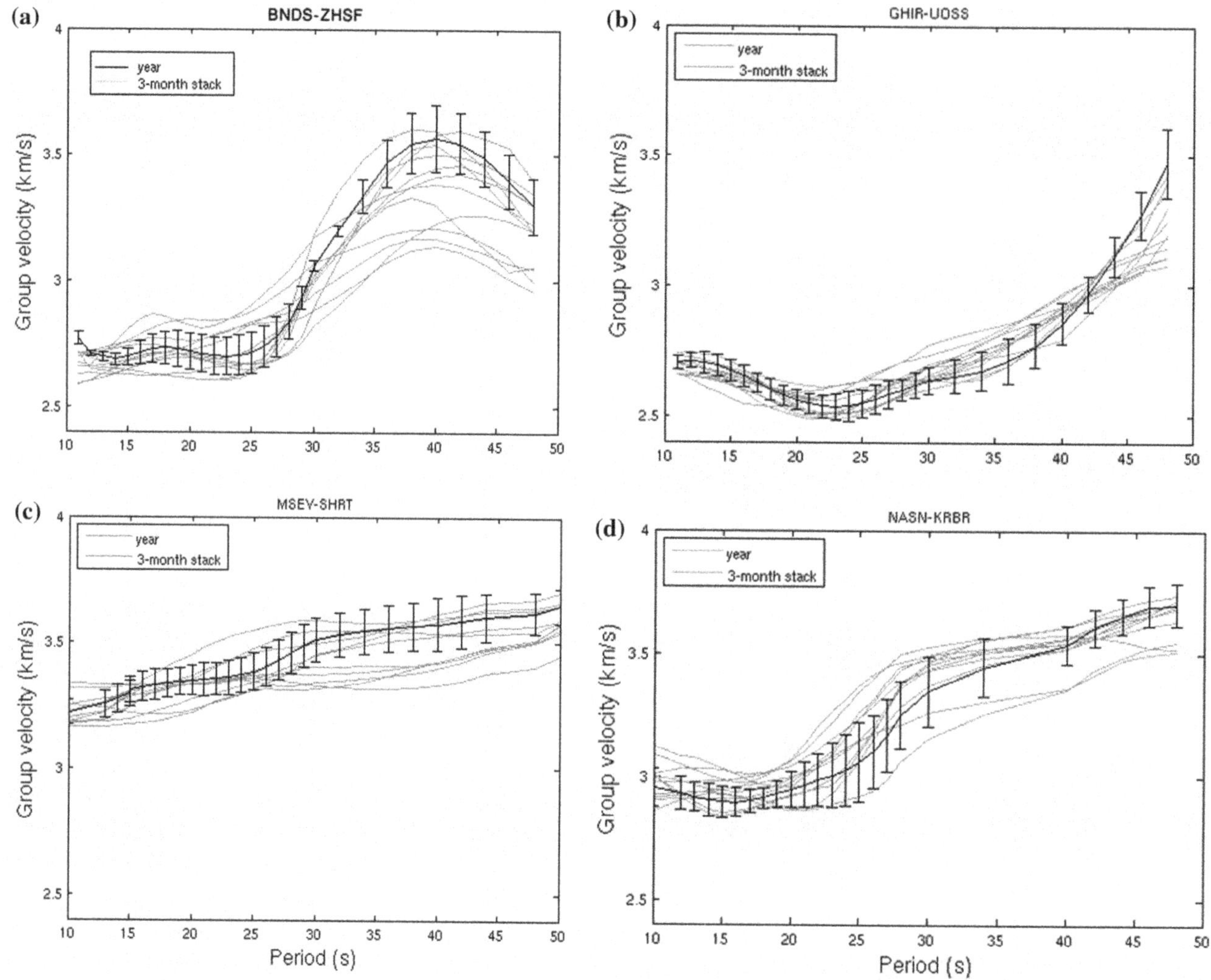

Fig. 6 Uncertainties estimates based on seasonal variability of the dispersion measurements for four station pairs (BNDS-ZHSF, GHIR-UOSS, MSEY-SHRT, NASN-KRBR). The *gray curves* are group velocity measurements obtained on twelve three-month cross-correlations. The 12-month measurement is plotted as *black line* and the *error bars* indicate the computed standard deviation. The paths for station pairs are depicted in Fig. 1a, b

layers from surface to 240 km and then a half space below 240 km. Group velocity dispersion curves were input data for the inversion procedure. The reference v_S model refers to the AK135 model (Kennett 1995). For the surf96 package, during the v_S structure inversion, it was also necessary to input other parameters. The input density was assigned using the parameter of the background model (AK135), which gradually increases from 2.3 g/cm^3 in the shallow crust to 3.3 g/cm^3 in the deeper crust and is set to 3.3 $\pm$ 0.1 g/cm^3 in the upper mantle. In addition to the shear wave speed (v_S), the P-wave speed (v_P) and density (ρ) are also required for inversion. These parameters were assigned based on parameters of the background model (AK135). An example of the 1-D v_S velocity models calculated between two station pairs (GHIR-ZHSF and KRBR-ZHSF) from the current dataset is presented in Fig. 7a next to the observed and synthetic dispersion curves (Fig. 7c). The paths related to two station pairs are depicted in Fig. 1b. The sensitivity kernels for different periods were also calculated from the AK135 velocity model (Fig. 7b). Because Rayleigh waves are more sensitive to shear wave velocities than compressional velocities, except in the uppermost crust (Bensen et al. 2009; Lin et al. 2007), we fixed the v_P/v_S ratio of each layer in accordance with the Ak135 velocity model, and the P-wave velocity was subsequently updated by the v_P/v_S ratio of the initial model in successive iterations.

The parametric test was used to determine the influence of the initial velocity model. The initial v_S models were perturbed using a normal random distribution with a standard deviation of 0.1, and 200 v_S models were produced. The Rayleigh dispersion curves were inverted using 200 perturbed initial models. Selected v_S models for the station pair BNDS-

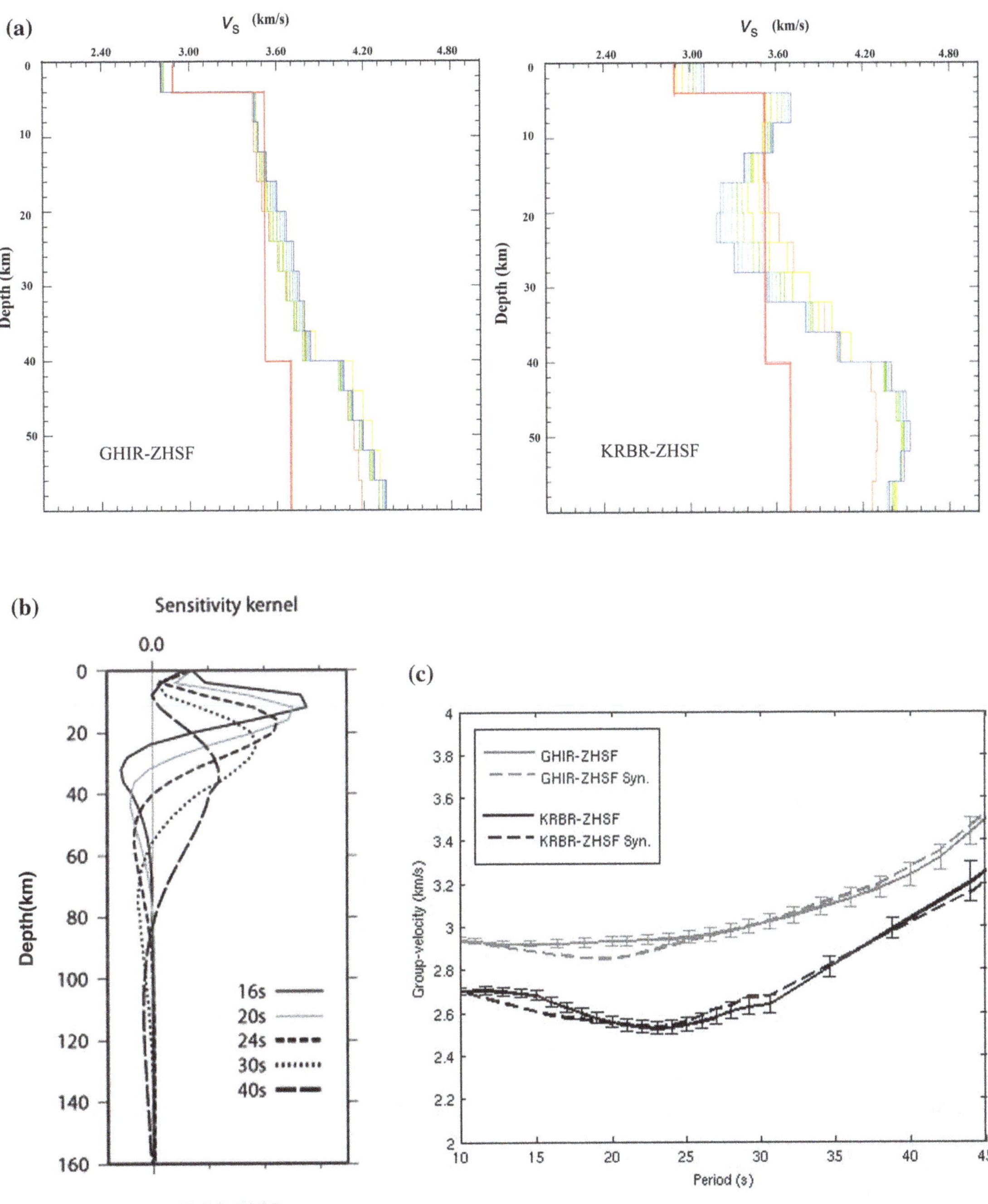

Fig. 7 **a** S-wave velocity model for two station pairs GHIR-ZHSF and KRBR-ZHSF. The starting v_S model is shown by *red lines*. The *blue line* is the final v_S model. **b** Sensitivity kernels of group velocity as a function of period for the v_S velocity measurements. **c** Dispersion curves related to the two station pairs mentioned above with their predicted dispersion *curves* (syn.) and their measured dispersion *curves*. The paths for GHIR-ZHSF and KRBR-ZHSF are depicted in Fig. 1b

GHIR from 200 runs at depths of 10, 20, 30, 40, 60, and 80 km are presented in Fig. 8. The parameter setting did not change. It can be observed that the histograms of inverted results tend to have a near normal distribution that congregates at the mean value with some deviation. The distributions reflect how well v_S is constrained at each depth (Fig. 8).

4 Rayleigh wave tomography

A 2-D tomographic inversion technique was applied for the calculation of the group velocity variations derived from the dispersion measurements of Rayleigh waves from the one-year cross-correlations recorded from January 1, 2009

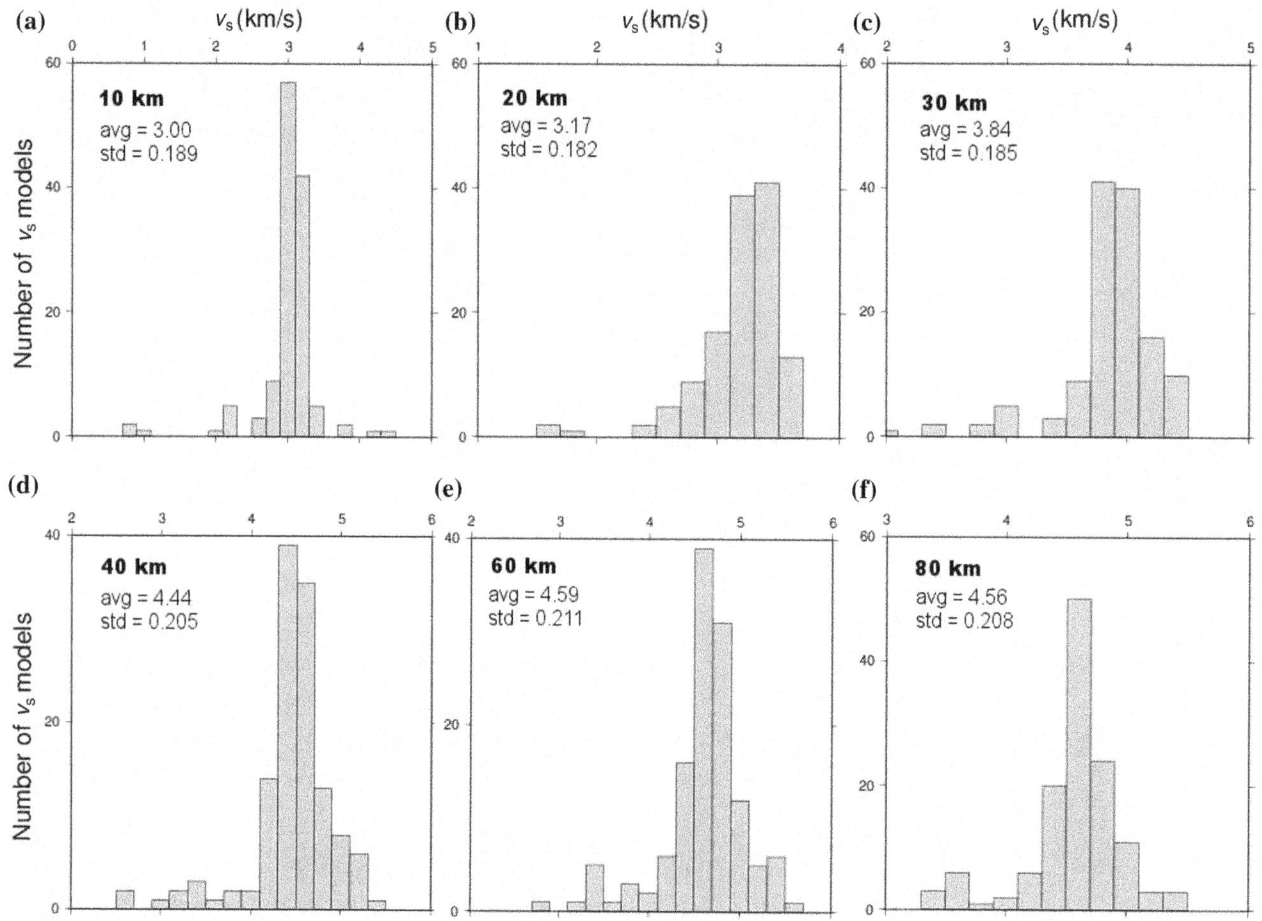

Fig. 8 The influence of the initial velocity model on the non-linear iterative damped least squares inversion was investigated using the normal randomly distribution with standard deviation 0.1 at 200 runs. Selected v_S models for station pair BNDS-GHIR and at 200 runs and at depths of **a** 10 km, **b** 20 km, **c** 30 km, **d** 40 km, **e** 60 km, and **f** 80 km are shown. The path for BNDS-GHIR is depicted in Fig. 1b

to January 1, 2010. Before conducting ANT, we checked the data availability of all broadband stations inside and around Iran and finally decided to use data from January 1, 2009 to January 1, 2010 because of its continuous data from most stations existed during this time period.

Fast marching surface wave tomography (FMST), the iterative non-linear inversion package developed by Rawlinson (2005) and Rawlinson and Sambridge (2005), was used for analysis. This method includes the forward calculation and inversion procedures. Subspace method was used in the inversion, which is based on an assumption of local linearity that can reduce the perturbation of the model parameters and then make the inversion result approach to the current model. The inversion step allows both smoothing and damping regularization to limit the non-uniqueness of the solution. The Fast Marching Method (FMM) is a grid-based numerical algorithm based on the eikonal equation that is formulated to determine the first arrival phase of surface waves rather than the group time. However, to describe the dissipation of the group energy, an eikonal solver can be used if multi-pathing is not included. In this case, the interfering waves cause the group energy to follow notably different paths. Therefore, if the phase and group velocities have a similar geographic pattern, comparable results can be obtained (Arroucau et al. 2010; Saygin and Kennett 2010; Young et al. 2011; Saygin and Kennett 2012). Young et al. (2011) obtained similar group and phase velocity maps using FMM in southeastern Australia. Generally, the FMM method is an iterative non-linear approach with an assumption of local linearity in the inversion step. The non-linear relationship between the travel time and the group velocity could have been explained by repeated applications of FMM and subspace inversions (Rawlinson 2005; Rawlinson and Sambridge 2005). The damping value was estimated based on a trade-off curve between the data misfit and model roughness. Combination of the FMM for calculation of the forward problem and the subspace method for inversion provides tomographic imaging.

The potential resolution of the tomographic results was evaluated using the checkerboard synthetic tests for 16, 20, 30, and 40 s periods that used actual path distribution. Two sets of tests with pattern sizes of 2° × 2° and 1° × 1° were conducted. The general recoveries for both tests were good; however, higher resolution for central Iran was attained for the 1° × 1° cell size (see the Appendix Fig. 12). The maximum error was 5 % noise signal with a perturbing velocity of 2.8 ± 0.1 km/s and superimposed alternating high- and low-velocity anomalies (as shown in the Appendix, Fig. 12). The number of iterations in the subspace inversion depends on the frequency value, which varies from 2 for the higher frequencies to 5 for the lower frequencies. The checkerboard test results of the observed data and the ray-path coverage for periods of 16, 20, 30, and 40 s are shown in Fig. 9a–d, respectively. The checkerboard results suggest that the resolution is fairly good in most periods in central Iran, indicating that the pattern and absolute amplitude values were recovered well. However, for the eastern and southeastern locations of the region, due to rare distribution of stations, the path coverage is not dense, and most waves travel in parallel; therefore, the resolution is so low that smearing effects are apparent in the eastern region, which make the inversion process not recover the absolute amplitudes well. The tomography maps at

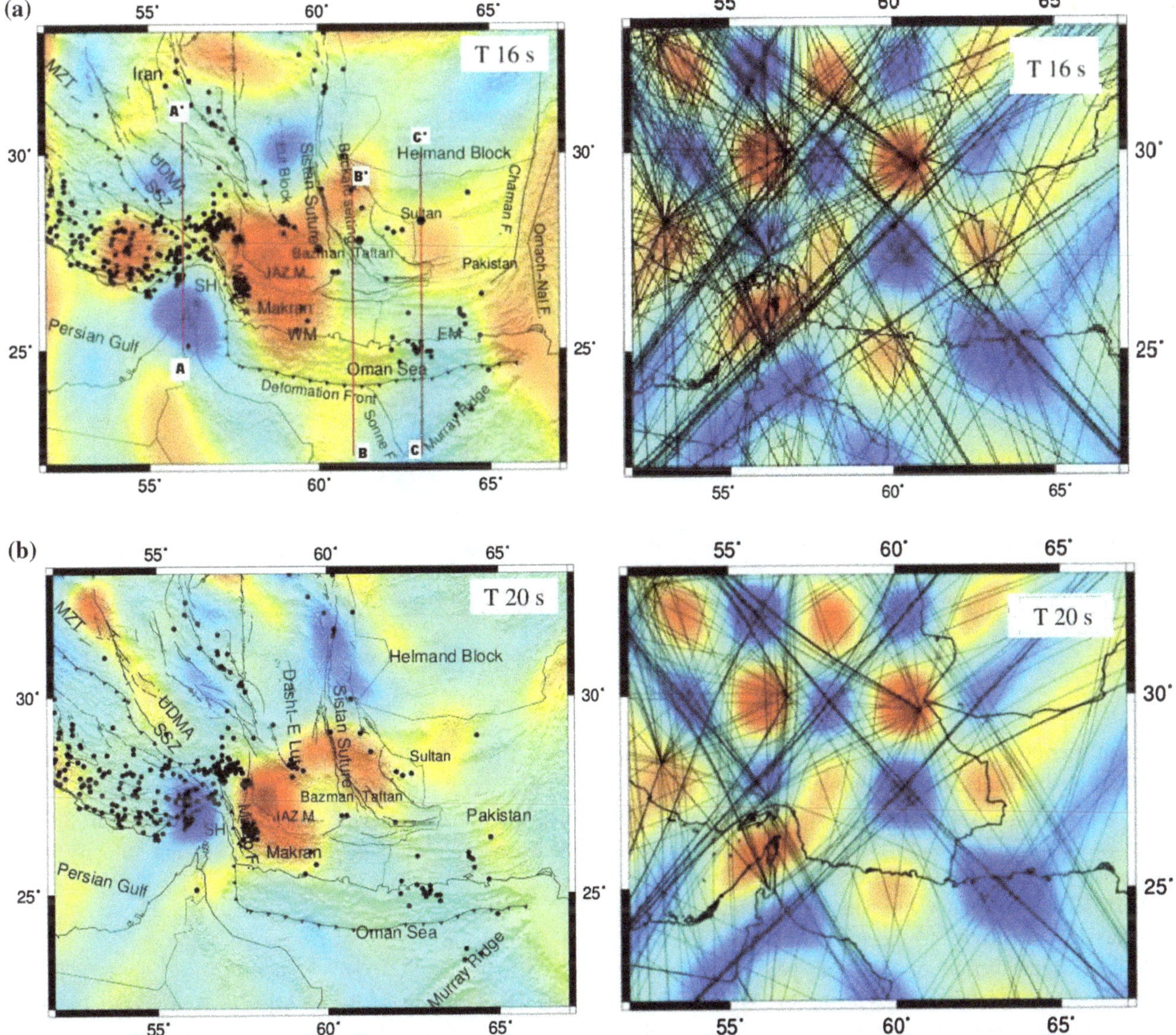

Fig. 9 Rayleigh wave group velocity tomography results for period 16 s (**a**), 20 s (**b**), 30 s (**c**), and 40 s (**d**). Corresponding checkerboards with ray-path coverage are depicted next to each map. Events marked as *black dots* with focal depths between 15 and 30 km are plotted on 16 s (**a**) and 20 s (**b**) group velocity maps, and events with focal depths between 30 and 55 km were plotted on 30 s (**c**) and 40 s (**d**) group-velocity maps. Seismicity with magnitude >4.0, during 1974–2008, is from EHB catalog (Engdahl et al. 1998)

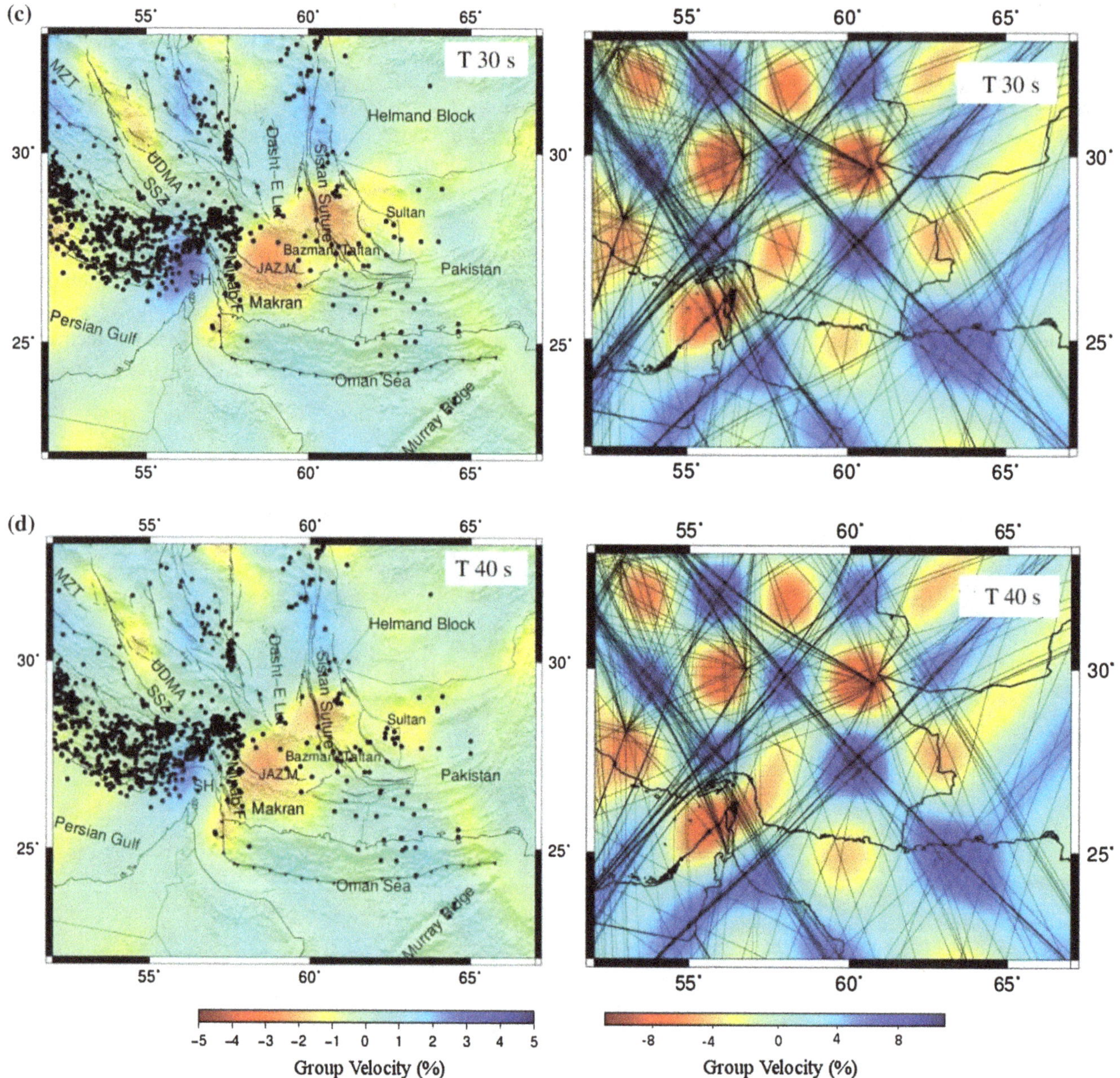

Fig. 9 continued

different periods indicate the general features of the structure at different depths is influenced by their sensitivity kernels (e.g., Yang et al. 2007; Huang et al. 2010; Tibuleac et al. 2011) (Fig. 9). To guide the interpretation, the sensitivity kernels for different periods were used (Fig. 7b). The Rayleigh wave with the shortest period of 16 s has fair sensitivity to the top 10 km, and the wave with the longest period of 40 s has peak sensitivity at ~60 km depth and fair sensitivity up to ~80 km. Thus, the use of the dispersion curves from the 16 to 40 s periods allows us to constrain the shear velocities at the depth range of 10 to 80 km. Each tomography map is illustrated with the corresponding seismicity in each period (Fig. 9). Events with focal depths between 15 and 30 km were plotted on the 16 and 20 s group velocity maps, and events with focal depths between 30 and 55 km were plotted on 30 and 40 s group velocity maps.

5 Discussion

The level of seismicity in Makran is low and increases from west to east. The western and eastern portions of the subduction zone of Makran have different seismic and tectonic characteristics (Byrne et al. 1992; Zarifi 2006). Three N-S profiles (AA', BB', and CC'; profile locations are

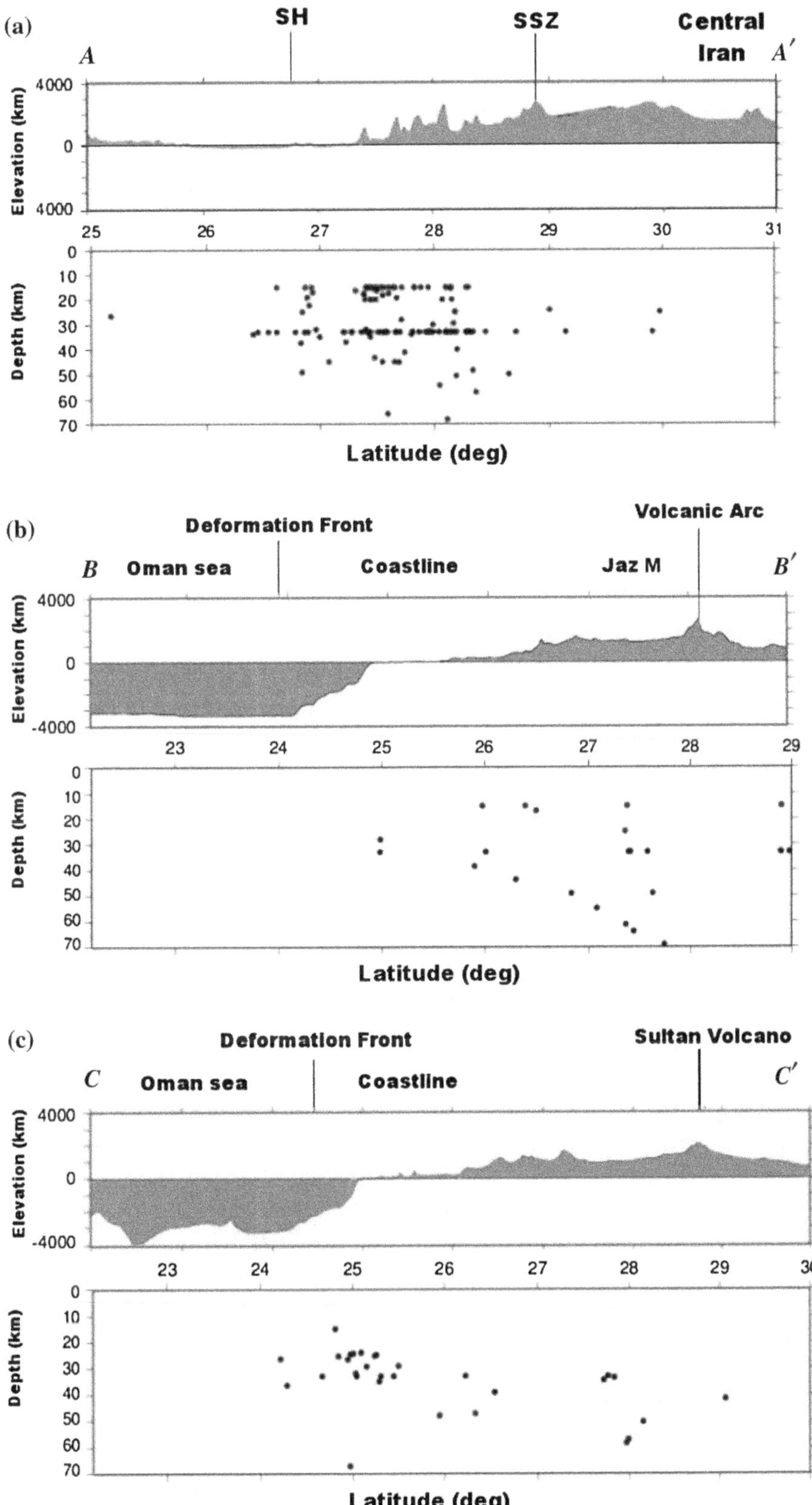

Fig. 10 The vertical cross section shows the depth distribution of earthquakes in upper 70 km along the profiles *AA′* (**a**), *BB′* (**b**), and *CC′* (**c**) (The positions of the profiles are depicted in Fig. 9a). The topography along the profiles is also shown in the *top panels*. The *black dots* are the earthquakes (with magnitude ≥ 4) from EHB catalog (Engdahl et al. 1998) occurred along the profiles with width 0.5° in each side

shown in Fig. 9a) are illustrated in Fig. 10, in which the focal depth distribution is shown. The western portion of Makran has an abnormally low level of deep seismicity. The intermediate depth seismicity characteristics related to western Makran within the downgoing plate are different from the dominant shallower seismicity of the Zagros region (Fig. 10). Across the Sistan Suture Zone, this seismicity pattern changes to a low seismicity condition compared with the Zagros region. Recent and active deformation in Sistan is dominated by right-lateral strike-slip and thrust faults related to the indentation of Iran by the Arabian shield (Berberian et al. 2000). The seismic activity in the mountain ranges, including the Taftan-Bazman volcanic arc, is quite weak (Fig. 9). In 1979, several right-lateral moderate-sized earthquakes occurred between the Lut and Helmand blocks, and inside these two blocks, there is little seismicity. This seismic activity increases the possibility that the Sistan Suture Zone plays a role in the segmentation between eastern and western Makran, and therefore, the continuity of this structure could be defined as a boundary between western and eastern Makran (Byrne et al. 1992). To the east, the distance of the volcanic arc and fore-arc setting increases, and this suggests that the slab dips shallower as it moves eastward (Byrne et al. 1992; Zarifi 2006; Shad Manaman et al. 2011). The eastern portion of Makran has relatively lower dips compared with the western portion (Zarifi 2006) (see Fig. 10b, c). Eastern Makran has experienced most of its seismic activity near Chaman and Ornach-Nal Faults (Zarifi 2006).

According to our tomographic maps, shallow earthquakes are expected to occur in the location of high-velocity anomaly, where Arabian Plate begins to subduct beneath the Straits of Hormuz (Fig. 9a, b). This high-velocity anomaly far beneath the Straits of Hormuz emerges more clearly at shorter periods, reflecting the thin crust under the Oman sea floor and Makran fore-arc setting (25–30 km). Most of the earthquakes that occur in this region are expected to be shallow, and as a confirmation of our outcome, nearly all of the seismicity associated with this region occurs at depths of <30 km (Jackson and McKenzie 1984), see Fig. 10a. Within the downgoing plate towards the north, where the low-velocity anomaly is located, we expect events would occur at intermediate depths due to the down-dip elongation of the subducting slab (see Fig. 10b, c). The deeper events occur along the downgoing slab where the subducting plate bends below the Taftan-Bazman volcanic arc (see Fig. 10b, c). The deepest earthquakes of the Makran region concentrate near the Taftan volcano due to the accommodation of the final component of the motion between Arabia and Eurasia (Byrne et al. 1992). We investigated the crustal thickness using the latest Moho map obtained for the same area using

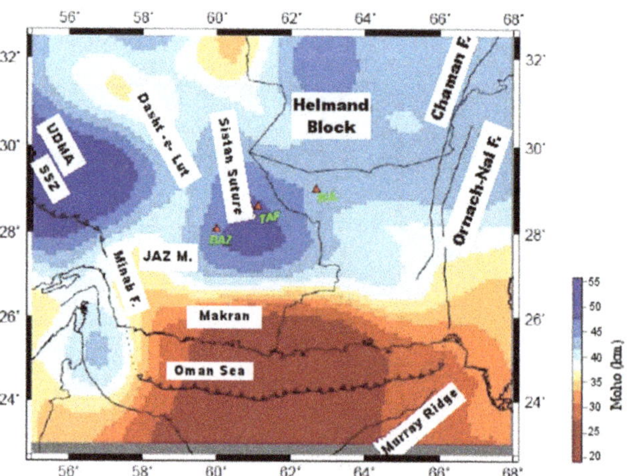

Fig. 11 The Moho map across the Makran region by Shad Manaman et al. (2011)

a different approach and the data reported by Shad Manaman et al. (2011). For better accuracy in the analysis, a high-resolution version of the Moho map in Shad Manaman et al. (2011) was used, as illustrated in Fig. 11.

The 16 and 20 s maps are sensitive to the upper crust at ~25 to 30-km depth based on the sensitivity curves in Fig. 7b. A low-velocity anomaly in the Oman Sea south of the Makran region is observable (Fig. 9a). Although low-velocity sediments with 6- to 7-km thickness in front of the Makran deformation front (White 1982; Fowler et al. 1985) might not be visible at this period, the Rayleigh waves of the 16 s period are still likely to sample the accretionary prism. At the same time, these waves might be influenced by the uppermost section of the subducted crust. The northward subducting Arabian Plate is shown as the high-velocity anomaly along the Straits of Hormuz in the 16 and 20 s map (see Fig. 9a, b). This high-velocity anomaly reflects the influence of subduction of the descending slab that is older, denser, and colder than the continental crust next to it. The Strait of Hormuz is considered a transition between the Zagros collision and the Makran oceanic subduction (Regard et al. 2010). A sharp transition boundary between the low- and high-velocity zones with a northwest trend is clearly depicted at the Minab fault system (see Fig. 9a, b). This transition indicates the boundary between the Zagros region with high seismicity in the northwest and the Makran region with low seismicity to the east. The earthquake distribution surrounding the Minab fault is restricted to the west of the Jaz Murian depression (Fig. 9a, b). The Jaz Murian depression has been interpreted as a fore-arc basin (Farhoudi and Karig 1977). The large trench-arc distance (~400–600 km) suggests that the angle of subduction is notably low and is consistent with thermal modeling in this region (e.g., Smith et al. 2013). An aseismic region extends from the deformation front for nearly 200 km in the western Makran (Fig. 9a, b). Most of the shallow seismicity is related to the Zagros Mountain.

A pronounced low-velocity anomaly extends to the SW-NE in the east of Minab fault, which is attributable to volcanic arc and back-arc settings of the Makran region and the Bazman and Taftan volcanoes (see Fig. 9a, b). This low-velocity anomaly has its origin in thicker crust caused by a warm lithosphere wedge overlying the subducting Arabian Plate and might be coincident with the source of the volcanic magmas. The subducting slab could cause partial melting when volatiles are released and rise into the overlying upper mantle. This effect is consistent with previous seismic observations in this region (e.g., Shad Manaman et al. 2011). Another low-velocity anomaly is observable at the Sultan volcanic setting; however, this anomaly occurs at the edges of the area with acceptable resolution (see Fig. 9a). The trend of low-velocity anomaly indirectly suggests geological and geophysical evidence for the geometry of slab. A transition from low to high velocity is observable in central Makran between the Sistan Suture Zone and the Lut block. Although Byrne et al. (1992) assumed that this suture zone separates the Lut and Helmand blocks, our results show that this suture does not appear to segment different blocks.

The 30 and 40 s maps are most sensitive to a depth of ~25–55 km based on the sensitivity curves in Fig. 7b; although these maps are of lower resolution than others, resolution within the area is still reasonable. The 40 s tomographic map displays its maximum sensitivities at a depth of ~55 km, as Fig. 7b illustrates. The low-velocity anomalies beneath the volcanic arc on the maps are similar to those at 16 and 20 s and reveal that the crustal thickness below the Taftan volcano is at least 50 km deep, which is compatible with the latest Moho Map obtained for the same area using a different approach and data by Shad Manaman et al. (2011) (Fig. 9d). The Moho map was produced using a surface wave tomography method to image the S-velocity structure of the upper mantle and Moho depth. The earthquake distributions surrounding these volcanoes and illustrated in the 30 and 40 s maps (Fig. 9c, d) indicate that the number of intermediate earthquakes increases with the downgoing plate. These earthquakes have normal faulting focal mechanisms (Jackson and McKenzie 1984; Laane and Chen 1989) with predominantly down-dipping T-axes, which indicate that the subducted slab is in tension. The high-velocity anomaly compatible with the descending slab moves northward in these two maps, which demonstrates the northward direction of Arabian Plate subduction.

Because few studies exist on the deep structure of the upper mantle in the Makran region, the use of ANT within this area provides new images on the crust and uppermost mantle. Results obtained from recent studies in this region have poor resolution compared with the results of the checkerboard test with input models consisting of anomalies of 4° spike size (e.g., Shad Manaman et al. 2011); however, in this study, the results demonstrate the higher resolution of ~1° × 1° for central Iran and limited resolution on the order of 2° × 2° towards the southeastern region of the study area. Many of the prominent features in our results are consistent with the known geological structures. The lateral resolution of the tomographic maps obtained by ANT greatly depends on various parameters, including the path coverage and inter-station distances. In this study, for the period range of 10–50 s, we retained only paths longer than 250 km (>3 wavelengths at 10 s). Therefore, we selected a grid spacing of ~110 km. Additionally, ANT is most powerful when ambient noise exists over a broad azimuthal range. Thus, we demonstrated that the fraction of high SNR paths in all azimuths are greater 50 % at four period bands of 10–20 s, 20–30 s, 30–40 s, and 40–50 s and were approximately 0.53, 0.64, 0.69, and 0.51, respectively. Consequently, the distribution of the useful amount of ambient noise is sufficient for use in tomography. Fig 7a shows the v_S velocity models obtained for two different inter-station distances at the study area. The sensitivity kernels as a function of period are also presented in Fig. 7b. The v_S velocity models presented in Fig. 7 demonstrate significant agreement with tomography results according to the sensitivity curves (see Fig. 7b). The v_S begins to increase gradually with depth. Comparing the v_S structures of the two station pairs, the most obvious v_S discrepancy is the low velocity of KRBR-ZHSF at a depth of 18–30 km. This low velocity is quite similar to the tomographic results along the path of KRBR-ZHSF at 16–20 s (Fig. 9a, b), which are sensitive down to a depth of ~20–30 km, according to the sensitivity curves in Fig. 7b.

6 Conclusions

The following results were obtained from this study:

(1) In this research, we explained how sufficient noise energy can be recorded in the Makran region at periods of 10–50 s from which the empirical Green's functions were extracted. It was also shown that the Rayleigh wave Green's functions were extracted by computing cross-correlations between records using observations collected over 12 months at pairs of seismic stations.

(2) Our results exhibit seasonal variability in the study area. This seasonal variation indicates that the Green's functions reconstructed using cross-correlation can show differences in quality during the summer and winter. Although we showed that coherent Rayleigh wave signals exist in all periods and most azimuths across the Makran region, they are sufficiently isotropically distributed in azimuth to deliver accuracy in dispersion measurements if integrated over a long time period, such as a year.

(3) In conclusion, various resolution tests showed that our data and methods are sufficient to provide high-resolution tomographic images of surface wave group velocities in the region. The group velocity maps show low-velocity anomalies beneath the volcanic arc correlated with large Moho depths, a finding compatible with the Moho Map obtained by Shad Manaman et al. (2011) (Fig. 11). The high-velocity anomalies along the Straits of Hormuz in the group velocity maps determine the northward subducting Arabian Plate. Finally, these group velocity maps are significantly improved in lateral resolution over those of previous studies, which have relied on traditional earthquake-based surface wave tomography.

Acknowledgments Seismicity of each period map is plotted from EHB catalog (Engdahl et al. 1998). The seismic data used in this study were obtained from the GSN/IRIS Global Seismographic Network (http://www.iris.edu/hq/programs/gsn), the Virtual European Broadband seismic Network (VEBSN) and international Institute of Earthquake Engineering and Seismology (IIEES). Many of the figures in this paper were prepared using GMT (Wessel and Smith 1998; www.soest.hawaii.edu/gmt, last accessed April 2015). We would like to thank Dr. N. Shad Manaman (from Sahand University of Technology, Iran) for generously providing us with his research results on the Moho depth and also Dr. N. Mirzaei (from University of Tehran, Iran) for his valuable suggestions in this study. A special note of thanks goes to Dr. T. Shirzad (from Islamic Azad University, Tehran, Iran) for his help and support. We also appreciate the research council of Tehran University for their support of this research.

Appendix

See Fig. 12.

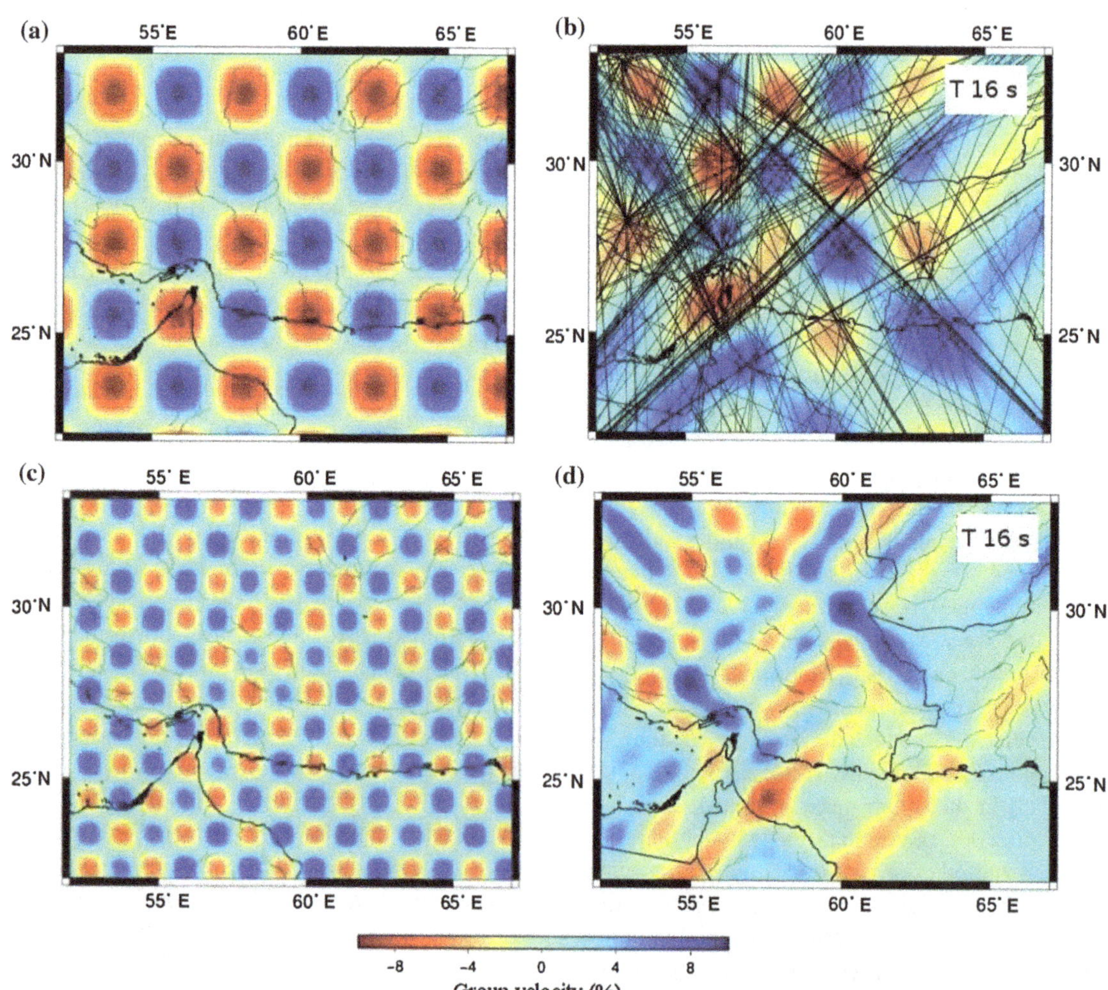

Fig. 12 Input checkerboard test models with velocity perturbation of about 2.8 ± 0.3 km/s and grid size 2° × 2° (**a**) and 1° × 1° (**c**), and the corresponding recovered models are given in (**b**) and (**d**)

References

Arroucau P, Rawlinson N, Sambridge M (2010) New insight into Cainozoic sedimentary basins and Palaeozoic suture zones in southeast Australia from ambient noise surface wave tomography. Geophys Res Lett 37(7):L07303. doi:10.1029/2009GL041974

Bayer R, Shabanian E, Regard V, Doerflinger E, Abbassi M, Chery J, Nilforoushan F, Tatar M, Vernant P, Bellier O (2003) Active deformation in the Zagros-Makran transition zone inferred from GPS measurements in the interval 2000–2002. Geophys Res Abstr 5:05891

Bensen GD, Ritzwoller MH, Barmin MP, Levshin AL, Lin F, Moschetti MP, Shapiro NM, Yang Y (2007) Processing seismic ambient noise data to obtain reliable broadband surface wave dispersion measurements. Geophys J Int 169:1239–1260

Bensen GD, Ritzwoller MH, Shapiro NM (2008) Broadband ambient noise surface wave tomography across the United States. J Geophys Res 113:B05306. doi:10.1029/2007JB005248

Bensen GD, Ritzwoller MH, Yang Y (2009) A 3-D shear velocity model of the crust and uppermost mantle beneath the United States from ambient seismic noise. Geophys J Int 177(3):1177–1196

Berberian M, Jackson JA, Qorashi M, Talebian M, Khatib M, Priestley K (2000) The 1994 Sefidabeh earthquakes in eastern Iran: blind thrusting and bedding-plane slip on a growing anticline, and active tectonics of the Sistan suture zone. Geophys J Int 142:283–299

Byrne DE, Sykes AR, Davis DM (1992) Great thrust earthquakes and aseismic slip along the plate boundary of the Makran subduction zone. J Geophys Res 97:449–478

Cho KH, Herrmann RB, Ammon CJ, Lee K (2007) Imaging the upper crust of the Korean peninsula by surface-wave tomography. Bull Seismol Soc Am 97:198–207

Dehghani G, Makris J (1984) The gravity field and crustal structure of Iran. Neues Jahrb Geol P-A 168:215–229

DeMets C, Gordon RG, Argus DF, Stein S (1990) Current plate motions. Geophys J Int 101:425–478

Derode A, Larose E, Tanter M, de Rosny J, Tourim A, Campillo M, Fink M (2003) Recovering the Green's function from field-field correlations in an open scattering medium. J Acoust Soc Am 113:2973–2976

Drewes H (2009) The actual plate kinematic and crustal deformation model APKIM2005 as basis for a non-rotating ITRF. In: Drewes H (ed) Geodetic reference frames, IAG Symposia, vol 134. Springer, Dordrecht pp 95–99

Engdahl ER, Van Der Hilst R, Buland R (1998) Global teleseismic earthquake relocation with improved travel times and procedures for depth determination. Bull Seismol Soc Am 88:722–743

Farhoudi G, Karig DE (1977) Makran of Iran and Pakistan as an active arc system. Geology 5:664–668

Fowler SR, White RS, Louden KE (1985) Sediment dewatering in the Makran accretionary prism. Earth Planet Sci Lett 75(4):427–438

Giese P, Makris J, Akashe B, Rower P, Letz H, Mostaanpour M (1984) Structure in southern Iran derived from seismic explosion data. Neues Jahrb Geol P-A 168:230–243

Gutenberg B (1951) Observation and theory of microseisms. In: Malone TF (ed) *Compendium of Meteorology*. American Meteorological Society, Providence, pp 1303–1311

Herrmann RB (1973) Some aspects of band-pass filtering of surface waves. Bull Seismol Soc Am 63:663–671

Herrmann RB, Ammon CJ (2013) Computer programs in seismology—surface waves, receiver functions and crustal structure. Saint Louis University, Available at http://www.eas.slu.edu/People/RBHerrmann/ComputerPrograms.html. Accessed 21 Dec 2013

Huang YC, Yao H, Huang BS, Van der Hilst RD, Wen KL, Huang WG, Chen CH (2010) Phase velocity variation at periods 0.5–3 s in the Taipei Basin of Taiwan from correlation of ambient seismic noise. Bull Seismol Soc Am 100:2250–2263

Jackson J, McKenzie D (1984) Active tectonics of the Alpine—Himalayan Belt between western Turkey and Pakistan. Geophys J Int 77:185–264

Kennett B (1995) Approximations for surface-wave propagation in laterally varying media. Geophys J Int 122:470–478

Laane JL, Chen WP (1989) The Makran earthquake of 1983 April 18; a possible analogue to the Puget Sound earthquake of 1965. Geophys J R Astronom Soc 98(1):1–9

Levshin AL, Yanovskaya TB, Lander AV, Buckchin BG, Barmin MP, Ratnikova LI, Its EN (1989) In: Keilis-Borok VI (ed) Seismic surface waves in a laterally inhomogeneous earth. Kluwer, Norwell

Lin FC, Ritzwoller MH, Shapiro NM (2006) Is ambient noise tomography across ocean basins possible? Geophys Res Lett 33(14):L14304. doi:10.1029/2006GL026610

Lin FC, Ritzwoller MH, Townend J, Bannister S, Savage MK (2007) Ambient noise Rayleigh wave tomography of New Zealand. Geophys J Int 170(2):649–666

Lin FC, Moschetti MP, Ritzwoller MH (2008) Surface wave tomography of the western United States from ambient seismic noise: Rayleigh and Love wave phase velocity maps. Geophys J Int 173:281–298

Maggi A, Priestley K (2005) Surface waveform tomography of the Turkish-Iranian plateau. Geophys J Int 160:1068–1080

Masson F, Anvari M, Djamour Y, Walpersdorf A, Tavakoli F, Daignieres M, Nankali H, Van Gorp S (2007) Large-scale velocity field and strain tensor in Iran inferred from GPS measurements: new insight for the present-day deformation pattern within NE Iran. Geophys J Int 170:436–440

McClusky S, Reilinger R, Mahmoud S, Ben Sari D, Tealeb A (2003) GPS constraints on Africa (Nubia) and Arabia plate motions. Geophys J Int 155:126–138

Nishida K, Kawakatsu H, Fukao Y, Obara K (2008) Background Love and Rayleigh waves simultaneously generated at the Pacific Ocean floors. Geophys Res Lett 35(16):L16307

Pedersen HA, Krüger F (2007) Influence of the seismic noise characteristics on noise correlations in the Baltic Shield. Geophys J Int 168:197–210

Rawlinson N (2005) FMST: fast marching surface tomography package. Research school of earth sciences. Australian National University, Canberra

Rawlinson N, Sambridge M (2005) The fast marching method: an effective tool for tomographic imaging and tracking multiple phases in complex layered media. Explor Geophys 36:341–350

Regard V, Hatzfeld D, Molinaro M, Aubourg C, Bayer R, Bellier O, Yamini-Fard F, Peyret M, Abbassi M (2010) The transition between Makran subduction and the Zagros collision: recent advances in its structure and active deformation. Geol Soc Lond Special Publications 330(1):43–64

Sabra KG, Gerstoft P, Roux P, Kuperman WA, Fehler MC (2005) Extracting time-domain Green's function estimates from ambient seismic noise. Geophys Res Lett 32:L03310. doi:10.1029/2004GL021862

Saygin E, Kennett BLN (2010) Ambient seismic noise tomography of Australian continent. Tectonophysics 481(1):116–125

Saygin E, Kennett BLN (2012) Crustal structure of Australia from ambient seismic noise tomography (1978–2012). J Geophys Res: Solid Earth. 117(B1):B01304. doi:10.1029/2011JB008403

Shad Manaman N, Shomali ZH, Koyi H (2011) New constraints on upper-mantle S-velocity structure and crustal thickness of the Iranian plateau using partitioned waveform inversion. Geophys J Int 184:247–267

Shapiro NM, Campillo M (2004) Emergence of broadband Rayleigh waves from correlations of the ambient seismic noise. Geophys Res Lett 31:L07614. doi:10.1029/2004GL019491

Shapiro NM, Campillo M, Stehly L, Ritzwoller MH (2005) High-resolution surface-wave tomography from ambient seismic noise. Science 307:1615–1618

Smith GL, McNeill LC, Wang K, He J, Henstock TJ (2013) Thermal structure and megathrust seismogenic potential of the Makran subduction zone. Geophys Res Lett 40(8):1528–1533

Snieder R (2004) Extracting the Green's function from the correlation of coda waves: a derivation based on stationary phase. Phys Rev E 69:046610. doi:10.1103/PhysRevE.69.046610

Snyder DB, Barazangi M (1986) Deep crustal structure and flexure of the Arabian plate beneath the Zagros collisional mountain belt as inferred from gravity observations. Tectonics 5:361–373

Stehly L, Campillo M, Shapiro N (2006) A study of seismic noise from its long-range correlation properties. J Geophys Res 111:B10306. doi:10.1029/2005JB004237

Sun X, Song X, Zheng S, Yang Y, Ritzwoller MH (2010) Three dimensional shear velocity structure of the crust and upper mantle beneath China from ambient noise surface wave tomography. Earthq Sci 23:449–463. doi:10.1007/s11589-010-0744-4

Tibuleac IM, Von Seggern DH, Anderson JG, Louie JN (2011) Computing Green's functions rom ambient noise recorded by accelerometers and analog, broadband, and narrow-band seismometers. Seismol Res Lett 82:661–675

Vernant C, Nilforoushan F, Masson F, Vigny P, Martinod J, Abbassi M, Nankali H, Hatzfeld D, Bayer R, Tavakoli F, Ashtiani A, Doerflinger E, Daignières M, Collard P, Chéry J (2003) GPS network monitors the Arabia-Eurasia collision deformation in Iran. J Geod 77:411–422

Weaver RL, Lobkis OI (2001) Ultrasonics without a source: thermal fluctuation correlation at MHz frequencies. Phys Rev Lett 87:134301–134304

Webb S, Zhang X, Crawford W (1991) Infragravity waves in the deep ocean. J Geophys Res 96:2723–2736

Wessel P, Smith WHF (1998) New, improved version of generic mapping tools released. EOS Trans Am Geophys Union 79:579

White RS (1982) Deformation of the Makran accretionary sediment prism in the Gulf of Oman (north-west Indian Ocean). Geol Soc Lond Spec Publ 10(1):357–372

Yang Y, Ritzwoller MH, Levshin AL, Shapiro NM (2007) Ambient noise Rayleigh wave tomography across Europe. Geophys J Int 168:259–274

Yang YJ, Li AB, Ritzwoller MH (2008) Crustal and uppermost mantle structure in southern Africa revealed from ambient noise and teleseismic tomography. Geophys J Int 174:235–248

Yao H (2012) Lithospheric structure and deformation in SE Tibet revealed by ambient noise and earthquake surface wave tomography: recent advances and perspectives. Earthq Sci 25(5–6):371–383

Young MK, Rawlinson N, Arroucau P, Reading AM, Tkalčić H (2011) High-frequency ambient noise tomography of southeast Australia: new constraints on Tasmania's tectonic past. Geophys Res Lett 38:L13313. doi:10.1029/2011GL047971

Zarifi Z (2006) Unusual subduction zones: case studies in Colombia and Iran. Dissertation, University of Bergen, Norway

Zheng Y, Yang Y, Ritzwoller MH, Zheng X, Xiong X, Li Z (2010) Crustal structure of the northeastern Tibetan plateau, the Ordos block and the Sichuan basin from ambient noise tomography. Earthq Sci 23(5):465–476

18

Dynamic interaction of twin vertically overlapping lined tunnels in an elastic half space subjected to incident plane waves

Zhongxian Liu · Yirui Wang · Jianwen Liang

Abstract The scattering of plane harmonic P and SV waves by a pair of vertically overlapping lined tunnels buried in an elastic half space is solved using a semi-analytic indirect boundary integration equation method. Then the effect of the distance between the two tunnels, the stiffness and density of the lining material, and the incident frequency on the seismic response of the tunnels is investigated. Numerical results demonstrate that the dynamic interaction between the twin tunnels cannot be ignored and the lower tunnel has a significant shielding effect on the upper tunnel for high-frequency incident waves, resulting in great decrease of the dynamic hoop stress in the upper tunnel; for the low-frequency incident waves, in contrast, the lower tunnel can lead to amplification effect on the upper tunnel. It also reveals that the frequency-spectrum characteristics of dynamic stress of the lower tunnel are significantly different from those of the upper tunnel. In addition, for incident P waves in low-frequency region, the soft lining tunnels have significant amplification effect on the surface displacement amplitude, which is slightly larger than that of the corresponding single tunnel.

Keywords Vertically overlapping lined tunnels · Scattering · Indirect boundary integration equation method (IBIEM) · Soil tunnel dynamic interaction

Z. Liu (✉) · Y. Wang
Tianjin Key laboratory of civil structure protection and reiforcement, Tianjin Chengjian University, Tianjin 300384, China
e-mail: zhongxian1212@163.com

J. Liang
Department of Civil Engineering, Tianjin University, Tianjin 300372, China

1 Introduction

With the rapid developing of underground space and the improvement of underground engineering technology, vertically overlapping tunnels (with vertical alignment) have been widely used for the urban subway and other underground transportation system in many large cities, for instance, the twin metro tunnels in Ankara, Turkey (Karakus et al. 2007), in Tehran, Iran (Chakeri et al. 2011), and in Wuhan, China (Wang et al. 2012). On the other hand, it has been observed that the underground tunnel may suffer from serious damage in great earthquakes, such as Taiwan Chi-Chi (Wang et al. 2001) earthquakes. Hence, in the last decades, the seismic response of underground tunnel has become an attractive research topic in earthquake engineering and has been intensively investigated by numerous researchers analytically or numerically.

The analytical method such as wave function expansion method (WFEM) has been widely used to solve the scattering of seismic waves by a lined tunnel (Lee and Trifunac 1979; Liang and Ji 2006; Liu et al. 2013; Yi et al. 2014). Due to the fact that the analytical methods are usually restricted to simple calculation models, for complex geometrical and material characteristics, it is necessary to develop numerical methods, such as the finite element method (FEM), finite difference method (FDM), the boundary element method (BEM), etc. Kobayashi and Nishimura (1983) used the BEM to solve the dynamic response of an underground tunnel. Stamos and Beskos (1996) solved the three-dimensional seismic response of long lined tunnels in a half space by BEM. Kattis et al. (2003) used the BEM to study the harmonic body waves scattering by lined and unlined tunnels in an infinite poroelastic saturated soil. Rodriguez-Castellanos et al.

(2006) studied the scattering of P and SV waves by cracks and underground cavities using the indirect boundary element method (IBEM). Esmaeili et al. (2006) analyzed the dynamic response of plane harmonic waves by a lined circular tunnel using the hybrid boundary and FEM. Yiouta-Mitra et al. (2007) adopted the FDM to study the dynamic response of a circular tunnel in a half space subjected to harmonic SV waves. Yu and Dravinski (2009) investigated the scattering of plane P, SV, and Rayleigh waves by a cavity embedded in an isotropic half space by BEM. Liu et al. (2010) further discussed the diffraction of P, SV waves by a tunnel in an elastic half space using a special BEM. Panji et al. (2013) studied the displacement response of an unlined truncated circular cavity in a homogenous isotropic medium under SH waves by BEM. Recently, Parvanova et al. (2014) investigated the seismic response of a lined tunnel in the half-plane with surface-traction relief. Pitilakis et al. (2014) presented a series of numerical analysis to investigate the dynamic response of shallow circular tunnels. Alielahi et al. (2015) utilized the BEM to study the seismic ground amplification by unlined tunnels subject to vertically P and SV wave propagation.

Note that the above-mentioned studies are mainly focused on the single-tunnel model. As for twin tunnels or tunnel group, the dynamic interaction between closely-spaced tunnels should be taken into account (Hasheminejad and Avazmohammadi 2007; Smerzini et al. 2009; Chen et al. 2010; Wang et al. 2012; An et al. 2015; Fang et al. 2015; Alielahi and Adampira 2016a). Fotieva (1980) studied the effect of the compressional and the shear waves by twin-parallel tunnels. Balendra et al. (1984) solved the dynamic response of a pair of circular tunnels under SH waves. Okumura et al. (1992) studied the seismic response of the twin circular tunnels by FEM. Moore and Guan (1996) investigated the three-dimensional seismic response of twin lined tunnels in full space using the successive reflection method. Liang et al. (2003, 2004) discussed a series of solutions for surface motion amplification of the underground twin tunnels under P, SV waves. Chen et al. (2006) presented the Null-field integral equations for stress field around circular holes under anti-plane shear waves. Zhou et al. (2009) applied a semi-analytical method to discuss the dynamic response of twin-parallel elliptic tunnels embedded in an infinite poroelastic medium. Alielahi and Adampira (2016b) studied the seismic ground response under vertically in-plane waves by the twin-parallel tunnels.

As stated above, there have been several studies on the dynamic models of twin tunnels or tunnels group. However, to the author's best knowledge, for the vertically overlapping tunnels (coincidence of plane projection) shallowly buried in a half space, available theoretical analysis is extremely limited. Note that the investigations of Liang et al. (2003, 2004) only considered horizontally twin tunnels by an approximate analytic method. In fact, in Liu and Wang (2012), it has been illustrated that the seismic response of vertically overlapping tunnels is significantly different from that of horizontally flat twin tunnels in a full space. However, it is restricted to deep-buried tunnels. In order to improve the qualitative level of seismic design of vertically overlapping tunnels in a half space, it is necessary to calculate and reveal the dynamic interactions between the upper and the lower tunnels, and the influence of nearby ground surface.

In this paper, we focus on the dynamic interaction between these two vertically overlapping tunnels which are shallowly buried in an elastic half space, based on the indirect boundary integration equation method (IBIEM). It has been demonstrated that this method has several advantages such as reducing dimensions of problems, automatic satisfaction of boundary condition, and high calculation precision (Luco and De Barros 1994). Moreover, the IBIEM does not require element discretization, and it can thus be implemented more efficiently. The rest of this paper is organized as follows. The numerical procedure for IBIEM solution is present in section II. Then, the precision of the method is verified by the satisfaction extent of boundary conditions and the comparison between the degenerated and available solutions. Based on the IBIEM, the effects of key parameters, such as the distance between the two tunnels, the stiffness and density of the lining material, and the incident frequency on dynamic response are investigated in detail through numerical examples. Finally, several conclusions are drawn, which provide some useful insights for the seismic design of underground vertically overlapping tunnels. Due to the semi-analytical feature of the IBIEM, the numerical example in this study can also be regarded as a benchmark scheme for other numerical methods.

2 Model definition

As shown in Fig. 1, a pair of vertically overlapping lined tunnels are shallowly buried in the elastic half space with the depth d (to the upper tunnel center), the inner and outer radius of the upper tunnel and lower tunnel a_1 and a_2, a_1' and a_2', respectively. The distance between the centers of upper and lower tunnels is denoted as D. The domain of tunnels and the half space are assumed to be elastic, homogenous and isotropic. Let D_0, D_1, D_2 denote the domain of the half space, the upper tunnel and the lower tunnel, respectively. S and S_0 denote the outer and inner surface of the upper tunnel; correspondingly, S' and S_0' denote those of the lower tunnel. The shear modulus, Poisson ratio, and the density in D_0 are μ_1, v_1, and ρ_1,

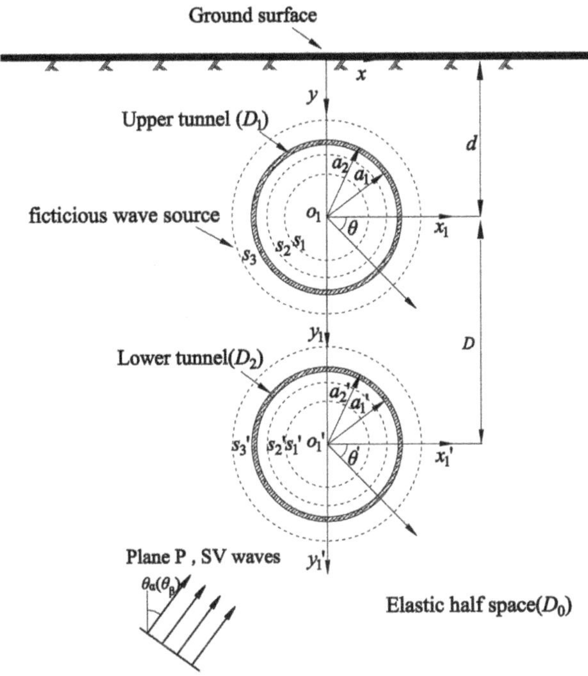

Fig. 1 The model of twin vertically overlapping lined tunnels shallowly buried in an elastic half space

respectively. Then the velocity of the P and SV waves in the half space can be defined by $c_{\beta 1} = \sqrt{\mu_1/\rho_1}$, $c_{\alpha 1} = c_{\beta 1}\sqrt{2(1-\nu_1)/(1-2\nu_1)}$, where, μ_2, ν_2, ρ_2, $c_{\alpha 2}$ and $c_{\beta 2}$ are the material parameters of the two lined tunnels. Suppose that the plane P and SV waves propagate in a half space, then the two-dimensional scattering of plane P and SV waves by the vertically overlapping lined tunnels in a half space needs to be studied. For simplicity, the following calculation is limited to the circular tunnel case, while the IBIEM is well suited for tunnels with arbitrary shapes.

3 Numerical solutions by IBIEM

IBIEM was first proposed by Wong (1982), since then it has applied extensively, among which Liang and Liu (2009) solved the wave motion problems. Based on the single-layer potential theory, the IBIEM solves wave motion problems in the following way. The whole wave field is divided into free field (without scattering) and scattered field, and the scattered waves are constructed by using a linear combination of fundamental solutions (fictitious wave sources). In order to avoid singularities, the fictitious wave sources are placed at some distance to the physical boundaries of scatters, and the densities of the fictitious wave sources are determined by the boundary conditions. In this paper, the compressional line source and shear line source in a half space are introduced as the fundamental solution.

Assume that the wave potential functions of P and SV waves in the medium are ϕ and ψ, respectively. The steady-state wave equation under two-dimensional plane strain state can be expressed as

$$\frac{\partial^2 \phi}{\partial x^2} + \frac{\partial^2 \phi}{\partial y^2} + k_P^2 \phi = 0, \quad (1)$$

$$\frac{\partial^2 \psi}{\partial x^2} + \frac{\partial^2 \psi}{\partial y^2} + k_S^2 \psi = 0, \quad (2)$$

where k_P, k_S are the wavenumber of the P and SV waves, respectively. Define the shear modulus of the medium as μ, the relationship between displacement, stress and the wave potential functions can be expressed as (Lamb 1904)

$$u_x = \frac{\partial \phi}{\partial x} + \frac{\partial \psi}{\partial y} \quad (3)$$

$$u_y = \frac{\partial \phi}{\partial y} - \frac{\partial \psi}{\partial x} \quad (4)$$

$$\sigma_{xx} = \mu\left(-k_S^2 \phi - 2\frac{\partial^2 \phi}{\partial y^2} + 2\frac{\partial^2 \psi}{\partial x \partial y}\right) \quad (5)$$

$$\sigma_{yy} = \mu\left(-k_S^2 \phi - 2\frac{\partial^2 \phi}{\partial x^2} - 2\frac{\partial^2 \psi}{\partial x \partial y}\right) \quad (6)$$

$$\sigma_{xy} = \mu\left(2\frac{\partial^2 \phi}{\partial x \partial y} - k_S^2 \psi - 2\frac{\partial^2 \psi}{\partial x^2}\right) \quad (7)$$

where u_x, u_y denote the horizontal and vertical displacement, respectively, σ_{xx}, σ_{yy} and σ_{xy} are the normal stress and shear stress, respectively.

3.1 Green's functions of compressional and shear wave sources buried in elastic half-space

It is known that the wave potential functions of compressional line source and shear line source in the whole space can be expressed by Hankel function of the second kind as $\phi_i(x,y) = \mathrm{H}_0^{(2)}(k_P r_2)$, $\psi_i(x,y) = \mathrm{H}_0^{(2)}(k_S r_2)$, with r_2 being the distance between the wave source at (x_S, y_S) and the observation point at (x,y). Note that the time factor $\exp(i\omega t)$ is omitted here and thereafter for simplicity, with ω and t being the excitation frequency and time variable, respectively.

According to the boundary condition of the free surface in the half space, and combining with the Fourier transform in the wave-number domain, the potential functions of total wave field can be derived (Lamb 1904).

(1) potential functions of total wave field in a half space under the compressional wave source can be expressed as

$$\phi(x,y) = \mathrm{H}_0^{(2)}\left(k_{\mathrm{P}}\sqrt{(x-x_{\mathrm{S}})^2+(y-y_{\mathrm{S}})^2}\right) + \mathrm{H}_2^{(2)}\left(k_{\mathrm{P}}\sqrt{(x-x_{\mathrm{S}})^2+(y+y_{\mathrm{S}})^2}\right) - \frac{4\mathrm{i}}{\pi}\int_0^{\infty}\frac{(2\xi^2-k_{\mathrm{P}}^2)^2}{\alpha F(\xi)}\mathrm{e}^{-\alpha(y+y_s)}\cos(\xi x)\mathrm{d}\xi, \quad (8)$$

$$\psi(x,y) = \frac{8\mathrm{i}}{\pi}\int_0^{\infty}\frac{\xi(2\xi^2-k_{\mathrm{S}}^2)^2}{F(\xi)}\mathrm{e}^{-\alpha y_{\mathrm{S}}-\beta y}\sin(\xi x)\mathrm{d}\xi; \quad (9)$$

(2) potential functions of total wave field in a half space under the shear wave source can be expressed as

$$\psi(x,y) = \mathrm{H}_0^{(2)}\left(k_{\mathrm{S}}\sqrt{(x-x_{\mathrm{S}})^2+(y-y_{\mathrm{S}})^2}\right) + \mathrm{H}_0^{(2)}\left(k_{\mathrm{S}}\sqrt{(x-x_{\mathrm{S}})^2+(y+y_{\mathrm{S}})^2}\right) - \frac{4\mathrm{i}}{\pi}\int_0^{\infty}\frac{(2\xi^2-k_{\mathrm{S}}^2)}{\beta F(\xi)}\mathrm{e}^{-\beta(y+y_{\mathrm{S}})}\cos(x\xi)\mathrm{d}\xi, \quad (10)$$

$$\phi(x,y) = \frac{-8\mathrm{i}}{\pi}\int_0^{\infty}\frac{\xi(2\xi^2-k_{\mathrm{S}}^2)}{F(\xi)}\mathrm{e}^{(-\beta y_{\mathrm{S}}-\alpha y)}\sin(\xi x)\mathrm{d}\xi, \quad (11)$$

where $\alpha=\sqrt{\xi^2-k_{\mathrm{P}}^2}$, $\beta=\sqrt{\xi^2-k_{\mathrm{S}}^2}$, $F(\xi)=(2\xi^2-k_{\mathrm{S}}^2)^2-4\xi^2\alpha\beta$.

3.2 Wave field construction

Define the inner and outer surfaces of lined upper tunnel are S and S_0, respectively, then introduce the fictitious wave source surface S_1 around S to construct the scattered field in the half space and the fictitious wave source surfaces S_2 and S_3 to construct the scattered field in the lined tunnel. Similarly, the inner and outer surfaces of the lower tunnel are denoted by S' and S'_0, and the fictitious wave source surfaces for the lower tunnel are S'_1, S'_2 and S'_3. Additionally, for convenience, suppose that the shapes of the fictitious wave source surfaces and the tunnel are identical.

The total wave field in the half space can be obtained as the superposition of the free field in the half space and the scattered field. First, we analyze the free field. Suppose that the plane P and SV waves with excitation frequency ω, and with incident angles θ_α and θ_β, respectively, then the wave potential function in the orthogonal coordinates can be expressed as

$$\varphi^{(\mathrm{i})}(x,y) = \exp[-\mathrm{i}k_{\alpha 1}(x\sin\theta_\alpha - y\cos\theta_\alpha)], \quad (12)$$

$$\psi^{(\mathrm{i})}(x,y) = \exp[-\mathrm{i}k_{\beta 1}(x\sin\theta_\beta - y\cos\theta_\beta)]. \quad (13)$$

Due to the existence of the half space surface, the incident P and SV waves will generate reflected P and SV waves as follows

$$\varphi^{(\mathrm{r})}(x,y) = A_2\exp[-\mathrm{i}k_{\alpha 1}(x\sin\theta_\alpha + y\cos\theta_\alpha)], \quad (14)$$

$$\psi^{(\mathrm{r})}(x,y) = B_2\exp[-\mathrm{i}k_{\beta 1}(x\sin\theta_\beta + y\cos\theta_\beta)], \quad (15)$$

where A_2, B_2 amplitude of the reflect waves, can be referred to Luco and De Barros (1994).

Once the lined tunnel exists, scattered field will appear in the half space and the inner of the lined tunnel. Based on the single-layer potential theory, the scattered field in the half space and the tunnel can be constructed by all the compressional line sources and shear line sources. Suppose that the scattered field in the half space is generated by the fictitious wave source surfaces S_1 and S'_1, then the displacement and stress can be expressed as

$$u_i(x) = \int_{S_1}[b(x_1)G_{i,1}^{(\mathrm{s})}(x,\ x_1) + c(x_1)G_{i,2}^{(\mathrm{s})}(x,\ x_1)]\mathrm{d}S_1 + \int_{S'_1}[b'(x'_1)G_{i,1}^{(\mathrm{s})}(x,\ x'_1) + c'(x'_1)G_{i,2}^{(\mathrm{s})}(x,\ x'_1)]\mathrm{d}S'_1 \quad (16)$$

$$\sigma_{ij}(x) = \int_{S_1}[b(x_1)T_{ij,1}^{(\mathrm{s})}(x,\ x_1) + c(x_1)T_{ij,2}^{(\mathrm{s})}(x,\ x_1)]\mathrm{d}S_1 + \int_{S'_1}[b'(x'_1)T_{ij,1}^{(\mathrm{s})}(x,\ x'_1) + c'(x'_1)T_{ij,2}^{(\mathrm{s})}(x,\ x'_1)]\mathrm{d}S'_1 \quad (17)$$

where $x \in D_1$, $x_1 \in S_1$, $x'_1 \in S'_1$. $b(x_1)$, $c(x_1)$, $b'(x'_1)$, $c'(x'_1)$ are the densities of the compressional line source and the shear line source at x_1 and x'_1 on fictitious wave source surfaces S_1 and S'_1, respectively. $G_{i,l}^{(\mathrm{s})}(x,x_1)$, $G_{i,l}^{(\mathrm{s})}(x,x'_1)$, $T_{ij,l}^{(\mathrm{s})}(x,x_1)$ and $T_{ij,l}^{(\mathrm{s})}(x,x'_1)$ are the Green's functions for the displacement and the traction in the half space (with the subscripts 1 and 2 corresponding to the compressional line source and the shear line source, respectively), which satisfy the wave equations and surface boundary conditions automatically. Note that subscripts $i, j = 1, 2$ denote the x, y directions, respectively.

The scattered field in the upper tunnel can be constructed by the superposition of the compressional line sources and shear line sources acted on the fictitious wave source surfaces S_2 and S_3, which can be expressed as

$$u_i(x) = \int_{S_2}[d(x_2)G_{i,1}^{(\mathrm{t})}(x,\ x_2) + e(x_2)G_{i,2}^{(\mathrm{t})}(x,\ x_2)]\mathrm{d}S_2 + \int_{S_3}[f(x_3)G_{i,1}^{(\mathrm{t})}(x,\ x_3) + g(x_3)G_{i,2}^{(\mathrm{t})}(x,\ x_3)]\mathrm{d}S_3 \quad (18)$$

$$\sigma_{ij}(x) = \int_{S_2}[d(x_2)T_{ij,1}^{(\mathrm{t})}(x,\ x_2) + e(x_2)T_{ij,2}^{(\mathrm{t})}(x,\ x_2)]\mathrm{d}S_2 + \int_{S_3}[f(x_3)T_{ij,1}^{(\mathrm{t})}(x,\ x_3) + g(x_3)T_{ij,2}^{(\mathrm{t})}(x,\ x_3)]\mathrm{d}S_3 \quad (19)$$

where $x \in D_2$, $x_2 \in S_2$, $x_3 \in S_3$, $d(x_2)$, $e(x_2)$ are the densities of the compressional line source and shear line source at x_2 on S_2, and $f(x_3)$, $g(x_3)$ denote those at x_3 on fictitious wave source S_3,$G^{(t)}_{i,l}$, $T^{(t)}_{ij,l}$ are the Green's function for the displacement and the traction in the lined tunnel, respectively.

Similarly, the stress and displacement of the lower lined tunnel can be obtained.

3.3 Boundary conditions and the numerical solutions

Due to the adoption of the fundamental solution for the elastic half space, the boundary condition of the free ground surface can be satisfied automatically. Thus, we only need to consider the continuity of displacement and traction on the interface between the lining and surrounding soil, and the zero traction condition on the inner surface of the lining, which are as follows:

$$u^{\mathrm{s}}_x = u^{\mathrm{t}}_x,\ u^{\mathrm{s}}_y = u^{\mathrm{t}}_y,\ \left(r = a_2,\ r' = a'_2\right) \tag{20}$$

$$\sigma^{\mathrm{s}}_{nn} = \sigma^{\mathrm{t}}_{nn},\ \sigma^{\mathrm{s}}_{nt} = \sigma^{\mathrm{t}}_{nt}\ \left(r = a_2,\ r' = a'_2\right) \tag{21}$$

$$\sigma^{\mathrm{t}}_{nn} = 0,\ \sigma^{\mathrm{t}}_{nt} = 0\ \left(r = a_1,\ r' = a'_1\right) \tag{22}$$

where the superscripts s, t denote the half space and the tunnel, respectively. For ease of numerical solution, we discrete the inner and outer surface of the tunnels and fictitious wave source surfaces. Suppose that the number of discrete points on the internal and external surface of the tunnels is N, and that of the fictitious wave source surface is N_1. Then the scattered displacement field and stress field in the half space can be expressed as

$$\begin{aligned} u_i(x_n) = &\sum_{n_1=1}^{N_1} b_{n_1} G^{(s)}_{i,1}(x_n, x_{n_1}) + c_{n_1} G^{(s)}_{i,2}(x_n, x_{n_1}) \\ &+ \sum_{n_1=1}^{N_1} b'_{n_1} G^{(s)}_{i,1}(x_n, x'_{n_1}) + c'_{n_1} G^{(s)}_{i,2}(x_n, x'_{n_1}) \end{aligned} \tag{23}$$

$$\begin{aligned} \sigma_{ij}(x_n) = &\sum_{n_1=1}^{N_1} b_{n_1} T^{(s)}_{ij,1}(x_n, x_{n_1}) + c_{n_1} T^{(s)}_{ij,2}(x_n, x_{n_1}) \\ &+ \sum_{n_1=1}^{N_1} b'_{n_1} T^{(s)}_{ij,1}(x_n, x'_{n_1}) + c'_{n_1} T^{(s)}_{ij,2}(x_n, x'_{n_1}) \end{aligned} \tag{24}$$

$x_n \in D_0$, $x_{n_1} \in S_1$, $x'_{n_1} \in S'_1$, $n = 1, 2, \cdots, N$,

$n_1 = 1, 2, \cdots, N_1$;

where b_{n_1}, c_{n_1}, and b'_{n_1}, c'_{n_1} are the source densities of P and SV waves for the n_1-th point on fictitious source surfaces S_1 and S'_1, respectively.

The scattered displacement field and stress field in the upper tunnel can be expressed as

$$\begin{aligned} u_i(x_n) = &\sum_{n_1=1}^{N_1} d_{n_1} G^{(s)}_{i,1}(x_n, x_{2,n_1}) + e_{n_1} G^{(s)}_{i,2}(x_n, x_{2,n_1}) \\ &+ \sum_{n_1=1}^{N_1} f_{n_1} G^{(s)}_{i,1}(x_n, x_{3,n_1}) + g_{n_1} G^{(s)}_{i,2}(x_n, x_{3,n_1}), \end{aligned} \tag{25}$$

$$\begin{aligned} \sigma_{ij}(x_n) = &\sum_{n_1=1}^{N_1} d_{n_1} T^{(s)}_{ij,1}(x_n, x_{2,n_1}) + e_{n_1} T^{(s)}_{ij,2}(x_n, x_{2,n_1}) \\ &+ \sum_{n_1=1}^{N_1} f_{n_1} T^{(s)}_{ij,1}(x_n, x_{3,n_1}) + g_{n_1} T^{(s)}_{ij,2}(x_n, x_{3,n_1}), \end{aligned} \tag{26}$$

$x_n \in D_1$, $x_{2,n_1} \in S_2$, $x_{3,n_1} \in S_3$, $n = 1, 2, \cdots, N$,

$n_1 = 1, 2, \cdots, N_1$

where d_{n_1}, e_{n_1} and f_{n_1}, g_{n_1} are the source densities of P and SV waves for the n_1-th point on fictitious source surfaces S_2 and S_3. Note that here we assume that the discrete numbers of S_2 and S_3 are also N_1.

Similarly, the scattered field in the lower tunnel can be constructed by the discrete wave source on S'_2 and S'_3. From Eqs. 20–22, we can obtain

$$\boldsymbol{H}_1\boldsymbol{Y}_1 + \boldsymbol{H}'_{11}\boldsymbol{Y}'_1 + \boldsymbol{F}_1 = \boldsymbol{H}_2\boldsymbol{Y}_2 + \boldsymbol{H}_3\boldsymbol{Y}_3, \tag{27}$$

$$\boldsymbol{H}_{11}\boldsymbol{Y}_1 + \boldsymbol{H}'_1\boldsymbol{Y}'_1 + \boldsymbol{F}'_1 = \boldsymbol{H}'_2\boldsymbol{Y}'_2 + \boldsymbol{H}'_3\boldsymbol{Y}'_3, \tag{28}$$

$$\boldsymbol{T}_2\boldsymbol{Y}_2 + \boldsymbol{T}_3\boldsymbol{Y}_3 = 0, \tag{29}$$

$$\boldsymbol{T}'_2\boldsymbol{Y}'_2 + \boldsymbol{T}'_3\boldsymbol{Y}'_3 = 0, \tag{30}$$

where H_1, H_2, H_3, H'_{11} are the Green's influence matrices relating to the displacements and tractions on the discrete points of the outer surface of the upper tunnel caused by the fictitious wave sources on S_1, S_2, S_3, S'_1; H'_1, H'_2, H'_3, H_{11} are the Green's influence matrices for the outer surface of the lower tunnel.T_2, T_3 are the Green's matrix (stress) relating to the traction on the discrete points of the inner surface of the upper tunnel caused by the fictitious wave sources on S_2, S_3 and T'_2, T'_3 are the corresponding Green's influence matrices for the internal surface of the lower tunnel. Y_1, Y_2, Y_3, Y'_1, Y'_2, Y'_3 are fictitious wave source densities vectors on the fictitious surfaces S_1,S_2, S_3,S'_1, S'_2,S'_3, respectively.F_1, F_2 are the free field vector related to the displacement and traction on the interfaces.

Equations 27–30 are written as a compact form as $\boldsymbol{HA} = \boldsymbol{B}$, and this overdetermined equation can be solved by least square (LS) method

$$\boldsymbol{A} = \left[\bar{\boldsymbol{H}}^{\mathrm{T}}\boldsymbol{H}\right]^{-1}\boldsymbol{H}^{\mathrm{T}}\boldsymbol{B} \tag{31}$$

where $\bar{\boldsymbol{H}}$, $\boldsymbol{H}^{\mathrm{T}}$ and $\bar{\boldsymbol{H}}^{\mathrm{T}}$ are the conjugate, transpose, and conjugate transpose matrices of $\boldsymbol{H}$, respectively. After solving the equations about the fictitious wave source densities from (31), we can obtain the scattered field. The total wave field is the superposition of scattered field and

Table 1 List of some important notations and symbols

Symbol	Description	Symbol	Description
d	The depth to the upper tunnel center	D	Distance between the centers of twin tunnels
a_1, a_2	Inner and outer radius of upper tunnel	D_0, D_1, D_2	Domain of the half space, upper, and lower tunnel
a_1', a_2'	Inner and outer radius of lower tunnel	ρ_1, ρ_2	Density of D_0 and tunnels
N	Discrete points	v_1, v_2	Poisson ratio of D_0 and tunnels
η	Non-dimensional frequency	μ_1, μ_2	Shear modulus of D_0 and tunnels
ζ	Damping ratio	$c_{\alpha 1}, c_{\beta 1}$	Velocity of the P and SV waves in D_0
$\theta_\alpha, \theta_\beta$	Incident angle of P and SV waves	$c_{\alpha 2}, c_{\beta 2}$	Velocity of the P and SV waves in tunnels

free field, and then we can calculate the displacement, stress at any location both in the half space and the tunnels.

4 Verification of accuracy and validation of the numerical solution

Until now, it is still a challenging task to obtain the accurate analytical solution for the scattering of plane P, SV waves by the lined tunnel in an elastic half space due to the difficulty in dealing with the traction-free boundary condition of the half-space surface. Thus, we verify the accuracy by the following steps: (1) test the satisfaction extent of the boundary conditions, (2) examine the numerical stability of the solutions, (3) and degenerate solutions to the single-tunnel case with well-known solutions.

Then define the non-dimensional frequency $\eta = \omega a_1/\pi c_{\beta 1}$, with $c_{\beta 1}$ being the shear wave velocity in a half space medium. To test the boundary conditions, plenty of calculation results show that with the increase of discrete points, the boundary residual value decreases gradually. When the incident frequency $\eta = 2.0$, the residual value can reach up to 10^{-4} for $N = 120$ and $N_1 = 80$. The notations and symbols are summarized in Table 1.

To verify the numerical stability of the solution, Tables 2, 3 and 4 illustrate the convergence of the surface displacement amplitudes and the hoop stress amplitudes on the inner surface of tunnels with the increase of discrete points. Parameters are set $a_1 = 1.0$, $d/a_1 = 4.0$, $D/a_1 = 3.0$, $a_{1/}a_{12} = 0.9$, $a_1'/a_2' = 0.9$, damping ratio $\zeta = 0.001$, $\eta = 1.0$, $v = 0.25$, $\rho_2/\rho_1 = 5/4$, $c_{\beta 2}/c_{\beta 1} = 5/1$(rigid lining), $N = 60$, 80, and 120, correspondingly $N_1 = 40$, 60, and 80. It is clearly shown that the surface displacement amplitudes and stress converge quite well with the increase of discrete points. This further validates the excellent numerical stability of this solution.

When the non-dimensional frequency is 0.5, the results indicate that as the depth of the lower tunnel is larger than $30a_1$, the impact to the upper tunnel or the ground surface can be ignored. To degenerate the solution to single-tunnel case, the following parameters are taken: $D/a_1 = 200$, $\rho_2/\rho_1 = 1.0$, $c_{\beta 2}/c_{\beta 1} = 1.0$, damping ratio $\zeta = 0.001$, $v = 1/3$ and $\eta = 0.5$. Figure 2 shows the surface displacement amplitudes of the half space and the dynamic stress concentration factors on the inner surface of the upper tunnel compared with the well-known results of the elastic half space by Luco and De Barros (1994) for vertically incident P waves and the buried depth of $d/a_1 = 1.5$ and 5.0, respectively.

Table 2 Numerical stability verification of surface displacement amplitudes under vertically incident P and SV waves ($\eta = 1.0$)

x/a_1	$N = 60, N_1 = 40$		$N = 80, N_1 = 60$		$N = 100, N_1 = 80$	
	$\lvert U_y/A_P \rvert$	$\lvert U_x/A_{SV} \rvert$	$\lvert U_y/A_P \rvert$	$\lvert U_x/A_{SV} \rvert$	$\lvert U_y/A_P \rvert$	$\lvert U_x/A_{SV} \rvert$
0.0	0.3983	0.5174	0.3904	0.5492	0.3904	0.5494
0.5	0.4199	0.4410	0.4163	0.4834	0.4164	0.4834
1.0	0.5316	0.3331	0.5404	0.3949	0.5402	0.3949
1.5	0.7875	0.6042	0.8021	0.6449	0.8021	0.6448
2.0	1.1534	1.1372	1.1626	1.1700	1.1625	1.1698
2.5	1.5637	1.7009	1.5645	1.7304	1.5645	1.7302
3.0	1.9592	2.1660	1.9602	2.1954	1.9602	2.1954
3.5	2.2826	2.4905	2.2925	2.5177	2.2926	2.5176
4.0	2.4745	2.6795	2.4855	2.6916	2.4854	2.6914

Table 3 Numerical stability verification of hoop stress amplitudes on the inner surface of the upper tunnel under vertically incident P and SV waves ($\eta = 1.0$)

θ	$N = 60, N_1 = 40$		$N = 80, N_1 = 60$		$N = 100, N_1 = 80$	
	$\lvert \sigma^*_{\theta\theta,P} \rvert$	$\lvert \sigma^*_{\theta\theta,SV} \rvert$	$\lvert \sigma^*_{\theta\theta,P} \rvert$	$\lvert \sigma^*_{\theta\theta,SV} \rvert$	$\lvert \sigma^*_{\theta\theta,P} \rvert$	$\lvert \sigma^*_{\theta\theta,SV} \rvert$
90°	5.1599	0.0000	5.5634	0.0000	5.6326	0.0000
60°	2.7410	4.1179	2.8433	4.4156	2.8458	4.4159
30°	0.6851	5.5921	1.2882	7.1904	1.2872	7.1895
0°	11.6011	3.5452	12.6431	6.1269	12.6434	6.1259
−30°	3.6969	13.1779	3.9972	17.3228	3.9970	17.3233
−60°	0.3768	7.0274	0.9150	9.2243	0.9142	9.2220
−90°	4.1719	0.0000	4.5758	0.0000	4.5754	0.0000

Table 4 Numerical stability verification of hoop stress amplitudes on the inner surface of the lower tunnel under vertically incident P and SV waves ($\eta = 1.0$)

θ	$N = 60, N_1 = 40$		$N = 80, N_1 = 60$		$N = 100, N_1 = 80$	
	$\lvert\sigma^*_{\theta\theta,P}\rvert$	$\lvert\sigma^*_{\theta\theta,SV}\rvert$	$\lvert\sigma^*_{\theta\theta,P}\rvert$	$\lvert\sigma^*_{\theta\theta,SV}\rvert$	$\lvert\sigma^*_{\theta\theta,P}\rvert$	$\lvert\sigma^*_{\theta\theta,SV}\rvert$
90°	3.8507	0.0000	3.7493	0.0000	3.7496	0.0000
60°	1.6496	13.3879	1.5015	15.5293	1.5012	15.5293
30°	7.8644	21.4634	7.6897	25.3231	7.6905	25.3227
0°	7.0338	15.5180	6.9707	17.5244	6.9710	17.5245
−30°	13.3862	8.8110	13.8012	10.8929	13.8006	10.8922
−60°	6.0876	12.2754	6.0504	12.7566	6.0502	12.7570
−90°	0.4880	0.0000	0.4305	0.0000	0.4306	0.0000

Figure 3 shows the surface displacement amplitudes of the half space compared with the well-known results of the elastic half space by Liu et al. (2013) for vertically incident P waves and the buried depth of $d/a_1 = 1.5$. The parameters are defined as $D/a_1 = 200$, $a_2/a_1 = a'_1/a'_1 = 1.1$, $\mu_2/\mu_1 = 0.8$, Poisson ratio $v = 1/3$ and $\eta = 0.5$. It is shown that the results of this study are in good agreement with the references.

5 Numerical results

A pair of vertically overlapping circular lined tunnels is shallowly buried in the elastic half space, with the buried depth of the upper tunnel $d/a_1 = 4.0$, and the inner and outer radius ratios $a_1/a_1 = 0.9$ and $a'_1/a'_2 = 0.9$. Considering the variation of the distance between the two tunnels, we choose $D/a_1 = 3.0$, 4.0, 5.0 (the normal range in practical engineering).

The practical parameters of the lining and the medium in the half-space are defined as follows: the radius and thickness of a real tunnel are $a_1 = 3$ and 0.33 m,

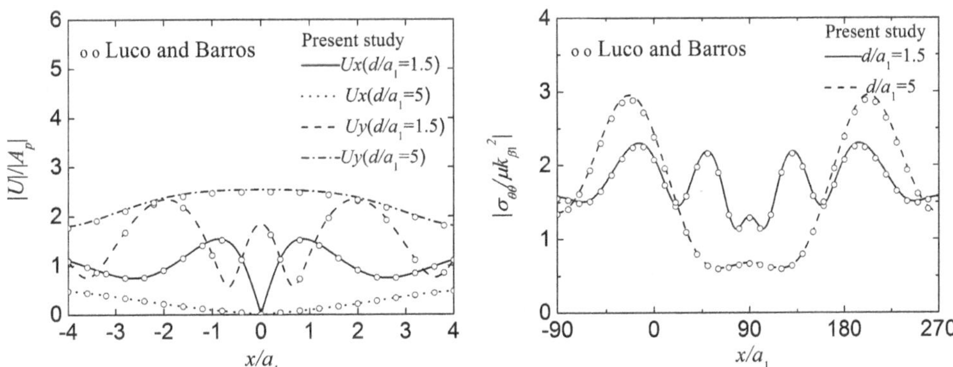

Fig. 2 Comparing present results for normalized displacement amplitude of the ground surface and normalized hoop stress amplitude of the *single tunnel* with the results of Luco and Barros (1994) for *vertically* incident P waves with the buried depth of the *upper tunnel* ($d/a_1 = 1.5$ and $d/a_1 = 5$), distance between the twin tunnels ($D/a_1 = 200$), frequency ($\eta = 0.5$), and Poisson ratio ($\upsilon = 1/3$)

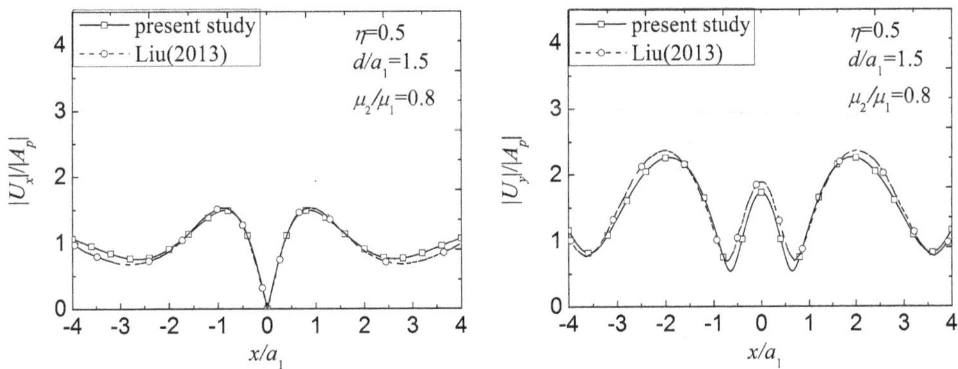

Fig. 3 Comparing present results for normalized displacement amplitude on the half space of the single tunnel with the results of Liu et al. (2013) for *vertically* incident P waves with the buried depth of the *upper tunnel* ($d/a_1 = 1.5$), distance between the twin tunnels ($D/a_1 = 200$), frequency ($\eta = 0.5$), and Poisson ratio ($\upsilon = 1/3$)

respectively. Considering the variation of the stiffness and density of the lining material, take the soft lining, homogenous lining, and rigid lining for parameter analysis. The stiff tunnel ($\rho_2/\rho_1 = 5/4$, $c_{\beta 2}/c_{\beta 1} = 5/1$) corresponds to the reinforced concrete tunnel in soft soil layer such as in Tianjin, Shanghai of China. The lining of the stiff twin tunnels are made of concrete and the density of tunnels lining is $\rho_2 = 2000$ kg/m^3. The shear wave velocity in the lining is $c_{\beta 2} = 2000$ m/s. Poisson ratio of half space and tunnels is $v_1 = v_2 = 0.25$, and damping ratio is $\zeta = 0.001$. In contrast, the soft tunnel ($\rho_2/\rho_1 = 4/5$, $c_{\beta 2}/c_{\beta 1} = 1/3$) denotes the case such that a shotcrete tunnel in granite. Moreover, the unlined tunnel is identical to the case of $\rho_2/\rho_1 = 1/1$, $c_{\beta 2}/c_{\beta 1} = 1/1$. For the convenience of analysis, the parameters used in follows are relative values, such as the non-dimensional frequency. For simplify, only the vertically incident P and SV waves are considered herein with $\theta_\alpha(\theta_\beta) = 0°$.

5.1 Dynamic hoop stress and deformation of vertically overlapping tunnels

First, define the dimensionless dynamic stress amplitudes as $\sigma^*_{\theta\theta} = |\sigma_{\theta\theta}/\sigma_0| = \left|\sigma_{\theta\theta}/\mu k^2_{\beta 1}\right|$, where $k_{\beta 1}$ is the wavenumber of SV waves in an elastic half space. Since the stress of the soft lining is not very large, we only study the hoop stress of the rigid lining in this paper. Figures 4 and 5 show the distribution of the hoop stress along the inner surface of the twin tunnels for vertically incident P and SV waves with distance $D/a_1 = 3.0, 4.0, 5.0$. For comparison, the case of single tunnel (identical with the upper tunnel) is also presented. The dimensionless frequency takes $\eta = 0.5, 1.0, 2.0$, respectively. It shows that the hoop stress under high frequency is quite different from those under low frequency. For low frequency ($\eta = 0.5$), the hoop stress of the upper tunnel under different intervals between tunnels ($D/a_1 = 3.0, 4.0, 5.0$) is similar to those of the single one. However, the hoop stress in the lower tunnel is significantly different among different intervals. For example, the max hoop stress value for $D/a_1 = 5.0$ is 19.6, while the max hoop stress is 10.7 for $D/a_1 = 3.0$ under P waves. From Figs. 4 and 5, we can observe that when the frequency is 0.5, the depth of the lower tunnel (the normal distance for practical subways) has little impact on the hoop stress, but the impact to the upper tunnel becomes significant when the frequency is 2.0, and the shielding effect decreases with the increase of the distance between the twin tunnels. The results coincide with the conclusion in Huang and Zhang (2013). At the high frequency ($\eta = 2.0$), the effect of the distance between the twin tunnels seems to be not so noticeable, but the hoop stress of the single tunnel is significantly larger than that of the upper tunnel. For example, the max hoop stress value of the single tunnel is 26.88, while that of the twin tunnels is only 9.35. It can be attributed to the strong scattering of high-frequency waves around the lower tunnel which leads to the significant attenuation of incident waves. This phenomenon implies that the lower tunnel may play a protective role for the seismic safety of the upper tunnel for high-frequency waves. Note that Chen and Chen (2008) have investigated the seismic response of vertical double-layered metro tunnels under near-fault strong ground motion by FEM, and similar conclusions have been obtained. While, the studies of Liang et al. (2003, 2004) indicate significant amplification effects due to the dynamic interaction of horizontally twin tunnels, which are substantially different from the model of vertically overlapping tunnels in this study.

Figures 6 and 7 illustrate the hoop stress amplitude spectrums on surface of twin tunnels under vertically incident P and SV waves for $D/a_1 = 3.0$. We select the observation points on the inner surface of the tunnel located at $\theta = 0°, \pm 30°, \pm 90°$ for P waves incidence, and $\theta = 0°, \pm 30°$ for SV waves. It is clearly shown that the stress amplitude highly depends on the incident frequency and the space location, and there are many peaks and troughs of the spectrum curve with the peak values usually appearing in low-frequency region $\eta < 0.5$. As for P waves, the hoop stress amplitude can reach up to 29.5 at $\theta = 0°$, but the amplification effects seem not remarkable at the apex of arch ($\theta = 90°$). In addition, both the peak value and the spectrum characteristic of the upper tunnel are significantly different from those of the lower tunnel. In general, for the lower tunnel, there are more peak frequencies and the peak value is slightly larger than that of the upper tunnel in this case. Compared with the case of P waves, the dynamic stress concentration seems more significant for SV waves and the peak value appears around $\theta = 30°$ for low-frequency waves and around $\theta = 0°$ for high-frequency waves. In addition, the spectrum for incident SV waves oscillates more violently than that for P waves.

Considering the incidence of P waves and the variation of the interval between the tunnels $D/a_1 = 3.0, 4.0, 5.0$, Figs. 8 and 9 illustrate the hoop stress amplitude spectrums at the points $\theta = 0°$ and $\theta = 30°$ on the inner surface of these twin tunnels and the corresponding single tunnel (only the upper tunnel or the lower tunnel exists). It shows that in low-frequency region ($\eta \leq 0.5$), the peak stress amplitude of the upper tunnel is close to the single-tunnel case. However, for the high frequencies, at $\theta = 0°$, the hoop stress amplitude of the single tunnel is much larger than the case of twin tunnels. For example, at $\eta = 1.56$, the hoop stress amplitudes of the single tunnel and the upper one of twin tunnels with $D/a_1 = 3.0$ are 13.80 and 7.25,

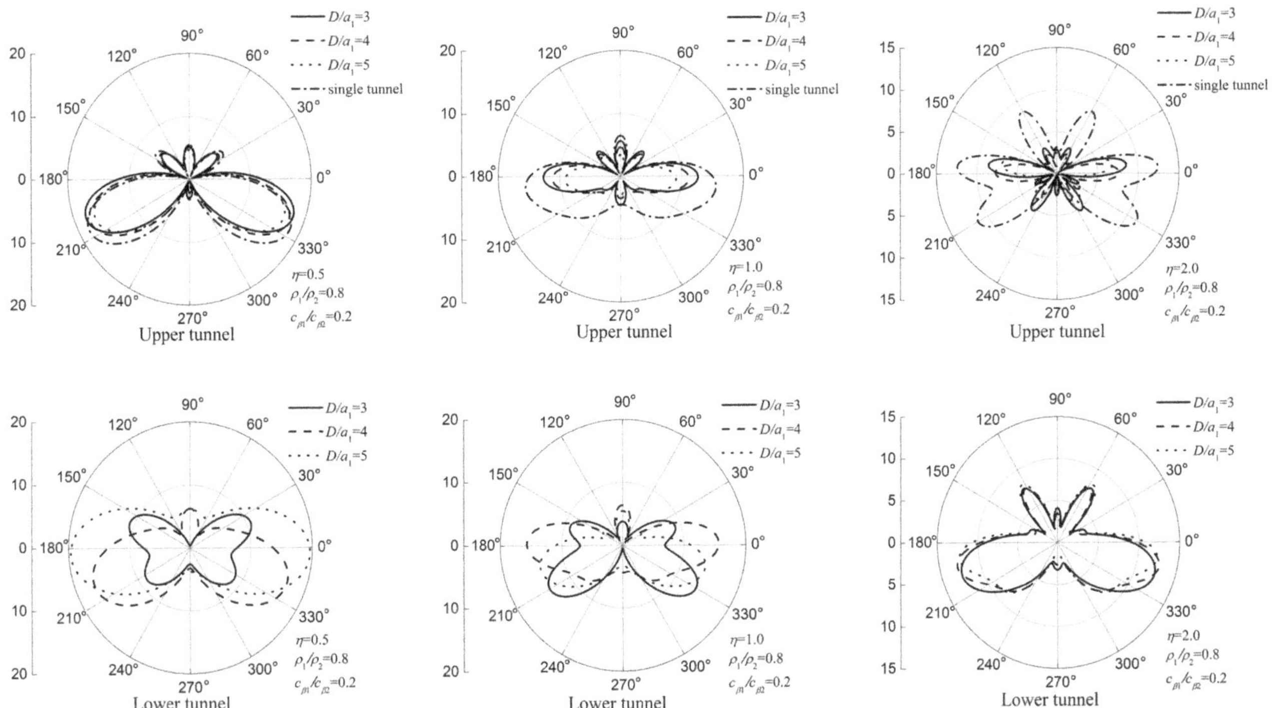

Fig. 4 Hoop stress amplitudes of the *inner wall* of *twin tunnels* and the *single tunnel* (identical with the *upper tunnel*) subject to vertically incident P waves, with frequency ($\eta = 0.5, 1.0, 2.0$), distance between the twin tunnels ($D/a_1 = 3, 4, 5$), and the rigid lining ($\rho_1/\rho_2 = 0.8$, $c_{\beta1}/c_{\beta2} = 0.2$)

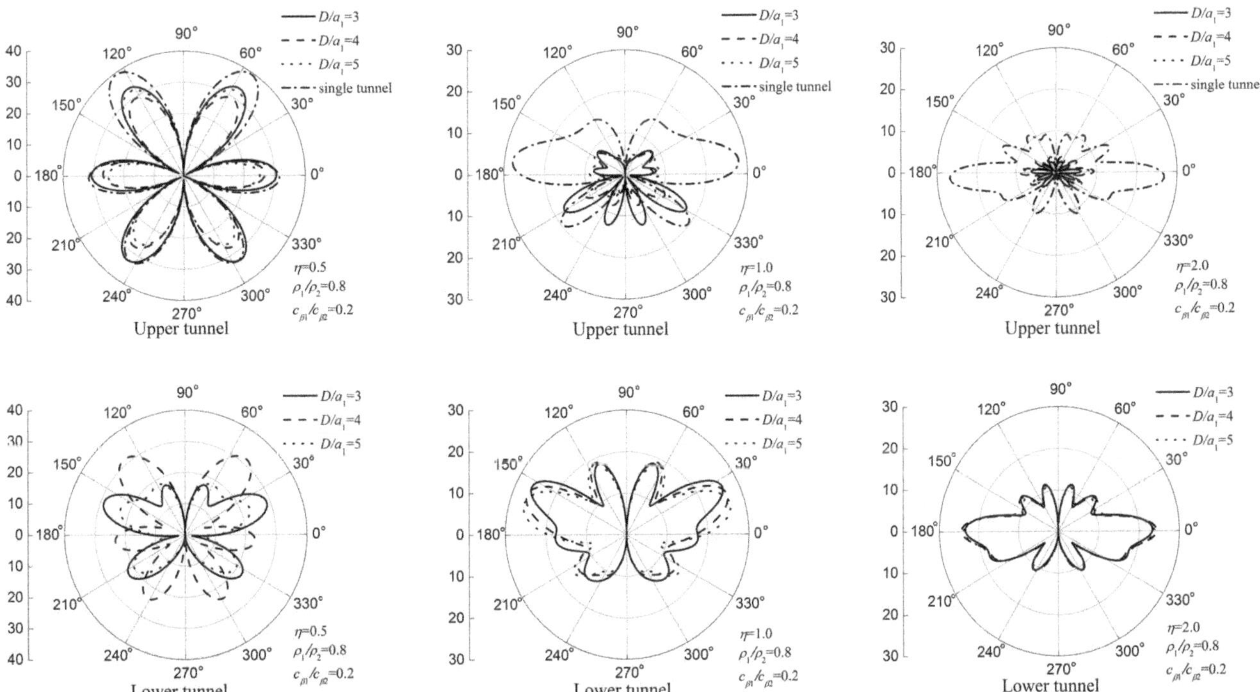

Fig. 5 Hoop stress amplitudes of the *inner wall* of *twin tunnels* and the *single tunnel* (identical with the upper tunnel) subject to vertically incident SV waves, with frequency ($\eta = 0.5, 1.0, 2.0$), distance between the twin tunnels ($D/a_1 = 3, 4, 5$), and the rigid lining ($\rho_1/\rho_2 = 0.8$, $c_{\beta1}/c_{\beta2} = 0.2$)

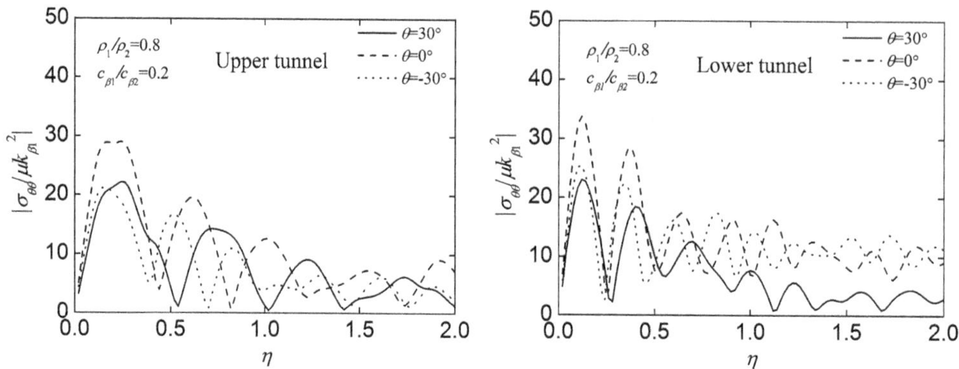

Fig. 6 Hoop stress amplitude spectrums at different points ($\theta = 0°, \pm 30°, \pm 90°$) on the *inner wall* of *twin tunnels*, with the distance between the *twin tunnel* ($D/a_1 = 3$), and the rigid lining ($\rho_1/\rho_2 = 0.8$, $c_{\beta 1}/c_{\beta 2} = 0.2$), subject to *vertically* incident P waves

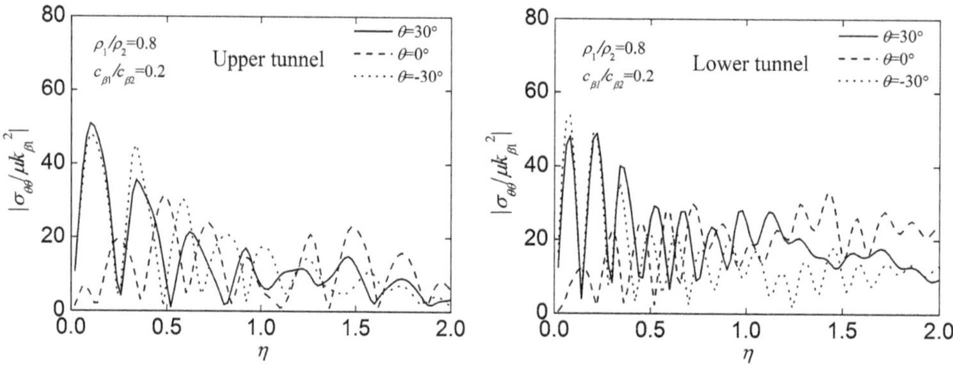

Fig. 7 Hoop stress amplitude spectrums at different points ($\theta = 0°, \pm 30°, \pm 90°$) on the *inner wall* of *twin tunnels*, with the distance between the *twin tunnel* ($D/a_1 = 3$), the rigid lining ($\rho_1/\rho_2 = 0.8$, $c_{\beta 1}/c_{\beta 2} = 0.2$), subject to *vertically* incident SV waves

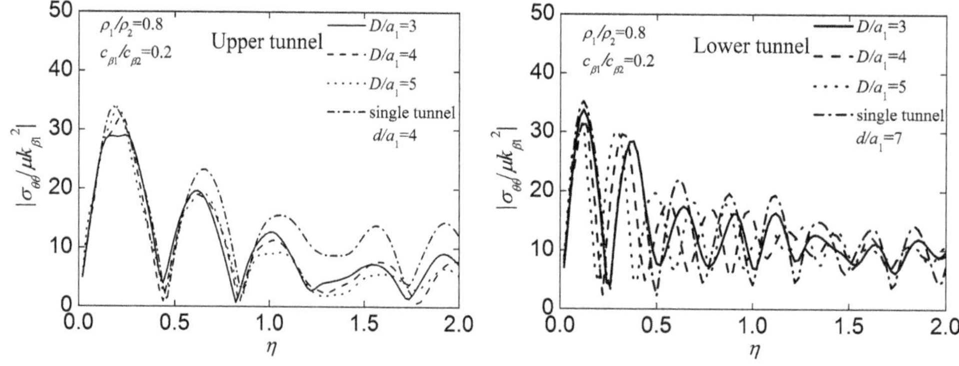

Fig. 8 Hoop stress amplitude spectrums at $\theta = 0°$ on the *inner wall* of *twin tunnels*, with the buried depth of the *single tunnel* ($d/a_1 = 4$ denotes the *single upper tunnel*, $d/a_1 = 7$ denotes the *single lower tunnel*), the distance between the twin tunnels ($D/a_1 = 3$, $D/a_1 = 4$, $D/a_1 = 5$), and the rigid lining ($\rho_1/\rho_2 = 0.8$, $c_{\beta 1}/c_{\beta 2} = 0.2$), subject to *vertically* incident P waves

respectively. But at the point $\theta = 30°$, the influence of the lower tunnel seems not so noticeable.

Figures 10 and 11 illustrate the hoop stress amplitude spectrums at points $\theta = 0°$ and $\theta = 30°$ on the inner surface of these twin tunnels and the corresponding single tunnel for incident SV waves. It shows that in low-frequency region ($\eta \leq 0.5$), the peak stress amplitude of the upper tunnel can be larger than the single-tunnel case. For high-frequency waves, in contrast, the shielding effect of the lower tunnel is very pronounced. For example, at $\eta = 1.5$, the stress amplitude at $\theta = 0°$ of the single tunnel is 38.4, while for the upper one of twin tunnels, the amplitudes are 22.7, 20.8, and 25.7 for $D/a_1 = 3.0$, 4.0 and 5.0 respectively. As for

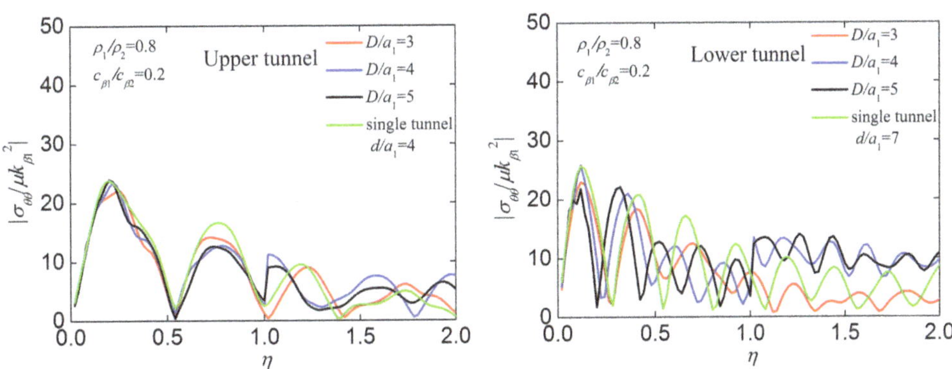

Fig. 9 Hoop stress amplitude spectrums at $\theta = 30°$ on the *inner wall* of *twin tunnels*, with the buried depth of the *single tunnel* ($d/a_1 = 4$ denotes the *single upper tunnel*, $d/a_1 = 7$ denotes the *single lower tunnel*), the distance between the *twin tunnels* ($D/a_1 = 3$, $D/a_1 = 4$, $D/a_1 = 5$), and the rigid lining ($\rho_1/\rho_2 = 0.8$, $c_{\beta1}/c_{\beta2} = 0.2$), subject to *vertically* incident P waves

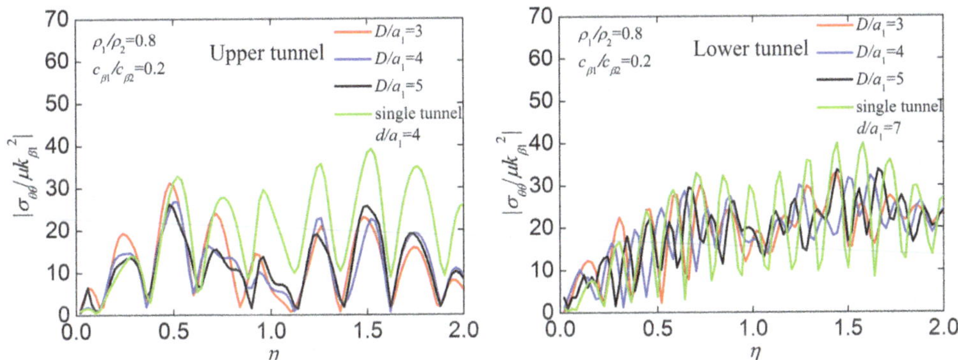

Fig. 10 Hoop stress amplitude spectrums at $\theta = 0°$ on the *inner wall* of the *twin tunnels* and the *single tunnel*, with the buried depth of the *single tunnel* ($d/a_1 = 4$ denotes the *single upper tunnel*, $d/a_1 = 7$ denotes the *single lower tunnel*), distance between the *twin tunnels* ($D/a_1 = 3$, $D/a_1 = 4$, $D/a_1 = 5$), and the rigid lining ($\rho_1/\rho_2 = 0.8$, $c_{\beta1}/c_{\beta2} = 0.2$), subjected to vertically incident SV waves

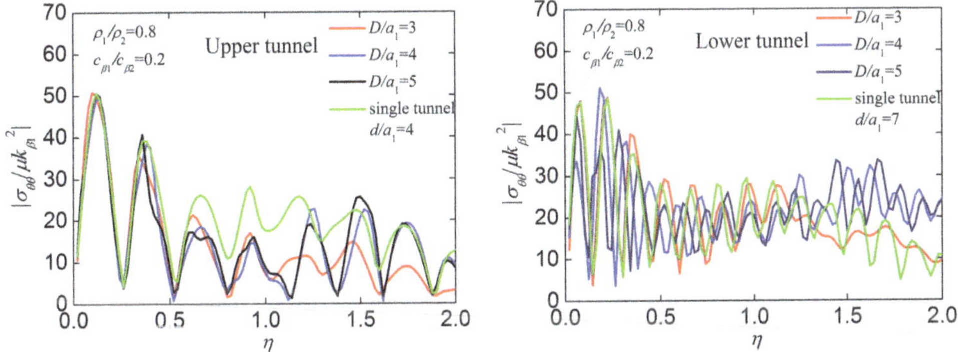

Fig. 11 Hoop stress amplitude spectrums at $\theta = 30°$ on the *inner wall* of the twin tunnels and the single tunnel, with the buried depth of the *single tunnel* ($d/a_1 = 4$ denotes the *single upper tunnel*, $d/a_1 = 7$ denotes the single *lower tunnel*), distance between the twin tunnels ($D/a_1 = 3$, $D/a_1 = 4$, $D/a_1 = 5$), and the rigid lining ($\rho_1/\rho_2 = 0.8$, $c_{\beta1}/c_{\beta2} = 0.2$), subject to *vertically* incident SV waves

the lower tunnel, the displacement spectrum characteristics become more complicated with the variation of D/a_1. As the buried depth increase, the spectrum curves oscillate more rapidly. Comparing the case $D/a_1 = 3.0$ with the corresponding single-tunnel case, we can observe that, for the high-frequency waves, the stress amplitude in the lower tunnel is decreased under the influence of the upper tunnel.

Figures 12 and 13 illustrate the imaginary and the real parts of the deformation of the inner wall of tunnels for vertically incident P and SV waves with $D/a_1 = 3.0$ and the non-dimensional frequency $\eta = 0.5$, 1.0, 2.0. It is

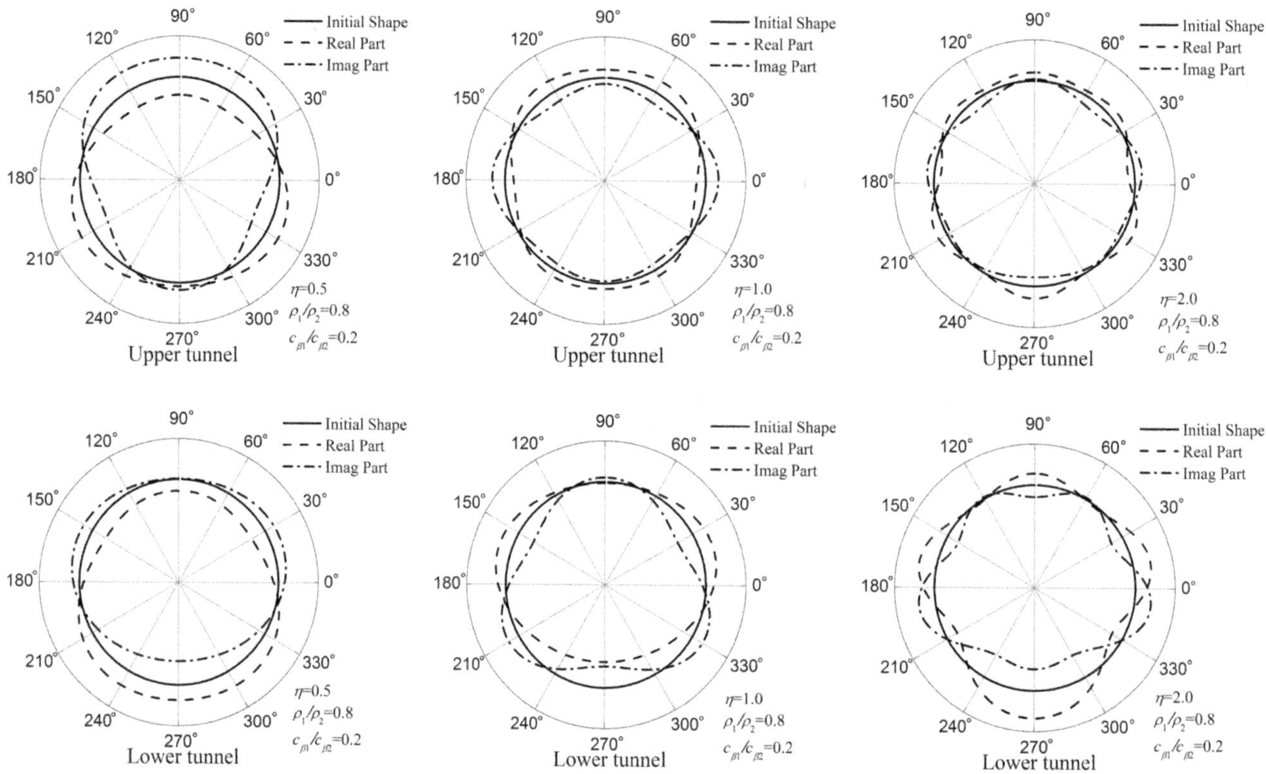

Fig. 12 Comparisons of the real part and the imaginary part deformation of the twin tunnels' *inner wall* by the *initial shape*, with frequency ($\eta = 0.5$, 1.0, 2.0), distance between the *twin tunnels* ($D/a_1 = 3$), and the rigid lining ($\rho_1/\rho_2 = 0.8$, $c_{\beta1}/c_{\beta2} = 0.2$), subject to *vertically* incident P waves

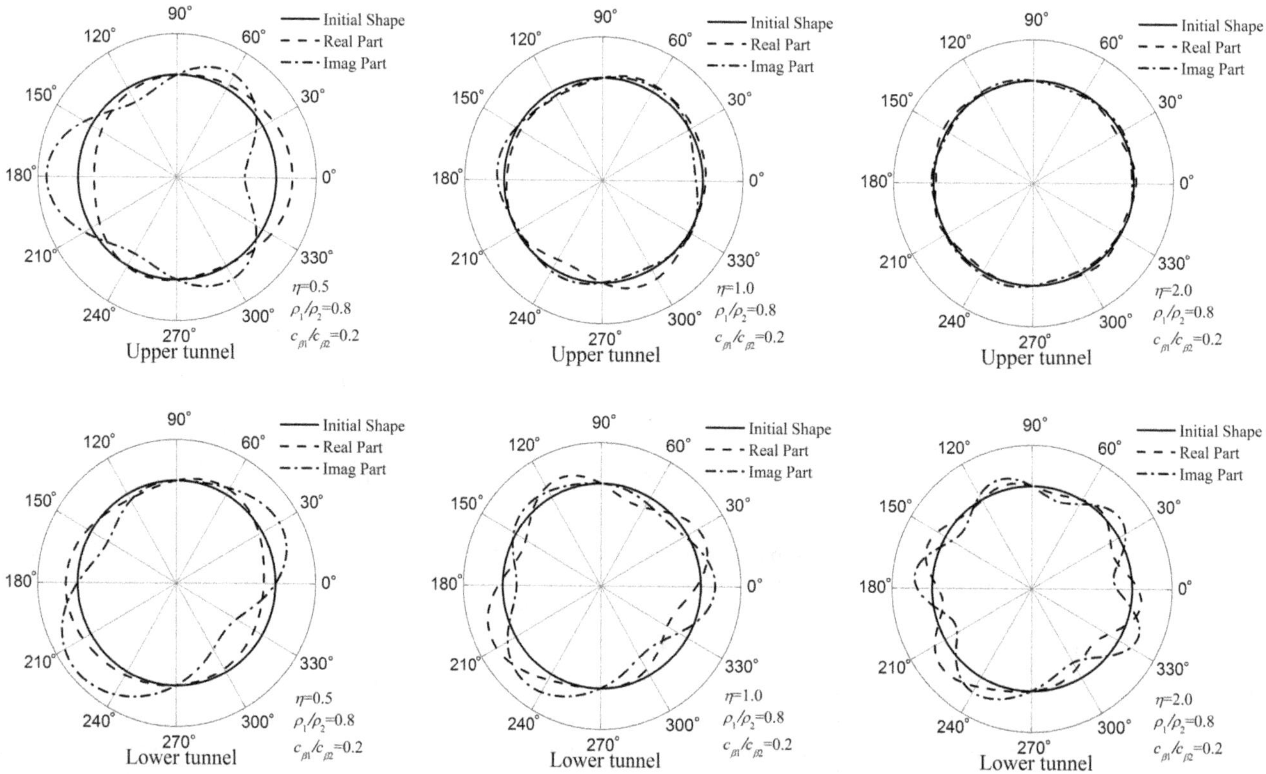

Fig. 13 Comparisons of the real part and the imaginary part deformation of the twin tunnels' inner wall by the *initial shape*, with frequency ($\eta = 0.5$, 1.0, 2.0), distance between the twin tunnels ($D/a_1 = 3$), and the rigid lining ($\rho_1/\rho_2 = 0.8$, $c_{\beta1}/c_{\beta2} = 0.2$), subject to vertically incident SV waves

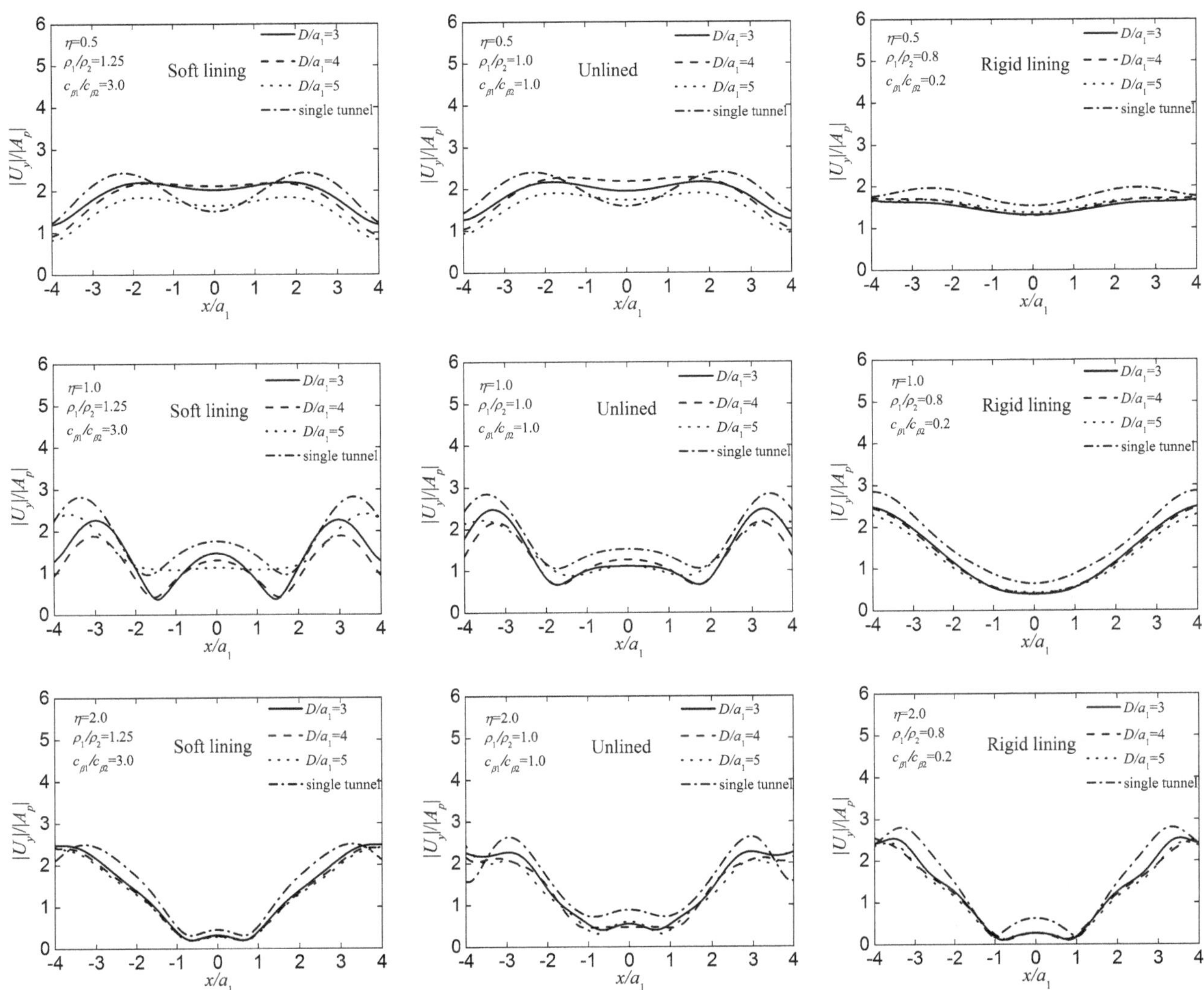

Fig. 14 The vertical displacement amplitudes of ground surface above the twin tunnels and the case of the single tunnel (identical with the *upper tunnel*), for soft lining ($\rho_1/\rho_2 = 1.25$, $c_{\beta1}/c_{\beta2} = 3.0$), homogenous lining ($\rho_1/\rho_2 = 1.0$, $c_{\beta1}/c_{\beta2} = 1.0$), rigid lining ($\rho_1/\rho_2 = 0.8$, $c_{\beta1}/c_{\beta2} = 0.2$), with incident frequency ($\eta = 0.5$, 1.0, 2.0), distance between the twin tunnels ($D/a_1 = 3$, $D/a_1 = 4$, $D/a_1 = 5$), subject to vertically incident P waves

shown that the deformation feature of the upper tunnel may be largely different from that of the bottom one, and for high-frequency waves, the shielding effect of the lower tunnel on the upper one can be also clearly seen for both P and SV waves.

5.2 Ground surface displacement response above the twin tunnels

Figures 14 and 15 show the vertical and horizontal surface displacement amplitudes on the ground surface above the tunnels for incident P and SV waves, considering the variation of the stiffness and the distance between these two tunnels. For comparison, the case of single tunnel (identical with the upper tunnel) is also presented. The non-dimensional frequency takes $\eta = 0.5$, 1.0 and 2.0, respectively. The surface displacement amplitudes are normalized by the displacement amplitudes of incident waves throughout the paper. For simplicity, only the vertically incident seismic waves are considered.

It can be seen that the incident frequency has large influence on the surface displacement response. For the soft tunnels, the amplification or deamplification effects strongly depend on the incident frequency. For the rigid tunnels, the shielding effect on nearby ground surface seems relatively more prominent. For example, at $\eta = 2.0$ the displacement amplitude at $x/a_1 = 0.0$ is only 0.27 and 0.42 for P, SV waves incidence respectively, while it is 2.0 for the free field. Furthermore, the displacement amplitudes and spatial distribution above the vertically overlapping tunnels are similar to those of the single one, and the influence of the distance between the twin tunnels on the

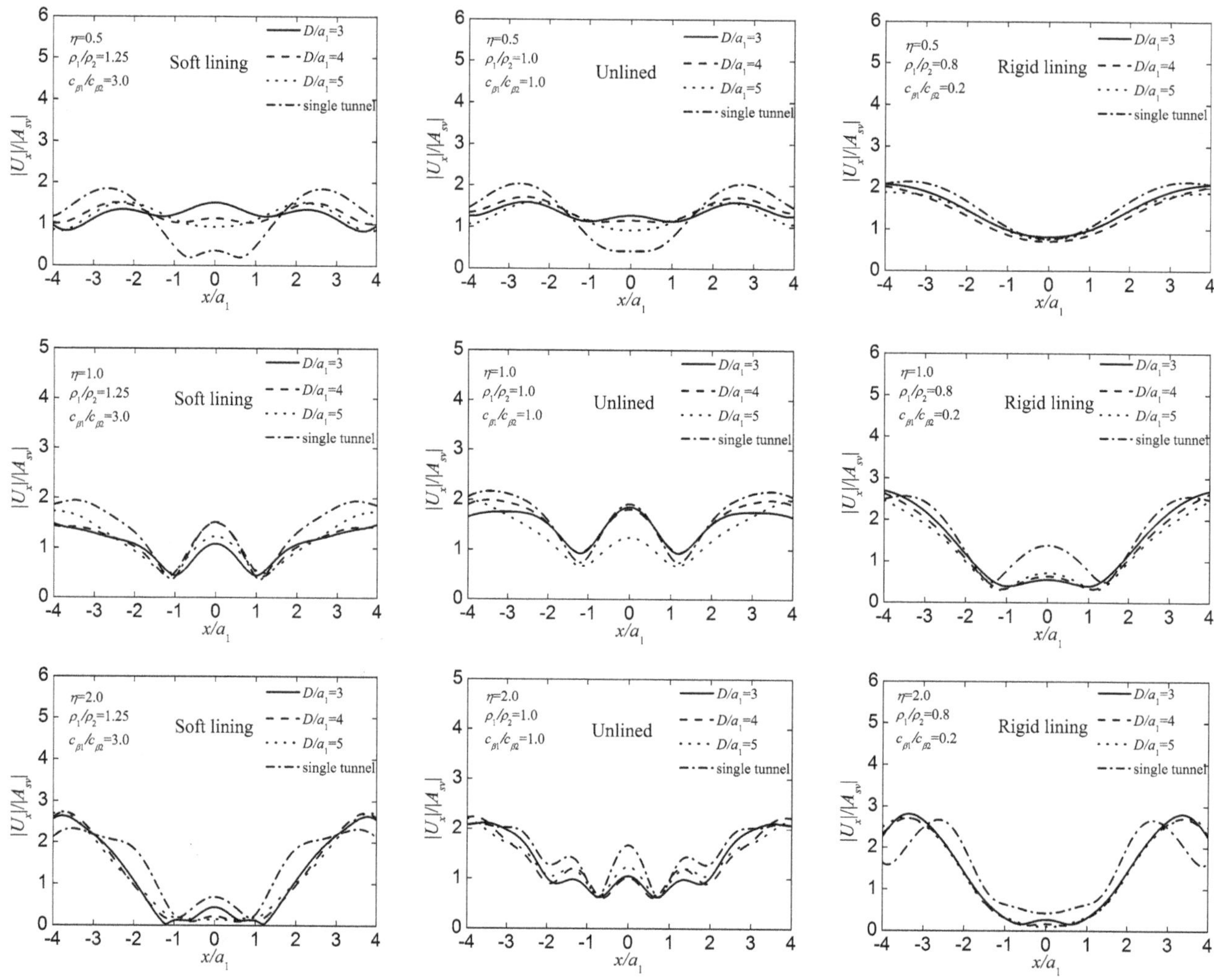

Fig. 15 The horizontal displacement amplitudes of ground surface above the twin tunnels and the case of the single tunnel (identical with the *upper tunnel*), for soft lining ($\rho_1/\rho_2 = 1.25$, $c_{\beta 1}/c_{\beta 2} = 3.0$), homogenous lining ($\rho_1/\rho_2 = 1.0$, $c_{\beta 1}/c_{\beta 2} = 1.0$), rigid lining ($\rho_1/\rho_2 = 0.8$, $c_{\beta 1}/c_{\beta 2} = 0.2$), with incident frequency ($\eta = 0.5, 1.0, 2.0$), distance between the twin tunnels ($D/a_1 = 3$, $D/a_1 = 4$, $D/a_1 = 5$), subject to vertically incident SV waves

surface displacement amplitudes is not so noticeable, especially for high-frequency cases.

Figures 16 and 17 show the surface displacement amplitude spectrums of the points ($x/a_1 = 0.0$, $x/a_1 = 2.0$) near the lined tunnels under vertically incident P and SV waves for both the vertically overlapping tunnels and the single tunnel (identical with the upper tunnel). It shows that the lining material of the tunnel has a significant effect on the surface displacement amplitudes and spatial distribution around the tunnels. In the case of soft lining and homogenous lining, the spectrum curves oscillate violently, while the spectra curves oscillate relatively smoothly for the rigid lining tunnel. As for incident P waves at low frequencies, the amplification effect is noticeable. For an example, at $\eta = 0.54$, the displacement amplitude at the point $x/a_1 = 0.0$ for the soft tunnel is up to 3.0 in the case of $D/a_1 = 3.0$, which is about 50 % larger than that of the free field. However, in the case of rigid tunnels, mainly the shielding effect of the tunnels can be observed, and the surface displacement amplitudes above the single tunnel are larger than those of the twin tunnels. Because under such circumstance, the twin tunnels have more pronounced shielding effect on the seismic waves. Moreover, the effect of the distance between twin tunnels on the surface displacement amplitudes seems not so significant. As for incident SV waves, it seems that the displacement spectrum curves oscillate more quickly, and note that in the case of rigid tunnels, the amplification effect can be observed above the tunnel ($x/a_1 = 0.0$) with the peak amplitude of horizontal displacement 2.36 for $D/a_1 = 3.0$ and $\eta = 0.32$, which is slightly larger than that of the single-tunnel case.

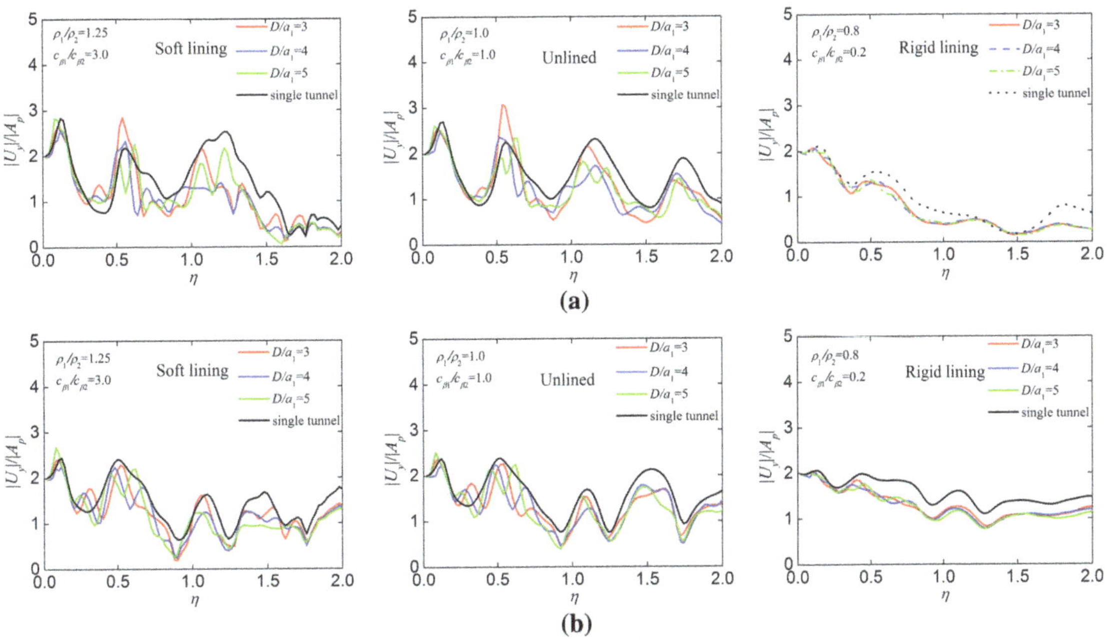

Fig. 16 The vertical displacement amplitudes spectrums of ground surface above the twin tunnels and the case of the single tunnel (identical with the *upper tunnel*), for soft lining ($\rho_1/\rho_2 = 1.25$, $c_{\beta 1}/c_{\beta 2} = 3.0$), homogenous lining ($\rho_1/\rho_2 = 1.0$, $c_{\beta 1}/c_{\beta 2} = 1.0$), rigid lining ($\rho_1/\rho_2 = 0.8$, $c_{\beta 1}/c_{\beta 2} = 0.2$), distance between the twin tunnels ($D/a_1 = 3$, $D/a_1 = 4$, $D/a_1 = 5$), subject to vertically incident P waves. **a** $x/a_1 = 0.0$, **b** $x/a_1 = 2.0$

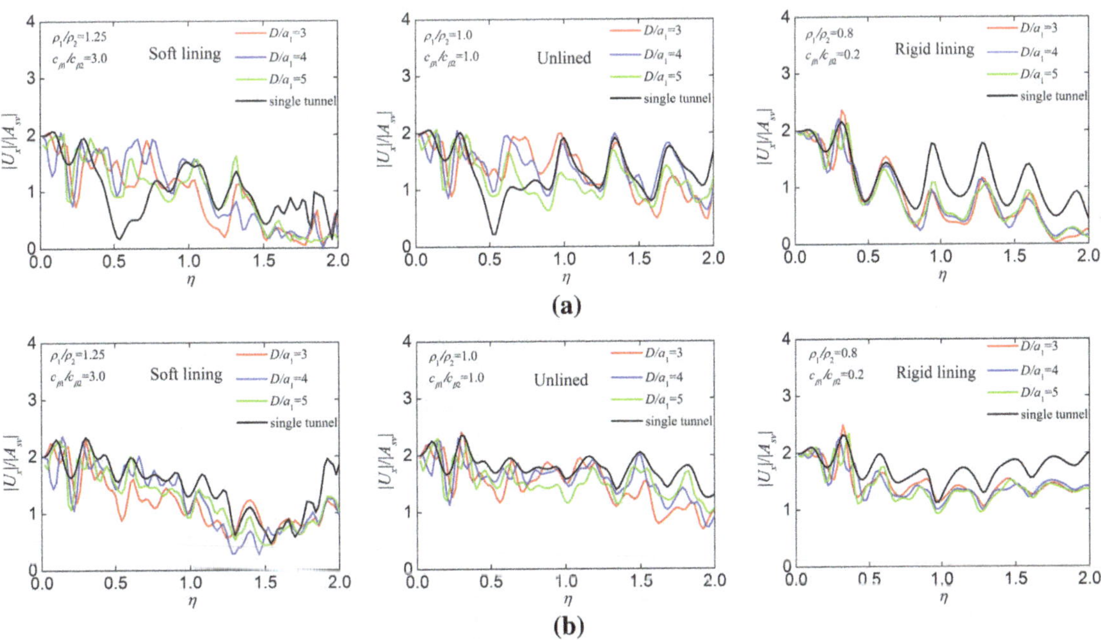

Fig. 17 The horizontal displacement amplitudes spectrums of ground surface above the twin tunnels and the case of the single tunnel (identical with the *upper tunnel*), for soft lining ($\rho_1/\rho_2 = 1.25$, $c_{\beta 1}/c_{\beta 2} = 3.0$), homogenous lining ($\rho_1/\rho_2 = 1.0$, $c_{\beta 1}/c_{\beta 2} = 1.0$), rigid lining ($\rho_1/\rho_2 = 0.8$, $c_{\beta 1}/c_{\beta 2} = 0.2$), distance between the twin tunnels ($D/a_1 = 3$, $D/a_1 = 4$, $D/a_1 = 5$), subject to vertically incident SV waves. **a** $x/a_1 = 0.0$, **b** $x/a_1 = 2.0$

It should be mentioned that in above examples we only considered the small damping case (the damping ratio is 0.001). We calculated the displacement amplitudes of ground surface with the damping ratio 0.001, 0.01, 0.05. The displacement amplitudes of ground surface decrease with the increase of damping ratio, and the decrease ratio is less than 10 % for the calculated frequency range $\eta \in [0, 2]$. Thus, for brevity, we did not analyze the impact of the damping ratio by the scattering of plane harmonic waves in detail.

6 Conclusions

The scattering of plane harmonic waves by a pair of vertically overlapping lined tunnels shallowly buried in an elastic half space is solved by a high-precision IBIEM. The convergence and numerical stability of the IBIEM for this model are verified. Through detailed numerical analysis, several beneficial conclusions can be drawn.

(1) Numerical results have been shown that the scattering of seismic wave strongly depends on the distance between these twin tunnels, the material properties of the lining and the non-dimensional frequency. The dynamic interaction between these twin tunnels cannot be neglected.
(2) It has been shown that the lower tunnel may play a protective role of isolating the P and SV waves of high frequency, leading to the great decrease of dynamic hoop stress amplitude of the upper tunnel, up to 50 % smaller than the case of the single tunnel at some frequencies. However, for the low-frequency SV waves, on the contrary, additional amplification effect between these twin tunnels can be observed at some locations in the upper tunnel.
(3) The soft tunnels may have significant amplification effect on the ground motion above the tunnels, while the rigid tunnels mainly have shielding effect on the nearby surface displacement response. The influence of the vertically overlapping tunnels seems slightly larger than that of the single-tunnel case.
(4) The deformation feature and the dynamic stress of the upper tunnel are significantly different from those of the lower tunnel, and there are more peaks and troughs of the spectrum curve of the lower tunnel with large buried depth.

This study was limited to the frequency domain analysis. But from the perspective of engineering, it is more attractive in the time domain. Moreover, the half-space model is suitable for the thick near-surface layer. For more general scenarios, the layered half-space model should be adopted. We leave these limitations for the future study.

Acknowledgments This work was supported by the Tianjin Research Program of Application Foundation Advanced Technology (14JCYBJC21900) and the National Natural Science Foundation of China under grants 51278327.

References

Alielahi H, Adampira M (2016a) Site-specific response spectra for seismic motions in half-plane with shallow cavities. Soil Dyn Earthq Eng 80:163–167

Alielahi H, Adampira M (2016b) Effect of twin-parallel tunnels on seismic ground response due to vertically in-plane waves. Int J Rock Mech Min 85:67–83

Alielahi H, Kamalian M, Adampira M (2015) Seismic ground amplification by unlined tunnels subjected to vertically propagating SV and P waves using BEM. Soil Dyn Earthq Eng 71:63–79

An JH, Tao LJ, Li JD, Bian J, Suo XA (2015) Nonlinear seismic response of double-decked intersecting metro tunnel. China Railw Sci 36(3):66–72 **(in Chinese with English abstract)**

Balendra T, Thambiratnam DP, Koh CG, Lee SL (1984) Dynamic response of twin circular tunnels due to incident SH-waves. Earthq Eng Struct Dyn 12(2):181–201

Chakeri H, Hasanpour R, Hindistan MA, Unver B (2011) Analysis of interaction between tunnels in soft ground by 3D numerical modeling. Bull Eng Geol Environ 70(3):439–448

Chen L, Chen GX (2008) Seismic response of vertically double-layered metro tunnels under near-fault strong ground motion. J Disaster Prev Mitig Eng 28(4):399–408 **(in Chinese)**

Chen JT, Shen WC, Wu AC (2006) Null-field integral equations for stress field around circular holes under anti-plane shear. Eng Anal Bound Elem 30(3):205–217

Chen L, Chen GX, Li LM (2010) Seismic response characteristics of the double-layer vertical overlapping metro tunnels under near-field and far-field ground motions. China Railw Sci 31(1):79–86 **(in Chinese)**

Esmaeili M, Vahdani S, Noorzad A (2006) Dynamic response of lined circular tunnel to plane harmonic waves. Tunn Undergr Sp Technol 21(5):511–519

Fang XQ, Jin HX, Wang BL (2015) Dynamic interaction of two circular lined tunnels with imperfect interfaces under cylindrical P-waves. Int J Rock Mech Min 79:172–182

Fotieva NN (1980) Determination of the minimum seismically safe distance between two parallel tunnels. Soil Mech Found Eng 17(3):111–116

Hasheminejad SM, Avazmohammadi R (2007) Harmonic wave diffraction by two circular cavities in a poroelastic formation. Soil Dyn Earthq Eng 27:29–41

Huang J, Zhang B (2013) Shaking table model test of subway coross-structure under near-field ground motion. Water Resour Power 31(2):120–122 **(in Chinese)**

Karakus M, Ozsan A, Basarir H (2007) Finite element analysis for twin metro tunnel constructed in Ankara clay, Turkey. Bull Eng Geol Environ 66(1):71–79

Kattis SE, Beskos DE, Cheng AHD (2003) 2-D dynamic response of unlined and lined tunnels in poroelastic soil to harmonic body waves. Earthq Eng Struct Dyn 32:97–110

Kobayashi S, Nishimura N (1983) Analysis of dynamic soil-structure interactions by boundary integral equation method. In: Proceedings of the Third ISNME, Paris, pp 353–362

Lamb H (1904) On the propagation of tremors over the surface of an elastic solid. Philos Trans R Soc London Ser A 203:1–42

Lee VW, Trifunac MD (1979) Response of tunnels to incident SH-waves. J Eng Mech 105(4):643–659

Liang JW, Ji XD (2006) Amplification of Rayleigh waves due to underground lined cavities. Earthq Eng Eng Vib 26(4):24–31 **(in Chinese with English abstract)**

Liang JW, Liu ZX (2009) Diffraction of plane SV waves by a cavity in poroelastic half-space. J Earthq Eng Eng Vib 8(1):29–46

Liang JW, Zhang H, Lee VW (2003) A series solution for surface motion amplification due to underground twin tunnels incident SV waves. Earthq Eng Eng Vib 2(2):289–298

Liang JW, Zhang H, Lee VW (2004) An analytical solution for dynamic stress concentration of underground Twin cavities due to incident SV waves. J Vib Eng 17(2):132–140 **(in Chinese)**

Liu QJ, Wang RY (2012) Dynamic response of twin closely-spaced circular tunnels to harmonic plane waves in a full space. Tunn Undergr Sp Technol 32(6):212–220

Liu ZX, Liang JW, Zhang H (2010) Scattering of plane P and SV waves by a lined tunnel in elastic half space(II): numerical results. J Nat Disasters 04:77–88 **(in Chinese)**

Liu QJ, Zhao MJ, Wang LH (2013) Scattering of plane P, SV or Rayleigh waves by a shallow lined tunnel in an elastic half space. Soil Dyn Earthq Eng 49(6):52–63

Luco JE, De Barros FCP (1994) Dynamic displacements and stresses in the vicinity of a cylindrical cavity embedded in a half-space. Earthq Eng Struct Dyn 23(3):321–340

Moore ID, Guan F (1996) Three-dimensional dynamic response of lined tunnels due to incident seismic waves. Earthq Eng Struct Dyn 25(4):357–369

Okumura T, Takewaki N, Shimizu K, Fukutake K (1992) Dynamic response of twin circular tunnels during earthquakes. NIST Spec Publ 840:181–191

Panji M, Kamalian M, Asgari MJ, Jafari MK (2013) Transient analysis of wave propagation problems by half-plane BEM. Geophys J Int 194(3):1849–1865

Parvanova SL, Dineva PS, Manolis GD, Wuttke F (2014) Seismic response of lined tunnels in the half-plane with surface topography. Bull Earthq Eng 12(2):981–1005

Pitilakis K, Tsinidis G, Leanza A, Maugeri M (2014) Seismic behaviour of circular tunnels accounting for above ground structures interaction effects. Soil Dyn Earthq Eng 67(67):1–15

Rodriguez-Castellanos A, Sanchez-Sesma FJ, Luzon F (2006) Martin R. Multiple scattering of elastic waves by subsurface fractures and cavities. Bull Seismol Soc Am 96(4):1359–1374

Smerzini C, Aviles J, Paolucci R, Sanchez-Sesma FJ (2009) Effect of underground cavities on surface earthquake ground motion under SH wave propagation. Earthq Eng Struct Dyn 38(12):1441–1460

Stamos AA, Beskos DE (1996) 3-D seismic response analysis of long lined tunnels in half-space. Soil Dyn Earthq Eng 15(2):111–118

Wang WL, Wang TT, Su JJ, Lin CH, Seng CR, Huang TK (2001) Assessment of damage in mountain tunnels due to the Taiwan Chi-Chi Earthquake. Tunn Undergr Sp Technol 16(3):133–150

Wang GB, Chen L, Xu HQ, Li P (2012) Study of seismic capability of adjacent overlapping multi-tunnels. Rock Soil Mech 33(8):2483–2490 **(in Chinese with English abstract)**

Wong HL (1982) Effect of surface topography on the diffraction of P, SV, and Rayleigh waves. J Bull Seismol Soc Am 72(4):1167–1183

Yi CP, Zhang P, Johansson D, Nyberg U (2014) Dynamic response of a circular lined tunnel with an imperfect interface subjected to cylindrical P-waves. Comput Geotech 55(1):165–171

Yiouta-Mitra P, Kouretzis G, Bouckovalas G, Sofianos A (2007) Effect of underground structures in earthquake resistant design of surface structures. Dynamic response and soil properties. Geo-Denver: New Peaks in Geotechnics

Yu CW, Dravinski M (2009) Scattering of plane harmonic P, SV or rayleigh waves by a completely embedded corrugated cavity. Geophys J Int 178(1):479–487

Zhou XL, Wang JH, Jiang LF (2009) Dynamic response of a pair of elliptic tunnels embedded in a poroelastic medium. J Sound Vib 325(4–5):816–834

Permissions

All chapters in this book were first published in ES, by Springer International Publishing AG.; hereby published with permission under the Creative Commons Attribution License or equivalent. Every chapter published in this book has been scrutinized by our experts. Their significance has been extensively debated. The topics covered herein carry significant findings which will fuel the growth of the discipline. They may even be implemented as practical applications or may be referred to as a beginning point for another development.

The contributors of this book come from diverse backgrounds, making this book a truly international effort. This book will bring forth new frontiers with its revolutionizing research information and detailed analysis of the nascent developments around the world.

We would like to thank all the contributing authors for lending their expertise to make the book truly unique. They have played a crucial role in the development of this book. Without their invaluable contributions this book wouldn't have been possible. They have made vital efforts to compile up to date information on the varied aspects of this subject to make this book a valuable addition to the collection of many professionals and students.

This book was conceptualized with the vision of imparting up-to-date information and advanced data in this field. To ensure the same, a matchless editorial board was set up. Every individual on the board went through rigorous rounds of assessment to prove their worth. After which they invested a large part of their time researching and compiling the most relevant data for our readers.

The editorial board has been involved in producing this book since its inception. They have spent rigorous hours researching and exploring the diverse topics which have resulted in the successful publishing of this book. They have passed on their knowledge of decades through this book. To expedite this challenging task, the publisher supported the team at every step. A small team of assistant editors was also appointed to further simplify the editing procedure and attain best results for the readers.

Apart from the editorial board, the designing team has also invested a significant amount of their time in understanding the subject and creating the most relevant covers. They scrutinized every image to scout for the most suitable representation of the subject and create an appropriate cover for the book.

The publishing team has been an ardent support to the editorial, designing and production team. Their endless efforts to recruit the best for this project, has resulted in the accomplishment of this book. They are a veteran in the field of academics and their pool of knowledge is as vast as their experience in printing. Their expertise and guidance has proved useful at every step. Their uncompromising quality standards have made this book an exceptional effort. Their encouragement from time to time has been an inspiration for everyone.

The publisher and the editorial board hope that this book will prove to be a valuable piece of knowledge for researchers, students, practitioners and scholars across the globe.

List of Contributors

Zhongxian Liu and Lei Liu
Key Laboratory of Soft Soils and Engineering Environmental of Tianjin, Tianjin Chengjian University, Tianjin 300384, China
Earthquake Engineering Research Institute of Tianjin, Tianjin 300384, China

Tianshi Liu and Haiming Zhang
Department of Geophysics, School of Earth and Space Sciences, Peking University, Beijing 100871, People's Republic of China

Yue Yang, Jian Wen and Xiaofei Chen
Laboratory of Seismology and Physics of Earth's Interior, University of Science and Technology of China, Hefei 230026, Anhui, China
Mengcheng National Geophysical Observatory, Hefei 230026, Anhui, China

Xizhu Guan, Li-Yun Fu and Weijia Sun
Institute of Geology and Geophysics, Chinese Academy of Sciences, Beijing 100029, China
CNOOC Research Center, Beijing 100027, China

Xueyan Li, Yanbin Wang and Yongshun John Chen
Department of Geophysics, School of Earth and Space Sciences, Peking University, Beijing 100871, China

Qiyan Yang and Yanrui Sheng
Earthquake Administration of Hebei Province, Shijiazhuang 050021, China

Qingju Wu and Fengxue Zhang
Institute of Geophysics, China Earthquake Administration, Beijing 100081, China

Xiaojun Ma
Earthquake Administration of Ningxia Autonomous Region, Yinchuan 751000, China

Yanyang Chen and Yanbin Wang
Department of Geophysics, School of Earth and Space Sciences, Peking University, Beijing 100871, China

Yuansheng Zhang
Lanzhou Institute of Seismology, China Earthquake Administration, Lanzhou 730000, China

Yan Cai, Weilai Wang, Lihua Fang, Liping Fan and Jianping Wu
Institute of Geophysics, China Earthquake Administration, Beijing 100081, China
Key Laboratory of Seismic Observation and Geophysical Imaging, Institute of Geophysics, China Earthquake Administration, Beijing 100081, China

Jie Dong and Wenke Sun
Key Laboratory of Computational Geodynamics, University of Chinese Academy of Sciences, Beijing 100049, China

Jian-Chang Zheng, Jin-Hua Zhao, Chang-Peng Xu and Peng Wang
Earthquake Administration of Shandong Province, Jinan 250014, China

Meiqing Song
Earthquake Prediction Center, Earthquake Administration of Shanxi Province, Taiyuan 030021, China

Yong Zheng
Institute of Geodesy and Geophysics, Chinese Academy of Sciences, Wuhan 430077, China

Chun Liu
Earthquake Administration of Shaanxi Province, Xi'an 710068, China

Li Li and Xia Wang
Earthquake Administration of Shanxi Province, Taiyuan 030021, China

Jia Fu and Lin Qin
Department of Civil Engineering, Tianjin University, Tianjin 300072, China

Jianwen Liang
Department of Civil Engineering, Tianjin University, Tianjin 300072, China
Tianjin Key Laboratory of Civil Engineering Structures & New Materials, Tianjin 300072, China

Muhammad Adeel Arshad
Department of Civil Engineering, University of Engineering & Technology, Peshawar, Peshawar 25120, KPK, Pakistan

Zehua Qiu, Lei Tang and Yanping Guo
Institute of Crustal Dynamics, China Earthquake Administration, Beijing 100085, China

Shunliang Chi
Earthquake Administration of Hebi, Hebi 458000, He'nan Province, China

Zhenming Wang and Seth Carpenter
Kentucky Geological Survey, University of Kentucky, Lexington, KY 40506, USA

Guang Yang
Guza Seismic Station, Earthquake Administration of Sichuan Province, Kangding 626001, Sichuan Province, China

Thang Le and Vincent W. Lee
Sonny Astani Civil & Environmental Engineering Department, University of Southern California, Los Angeles, CA 90089, USA

Hao Luo
HNTB Corporation, 200 E Sandpointe Ave #200, Santa Ana, CA 92707, USA

Xianghua Jiang and Yanbin Wang
Department of Geophysics, School of Earth and Space Sciences, Peking University, Beijing 100871, China

Yanfang Qin
Equipe de Géosciences Marines, Institut de Physique du Globe de Paris, 4 Place Jussieu, 75252 Paris Cedex 05, France

Hiroshi Takenaka
Department of Earth Sciences, Faculty of Science, Okayama University, 3-1-1 Tsushima-Naka, Kita-ku, Okayama 700-8530, Japan

Mahsa Abdetedal, Zaher Hossein Shomali and Mohammad Reza Gheitanchi
Institute of Geophysics, Tehran 14155-6466, Iran

Zhongxian Liu and Yirui Wang
Tianjin Key laboratory of civil structure protection and reiforcement, Tianjin Chengjian University, Tianjin 300384, China

Jianwen Liang
Department of Civil Engineering, Tianjin University, Tianjin 300372, China

Index

T

U

V

W

www.ingramcontent.com/pod-product-compliance
Lightning Source LLC
Chambersburg PA
CBHW082033190326
41458CB00010B/3353

9781641160902